Introduction to Environmental Assessment

Introduction to Environmental Assessment
A Guide to Principles and Practice

Bram F. Noble

FOURTH EDITION

UNIVERSITY PRESS

Oxford University Press is a department of the University of Oxford.
It furthers the University's objective of excellence in research, scholarship,
and education by publishing worldwide. Oxford is a registered trade mark of
Oxford University Press in the UK and in certain other countries.

Published in Canada by
Oxford University Press
8 Sampson Mews, Suite 204,
Don Mills, Ontario M3C 0H5 Canada

www.oupcanada.com

Copyright © Oxford University Press Canada 2021

The moral rights of the author have been asserted

Database right Oxford University Press (maker)

First Edition published in 2006
Second Edition published in 2010
Third Edition published in 2015

All rights reserved. No part of this publication may be reproduced, stored in
a retrieval system, or transmitted, in any form or by any means, without the
prior permission in writing of Oxford University Press, or as expressly permitted
by law, by licence, or under terms agreed with the appropriate reprographics
rights organization. Enquiries concerning reproduction outside the scope of the
above should be sent to the Permissions Department at the address above
or through the following url: www.oupcanada.com/permission/permission_request.php

Every effort has been made to determine and contact copyright holders.
In the case of any omissions, the publisher will be pleased to make
suitable acknowledgement in future editions.

Library and Archives Canada Cataloguing in Publication
Title: Introduction to environmental assessment :
a guide to principles and practice / Bram F. Noble.
Names: Noble, Bram F., 1975- author.
Description: Fourth edition. | Includes bibliographical references and index.
Identifiers: Canadiana (print) 20200245368 | Canadiana (ebook) 20200245465 |
ISBN 9780199028894 (softcover) | ISBN 9780199028917 (ebook)
Subjects: LCSH: Environmental impact analysis—Textbooks. | LCSH: Environmental impact analysis—
Canada—Textbooks. | LCGFT: Textbooks.
Classification: LCC TD194.6 .N62 2020 | DDC 333.71/4—dc23

Cover image: ©Joel Sartore
Cover and interior design: Sherill Chapman

Oxford University Press is committed to our environment.
This book is printed on Forest Stewardship Council® certified paper
and comes from responsible sources.

Printed and bound in Canada

2 3 4 — 23 22 21

Contents

List of Boxes, Boxed Features, Figures, and Tables   ix
Preface   xiii
Acknowledgements   xiii

1   Aims and Objectives of Environmental Assessment   1
　　Introduction   1
　　Environmental Assessment   2
　　The EA Process   6
　　Purpose and Objectives of EA   6
　　Who's Who in the EA Process   10
　　Getting the Big Picture   12
　　Key Terms   14
　　Review Questions and Exercises   14
　　References   15

2   Environmental Assessment in Canada   17
　　Overview of Environmental Assessment in Canada   17
　　Provincial EA Systems   17
　　Northern EA   23
　　Origins and Development of EA in Canada   25
　　Continuous Learning Process   33
　　Key Terms   34
　　Review Questions and Exercises   34
　　References   34

3   Pre-project Planning and Public Engagement   37
　　Introduction   37
　　Roles and Responsibilities   37
　　Project Need and Consideration of Alternatives   37
　　Public Engagement   47
　　Project Description   54
　　Key Terms   55
　　Review Questions and Exercises   55
　　References   56

4   Determining the Need for Assessment   58
　　Screening   58
　　Screening Approaches   58
　　Level of Assessment Required   65
　　Screening and the Precautionary Principle   68
　　Key Terms   68

Review Questions and Exercises   69
References   69

5   Scoping and Baseline Assessment   71

Scoping   71
Baseline Assessment   72
Knowledge to Support Baseline Assessments   88
Key Terms   90
Review Questions and Exercises   90
References   91

6   Impact Prediction and Characterization   93

Impact Prediction   93
Change and Project Effects   93
What to Predict   94
How to Predict   107
Characterizing Predicted Impacts   113
Addressing Uncertainty   121
Key Terms   123
Review Questions and Exercises   124
References   124

7   Managing Project Impacts   127

Impact Management   127
Mitigation Hierarchy   127
Checklist for Management Prescriptions   133
Adaptive Management   136
Creating and Enhancing Positive Impacts   139
Key Terms   144
Review Questions and Exercises   144
References   145

8   Significance Determination   147

Impact Significance   147
Measurement and Meaning: Components of Significance   149
Approaches to Significance Determination   162
Key Principles for Determining Significance   167
Key Terms   168
Review Questions and Exercises   168
References   169

Contents   vii

9 Follow-Up and Monitoring   172
Follow-Up   172
Rationale for Post-decision Monitoring   174
Effective Follow-Up and Monitoring   179
Monitoring Methods and Techniques   188
Key Terms   188
Review Questions and Exercises   189
References   189

10 Indigenous Consultation and Engagement   192
Indigenous Engagement   192
Duty to Consult   193
Indigenous and Local Knowledge Systems   196
Enduring Challenges to Indigenous Engagement   199
Toward Meaningful Indigenous Engagement in EA   202
Key Terms   208
Review Questions and Exercises   208
References   208

11 Cumulative Effects Assessment   211
Cumulative Effects   211
Assessing Cumulative Environmental Effects   211
Regional Assessment   219
Basic Science Components of a CEA Framework   226
Governance for Cumulative Effects Management   232
Key Terms   233
Review Questions and Exercises   234
References   234

12 Strategic Environmental Assessment   238
Higher-Order Assessment   238
Defining Strategic EA   238
Origins and Evolution   243
Foundational Principles of Strategic EA   249
SEA Benefits   258
SEA Design   259
Enduring Challenges   268
Key Terms   268
Review Questions and Exercises   269
References   269

13 Professional Practice and Ethics 272
Professional Practice 272
Ethical Conduct 278
Key Terms 282
Review Questions and Exercises 282
References 282

14 Environmental Assessment Prospects 283
References 285

Glossary 287
Index 295

List of Boxes, Boxed Features, Figures, and Tables

BOXES

1.1 The International Association for Impact Assessment   4
1.2 Definitions of environmental assessment   5
1.3 Components of the EA process   7
2.1 Oldman River dam project, Alberta   28
3.1 Alternative means and the James Bay Lithium Mine, Quebec   40
3.2 Peterson matrix for comparing project route alternatives   42
3.3 Participant funding for the Sisson Project, New Brunswick   51
4.1 Designated activities requiring assessment under the Impact Assessment Act   59
4.2 EA screenings and reviews in Nunavut   65
5.1 Principles for VC selection in the Elk Valley, British Columbia   73
5.2 Basic principles for spatial bounding   82
5.3 Network diagrams for mapping impact pathways   86
5.4 Shifting baseline syndrome   87
6.1 Climate considerations in EA practice in British Columbia's LNG sector   99
6.2 ALCES and MARXAN: Tools for scenario analysis   112
6.3 Actions, causal factors, effects, and impacts   114
6.4 Impacts of fly-in fly-out mining operations   117
7.1 Ecoducts for avoiding wildlife collisions, northern Sweden   128
7.2 The burgeoning cost of remediating Alberta's oil and gas well sites   131
7.3 LNG Canada export terminal wetland compensation strategy   132
7.4 Typical components of an environmental management plan   134
7.5 Tales of adaptive management from two mining projects   137
8.1 Debunking the "compared-to" approach for significance determination   153
8.2 Factors to be considered in determining public interest under the Impact Assessment Act   162
8.3 Issues of national significance   165
9.1 Types of EA audits   174
9.2 Monitoring of socio-economic agreements: Diavik diamond mine   175
9.3 Following up for impact management: Bison reintroduction and experimental grazing in Grasslands National Park, Saskatchewan   176
9.4 Follow-up program objectives for the Hardrock project, Ontario   180
10.1 The United Nations Declaration on the Rights of Indigenous Peoples   193
10.2 Duty to consult and the Haida case ruling   194
10.3 A water licence retracted: Clarity of roles and responsibilities in discharging the duty to consult   195
10.4 Indigenous and local knowledge in Keeyask hydroelectric generation project EIS   197
10.5 Ur-Energy Inc. Screech Lake uranium exploration project   200

10.6	Researchers' experiences working with environmental assessment registries	204
11.1	Cumulative effects on social systems—the orphan of CEA	212
11.2	Project EA and individually insignificant actions	215
11.3	Eastmain 1A project, Quebec, and the need for collaborative approaches to managing cumulative effects	218
11.4	Regional assessment of offshore oil and gas exploratory drilling east of Newfoundland and Labrador	220
11.5	British Columbia's cumulative effects framework	233
12.1	Canada's climate change strategic environmental assessment	241
12.2	SEA in Atlantic Canada's offshore oil and gas sector	242
12.3	International Atomic Energy Agency's nuclear power program SEA guidance	242
12.4	Characteristics of strategic environmental assessment	249
12.5	What is "strategic"?	251
12.6	Compliance-based SEA	253
12.7	SEA benefits and opportunities	259
12.8	Future scenarios and strategic options in SEA	263
13.1	International Association for Impact Assessment code of conduct	279

BOXED FEATURES

Evaluation of alternatives for managing impacts to wetlands for the Louis Riel Trail Highway Twinning Project, Saskatchewan   44
Screening out small projects in a grassland ecosystem   61
Predicting the impacts of the Lower Churchill Hydroelectric Generating Project on community health and well-being   97
Galore Creek Project and Athabasca uranium mining agreements   142
Sustainability-based review of the Kemess North copper-gold mine project   158
Wood Buffalo Environmental Association   186
Squamish Nation EA of the Woodfibre liquefied natural gas facility   206
Elk Valley Cumulative Effects Management Framework   220
Strategic futures SEA in Saskatchewan's Great Sand Hills   254
What would you do?   280

FIGURES

1.1	Human access and industrial concessions in Canada	2
1.2	Project EA as part of an integrated system of policy, land-use planning, and natural resource management	13
2.1	Provincial, territorial, and land-claims–based EA systems in Canada	18
2.2	Proportion of Canadian court decisions that include the term "environmental assessment," 2002 to 2017	31
2.3	Simplified overview of the basic steps involved in the federal impact assessment process under the Impact Assessment Act	32
3.1	Hierarchy of project alternatives	39
3.2	Arnstein's ladder of participation	48
3.3	Influence versus stake in outcome when identifying publics	53

List of Boxes, Boxed Features, Figures, and Tables   xi

4.1   Hybrid screening model   64
5.1   Past, present, and future baseline   72
5.2   Section of an impact identification matrix from the Tembec, Pine Falls Operations, Manitoba, Forest Stewardship Plan 2010–2029, depicting potential interactions of forest management activities with valued wildlife components   75
5.3   Section of a simple Leopold matrix to scope and communicate project impacts and characterize impact importance   76
5.4   Example of a valued component effect pathway   77
5.5   Right-of-way cleared for a liquefied natural gas pipeline development north of Kitimat, British Columbia   79
5.6   Cautionary, target, and critical thresholds   81
5.7   Spatial bounding based on local study area, regional study area, and broader context   83
6.1   Change in VC condition and predicted effect, with shaded area depicting range of normal variability   94
6.2   Determinants of impacts on human health and well-being   96
6.3   Predicting a project's climate risk   102
6.4   Permafrost thaw and slumping on Ellesmere Island, near the Eureka weather station   103
6.5   Sources of potential cumulative effects in a watershed   104
6.6   Predicted project and cumulative impacts   106
6.7   Scenarios in EA   111
6.8   Additive effects   115
6.9   Synergistic effects   116
6.10 Relationship between disturbance frequency, duration, magnitude, and VC recovery time and condition   119
6.11 Simple 3 x 3 risk classification matrix   120
6.12 Uncertainty matrix for communicating the location, level, and nature of uncertainty   122
7.1   Impact mitigation hierarchy   128
7.2   Adaptive management cycle   137
7.3   Timing of IBA negotiation in the EA process   140
8.1   Placer gold mine in the Indian River valley, Yukon, south of Dawson   152
8.2   Using range of normal or natural variability for setting acceptable thresholds   154
8.3   Vulnerability and irreplaceability as context for significance determination   155
8.4   Range of acceptability for significance determination   166
8.5   Decision tree for determining acceptability of impacts   167
9.1   Follow-up and monitoring in EA   172
9.2   Follow-up components   174
9.3   Control-impact monitoring   184
9.4   Gradient-to-background monitoring   184
11.1 Project- or activity-centred and valued component–centred approaches to effects   213
11.2 Components of a cumulative effects model   226
12.1 Example of assessment tiers for energy policies, plans, programs, and projects   240

12.2 Concentration of oil refineries, referred to locally as refinery low, east of the city of Edmonton, Alberta   240
12.3 Spectrum of approaches to SEA from less to more strategic   252
12.4 Generic SEA process components   259
12.5 Critical decision factors, objectives, and indicators   261

TABLES

1.1 Some of the many labels of impact assessment   4
1.2 Spectrum of EA philosophies and values   9
2.1 A brief history of federal EA in Canada   25
3.1 Key roles and responsibilities in pre-project planning   38
3.2 Example methods for comparing or evaluating alternatives   42
3.3 Basic characteristics of meaningful engagement in EA   49
3.4 Objectives of public engagement throughout the EA process   50
3.5 Selected techniques for public involvement and communication   54
4.1 Generic content of an EIS terms of reference   67
5.1 Example environmental components, key issues, VCs, and indicators from the Cold Lake oil sands project   78
5.2 Examples of condition- and stress-based indicators for terrestrial and aquatic VCs   78
5.3 Example of grizzly bear as a value and its assessment components, indicators, and management targets   81
6.1 Typical biophysical and human environment impacts of projects and examples of what to predict   95
6.2 Illustrative examples of the adverse impacts of climate change on projects in different sectors   101
6.3 Nature and sources of change that contribute to cumulative impacts or effects   107
6.4 Classification of environmental effects and impacts   113
8.1 Interpretations of significance in the EA process   148
8.2 Measurement and meaning of impact significance   149
9.1 Effectiveness and implementation monitoring based on risk and uncertainty about mitigation   178
9.2 Selected valued components and monitoring indicators for the Cold Lake oil sands project, Alberta   182
9.3 Selected biophysical monitoring components, parameters, and techniques   188
11.1 Characteristics of status quo thinking versus requirements for CEA   217
12.1 Definitions of strategic environmental assessment—past to present   239
12.2 Timeline of SEA development in Canada   244
12.3 Summary of 2015–18 federal SEA audits by the Auditor General of Canada, Commissioner of the Environment and Sustainable Development   247
13.1 Guideline standards for impact assessment practitioners   274
13.2 Guideline standards for impact assessment administrators   276

Preface

Environmental assessment (EA) in Canada is in a constant state of change. Although most changes seem to be incremental, they have largely been positive. Since the first edition of this book in 2006, federal EA legislation has been repealed and replaced twice: the Canadian Environmental Assessment Act, 2012 and the Impact Assessment Act, 2019. Several provincial and territorial jurisdictions have also introduced revisions to EA legislation and regulations, with British Columbia being the first to introduce legislation to implement commitments under the United Nations Declaration on the Rights of Indigenous Peoples. The period from 2006 to the present has also witnessed an increasing public awareness of the EA brand name—sometimes for good, as illustrated by increasing demand for more regional and strategic EA systems to tackle complex policy and cumulative effects challenges; sometimes not so good, as illustrated by the politicizing of EA and pipelines of misinformation about EA's impact on Canada's oil and gas sector.

Given this context, it can be quite challenging, if not impossible, to generate a text that captures the currency of EA legislation and regulations. This fourth edition was started as Bill C-69 was making its way to royal ascent and was completed in the months immediately following the 2019 federal election—both events meant much uncertainty about the future of federal EA. The first three editions of this text were tightly coupled with EA legislation. This fourth edition takes a slightly different direction, stepping back from legislation and focusing on the foundational principles and practices that characterize "good" EA—regardless of the prevailing legislation or jurisdiction of application. Of course, as a text about EA it is still important to introduce underlying legislative provisions, since this is what ultimately sets the minimum standard of practice. But the fourth edition does so only to illustrate the diversity of practice—not to critique or "teach" EA requirements.

This book is written for the student of EA—the current and future practitioner, regulator, and decision-maker.

Bram Noble
Professor, Department of Geography and Planning
University of Saskatchewan

Acknowledgements

I would like to acknowledge Oxford University Press for their assistance throughout the manuscript writing and editing process and the constructive input of peer reviewers. I am especially grateful to numerous friends and colleagues for their critical insights, reflections, and debates on EA over the years—you know who you are.

Aims and Objectives of Environmental Assessment

Introduction

The United Nations Environment Programme's International Resource Panel predicts that global material resource use in 2050 could be more than double that of 2015 levels (Bringezu et al., 2017). In Canada, four of the country's ecozones already are more than 50 per cent accessed (Cheng & Lee, 2014). **Human access** refers to the combined land surface anthropogenic disturbance caused by industrial activities, such as pipelines, petroleum and natural gas well sites, transmission lines, roads, mine sites, clear-cuts, and agricultural clearings. More than 25 per cent of Canada's land mass, or about 2.6 million km^2, is covered by at least one **industrial concession** or land tenure. This includes 58 million hectares of mineral claims, 2.1 million hectares under mineral lease, and nearly 500,000 km^2 of onshore petroleum and natural gas concessions (Cheng & Lee, 2014) (Figure 1.1).

Society derives significant benefit from natural resources. Canada's natural resource sector supports more than 1.82 million jobs, accounts for 17 per cent of the country's gross domestic product, and contributes more than $22 billion each year in government revenues. As of late 2018, there were 418 major resource development projects either under construction or planned over the next 10 years, valued at over $585 billion in investment (Natural Resources Canada, 2018). The *total* benefits derived from the natural environment, however, far exceed the economic benefits stemming from the extraction of natural resources—and include ecosystem provisioning, regulating, and cultural and supporting services that range from carbon sequestration and flood control to cultural value, recreation, and medicinal supply.

While the goal of land and resource use is typically positive change, it can also create significant and long-term adverse environmental and socio-economic consequences—often affecting the most vulnerable of communities and ecological systems. Since the early 1900s, for example, up to 70 per cent of wetlands in the Prairie region of Canada have been lost or degraded (Serran & Creed, 2016). Between 2000 and 2013, nearly 5 per cent of Canada's intact forest landscape was lost or degraded, with 92 per cent of this degradation occurring in areas containing species at risk (Smith & Cheng, 2016). The challenge facing governments, industry, Indigenous communities, and society at large is to find ways to support land use and development that enhances social and economic opportunity without adversely affecting the environment and the well-being of those who depend on it. The need for tools and processes to scrutinize development proposals, engage the public and affected communities, inform decision-makers, and manage impacts is of ever-increasing importance.

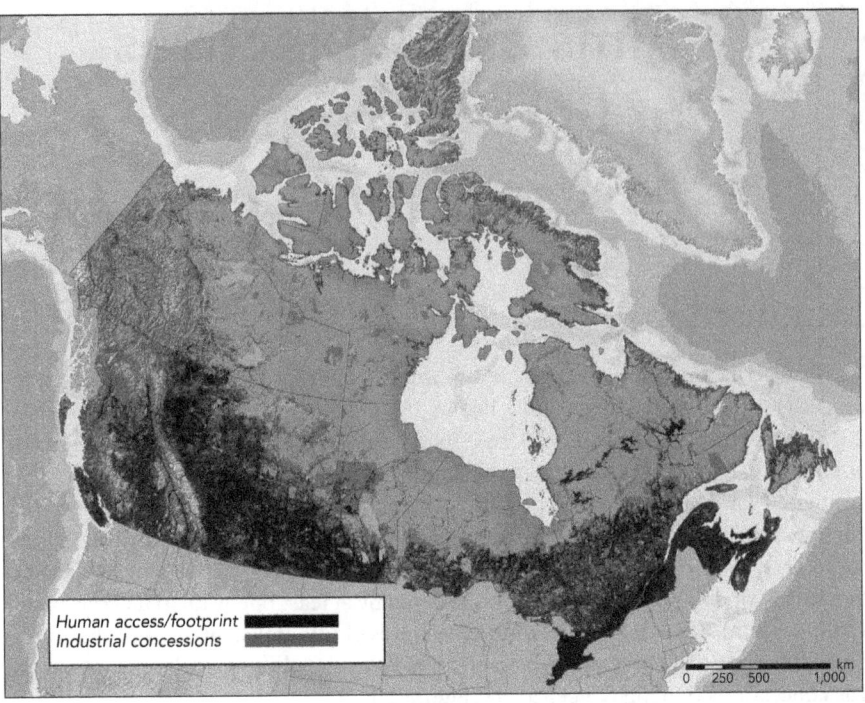

Figure 1.1 Human access and industrial concessions in Canada

Source: Compiled using Data Basin, Conservation Biology Institute, based on data provided by Global Forest Watch Canada, current to 2017

Environmental Assessment

Environmental assessment (EA), also referred to as impact assessment (IA) or environmental impact assessment (EIA), originated in the United States under the **National Environmental Policy Act** (**NEPA**) of 1969, which became law in 1970. Throughout North America and western Europe, the 1960s were characterized by a sudden growth in awareness of the relationship between an expanding industrial economy and local environmental change. While many characterize the 1960s as an era of idealism, triggered by a number of environmentally based legal challenges about pollution and pesticide use and sparked by such works as Rachel Carson's *Silent spring* (1962), the decade also led to increasing environmental awareness and public demand and pressure on central governments that environmental factors be explicitly considered in development decision-making. Because of such pressures, the 1960s and early 1970s witnessed the passage of several legislations concerning resource protection, hazardous waste management, and control of water and air pollution. Perhaps the most significant and influential legislation was NEPA. The term "environmental impact assessment" is derived from NEPA, which for the first time in the United States required by law that those proposing to undertake certain development projects had to demonstrate that the project would not significantly adversely affect the environment. To do so, those proposing to undertake development had to include in their proposals an **environmental impact statement (EIS)** describing the proposed undertaking, the affected environment, likely impacts, and actions to manage and monitor those impacts.

Canada was the first country to follow the US NEPA beginnings, formally implementing an EA system in 1973 as a guidelines order. It was not until 1995, however, that EA became entrenched in Canadian law, and numerous amendments have since taken place. Australia was next to formally adopt EA, through its Environmental Protection Act, formally implemented in 1975. As with Canada's provinces and territories, most states in Australia have opted for their own form of EA legislation, and projects of national significance are assessed under a joint state and federal system. EA was formally adopted in Europe through the 1985 European Union Directive 85/337/EEC, which established a legal basis for individual member states' EA regulations.

EA is also widely practised in many developing countries—either to comply with national EA provisions or to meet the requirements of international development aid agencies (Noble, 2011). EA was introduced early in some developing countries, including Colombia (1974) and the Philippines (1977). However, most developing countries established formal legislative bases for EA or introduced EA provisions into their existing environmental legislative frameworks later in the 1980s (e.g., Brazil, Indonesia, Mexico, Algeria, and Turkey) and particularly in the 1990s (e.g., Belize, Bolivia, Gambia, Mongolia, and Tunisia). EA legislation in many developing countries is high-quality, much like legislation in Canada or the United States. The main differences lie in implementation, enforcement, and government capacity. Since NEPA, EA has emerged as one of the most widely practised environmental management tools in the world (Box 1.1). Today, more than 190 member nations of the United Nations either have national legislation or have signed some form of international agreement that refers to the use of EA.

Defining EA

EA was initially conceived to ensure that the biophysical impacts of major development proposals were considered during decision-making. The scope of EA has broadened considerably over the years to include many other factors, such as social impacts, human health, culture, Indigenous rights, and gender-based analysis. Since NEPA, many different forms of EA have also emerged, including **social impact assessment (SIA)**, **health impact assessment (HIA)**, **strategic environmental assessment (SEA)**, and **sustainability assessment (SA)**, to name a few (Table 1.1). Morrison-Saunders et al. (2014) identified more than 40 different types of impact assessment. The emergence and development of EA as a key component of environmental management has coincided with the increasing recognition of the nature, magnitude, and implications of environmental and socio-economic change triggered by human actions, the distributions of costs and benefits associated with those actions, and the need to manage within ecological limits.

> **Box 1.1 The International Association for Impact Assessment**
>
> The International Association for Impact Assessment (IAIA), the leading global network on best practices in the use of impact assessment, has played a significant role in the development of EA globally. The IAIA was formed in 1980 to bring together practitioners, researchers, and other users of impact assessment from across the world. The IAIA has more than 1600 members from over 120 countries, with affiliate branches established in several countries, including Canada, Germany, New Zealand, Portugal, Zambia, Mozambique, and Ghana. The IAIA membership is diverse and includes planners, engineers, social and natural scientists from various disciplines, corporate managers, public interest advocates and environmental non-government organizations, government regulators and senior administrators, private consultants, and educators and students.
>
> The IAIA's mission is to provide an international forum for advancing innovation and communication of best practice in all forms of impact assessment to further the development of local, regional, and global capacity in impact assessment. This is achieved, in part, through annual international conferences where research and practice experience are shared and future directions for impact assessment discussed, a formal journal (*Impact Assessment and Project Appraisal*) to disseminate research focused on advancing impact assessment, and a variety of professional networking opportunities. The IAIA's website contains a variety of resources on different forms of impact assessment, many of which are available to non-members, including special publications on the principles of best practice in environmental, social, and health impact assessment and standards for impact assessment professionals. The website can be accessed at www.iaia.org.

There is no single, universally accepted definition of EA (Box 1.2). The International Association for Impact Assessment and the UK Institute of Environmental Assessment (IEA) define EA as: "The process of identifying, predicting, evaluating, and mitigating the biophysical, social, and other relevant effects of development proposals prior to major decisions being taken and commitments made" (IAIA & IEA, 1999). *Look before you leap* is the basic concept behind EA—it is an integral component of sound decision-making, serving both an information gathering and an analytical purpose to inform decision-makers and affected interests about the impacts of proposed developments and impact management solutions.

Table 1.1 Some of the Many Labels of Impact Assessment

sustainability assessment	risk assessment	technology impact assessment
life-cycle assessment	health impact assessment	economic impact assessment
poverty assessment	cultural impact assessment	ecological impact assessment
policy impact assessment	regulatory impact assessment	environmental impact assessment
social impact assessment	gender-based assessment	visual impact assessment

Box 1.2 Definitions of Environmental Assessment

- An activity designed to identify and predict the impact on human health and well-being of legislative proposals, policies, programs, and operational procedures and to interpret and communicate information about the impacts (Munn, 1979).
- The study of the full range of consequences, immediate and long-range, intended and unanticipated, of the introduction of a new technology, project, or program (Rossini & Porter, 1983).
- A planning tool whose main purpose is to give the environment its due place in the decision-making process by clearly evaluating the environmental consequences of a proposed activity before action is taken (Gilpin, 1995).
- The evaluation of the effects likely to arise from a major project (or other action) significantly affecting the environment. It is a systematic process for considering possible impacts prior to a decision being taken on whether or not a proposal should be given approval to proceed (Jay et al., 2007).
- A process to predict the environmental effects of proposed initiatives before they are carried out (CEAA, 2013).
- An assessment of the potential environmental, health, social, and economic impacts of proposed projects, including benefits, and potential impacts on Aboriginal treaty rights (IAAC, 2019).

EA is often described as an environmental protection tool, a methodology, and a regulatory requirement. Morrison-Saunders (2018) differentiates between two basic types of EA—*formal* and *informal*. Formal EA is grounded in a legal or regulatory process. There is a requirement that EA take place for certain development actions and that the process results in an approval decision that authorizes (or not) the proposed development and sets the terms and conditions under which the development may proceed. Informal EA, in contrast, may adopt the same procedural elements of formal EA, but its application is ad hoc and with no legal requirement that the process occur and no regulatory decision taken. This book is focused primarily on formal EA and its good-practice principles and guidance.

EA also has both a technical guise, providing and analyzing information about the potential impacts of development actions, and a regulatory or institutional guise, establishing the legal and procedural matters as they relate to engagement, laws, and project authorizations (IAIA, 2012). It is both an art and a science (Morrison-Saunders, 2018) through which concerns about the potential environmental or socio-economic consequences of proposed actions, public or private, are analyzed and incorporated into decisions regarding those actions. In this regard, EA not only informs decision-makers about whether a proposed development is in the public and environmental interest, it also strengthens the environmental management process that follows the consent decision for development (Morrison-Saunders & Bailey, 1999).

EA is not simply a means to secure a development permit, but it is also not a mechanism for preventing all development that might cause adverse impacts. If all projects with potentially adverse impacts were rejected, then few projects would proceed. Most projects subject to EA are approved—even in cases where the effects are deemed adverse and significant. Consider Teck Resources' Frontier oil sands mine, Alberta—one of the largest oil sands mines ever proposed, at more than 24,000 hectares and with a peak production capacity of 260,000 barrels of bitumen per day (Riley, 2019). The EA review panel recommended that the project was in the public interest and should be approved, notwithstanding the panel's conclusion that the project would cause irreversible environmental impacts, including permanent destruction of fish habitat, wetlands, and areas of high species biodiversity (JRP, 2019). Teck withdrew its application in early 2000, however, citing uncertainties around climate change policy and market conditions as impacting the project's viability. In a recent analysis of projects subject to review under the Canadian federal EA system, Orenstein (2018) reports that 95 per cent of projects that completed the EA process were approved; even for those projects where significant adverse effects were identified, 73 per cent were still approved.

The EA Process

Notwithstanding the diversity of EA definitions and its adaptations to different international contexts, the core elements of EA are generally agreed upon (Jay et al., 2007). From an applied perspective, EA systematically examines the potential environmental implications of development actions prior to their approval. While not all EA systems contain the exact same elements or design features, the general process emanating from NEPA and subsequently diffused throughout the world is typically comprised of a series of systematic steps (Box 1.3).

Presented here as a linear process, EA is often quite iterative. Public engagement occurs early and ideally at all stages of EA, and post-project evaluations and adaptive management continue to refine project design, impact predictions, and impact management strategies. The time required to complete an EA for any given project is highly context-specific, often depending on the regulatory requirements, the complexity of the proposed development, the uncertainty about appropriate mitigation measures, and the level of political interest or conflict that might surround the project. For example, an EA for a proposed nuclear power plant or major energy pipeline may take considerably longer and be far more complex than an EA for a new forest access road or even an expansion to an existing mine site operation. Reviewing federal EA processes, which typically include major resource development undertakings, Orenstein (2018) found that the EA process for most projects lasted on average 3.5 years before a decision was reached, though some lasted more than 10 years.

Purpose and Objectives of EA

EA is about identifying and evaluating the potential impacts of proposed development actions, proposing strategies for managing those impacts, and ensuring that development proceeds in a manner that is in the public interest. When applied to development projects, the intent of EA in the *short term* is to ensure that environmental

1 | Aims and Objectives of Environmental Assessment

Box 1.3 Components of the EA Process

Pre-project planning	The proponent consults with potentially affected communities about the project's purpose, potential benefits, and impacts, and develops a plan for community engagement. The proponent may negotiate private agreements with communities that include benefits-sharing arrangements. Governments engage in consultations with communities about land use, including land titles.
Project description	The proponent prepares a project description, including a description of the project's need. As part of the project description, the proponent may be required to show that alternatives to the project, and alternative means of carrying out the project (e.g., design, location) have been considered.
Screening	A determination is made as to whether an EA is required under the regulations or guidelines and, if so, who is responsible and what type of assessment is required. In some jurisdictions, EA is triggered only for large projects or for projects for which significant adverse effects are likely. In other jurisdictions, even small or routine developments are subject to assessment.
Scoping and baseline assessment	If an EA is required, the key issues to be included and the spatial and temporal boundaries of the assessment are determined. This is an iterative exercise involving the regulatory authorities, the proponent, and affected communities and other interests. Baseline data are collected to identify the condition of potentially affected components, and trends or changes in those components are assessed.
Impact assessment	The project's potential impacts are predicted and characterized, based on trends, scenarios, scientific and local knowledge, and previous projects and experience.
Identifying strategies for managing impacts	Strategies are identified to manage potential impacts, ranging from avoidance and mitigation to compensating for the impact. Environmental management or protection plans and adaptive management programs are formulated.
Significance determination	Potentially significant adverse impacts are identified after considering impact management strategies, uncertainties, the vulnerability of the affected component, and applicable thresholds, sustainability objectives, or regulatory standards.
Submission and review of the EIS	The EIS is prepared, including related technical documents and reports, and submitted for technical and public review. Typically, the EIS is prepared by the proponent or a team of consultants on behalf of the proponent, though in some jurisdictions, for small projects the EIS is prepared by government. The EIS presents the findings of the EA. The nature and transparency of the EIS review varies by jurisdiction and by the complexity of the proposed undertaking.

Consultation, participation, and engagement

continued

Recommendations and decision	Recommendations are made to a decision-maker, usually the responsible minister, and a decision is made as to whether the proposed undertaking should proceed and, if so, under what conditions. This involves some consideration of the project's net benefits and whether any adverse effects are justified. In some cases, public hearings are required or additional information gathered before a decision is issued.
Implementation and follow-up	The EA process does not end once a decision is made that a project is approved. Impact management strategies are applied and monitoring programs implemented to determine compliance with the conditions of approval, to determine the effectiveness of impact management, and to implement adaptive management measures where needed.

and socio-economic factors are considered in decision-making; to improve the design of proposed undertakings; to anticipate, avoid, minimize, and offset adverse effects; to ensure a proponent's accountability and compliance with laws and regulations; and to provide fair and meaningful opportunities for participation in the development process (Noble & Udofia, 2015). In the *long term*, EA is one of several public policy tools to regulate development and promote sustainable resource use.

At the most basic level, the immediate result of EA is a development proposal that is a well-thought-out environmental, social, economic, and technical design accompanied by a management plan to deal with potential risks and unanticipated outcomes (Leadem, 2013). Cashmore, Gwilliam, Morgan, Cobb, & Bond (2004) suggest that the potential for EA to contribute to sustainable development may be widely underestimated. Empirically and conclusively proving its contribution(s) to sustainable development is difficult, but Cashmore et al. argue that EA, over time, contributes to incremental changes in several areas that can ultimately direct the course of development to achieve more sustainable outcomes. These incremental changes or influences of EA cut across several domains, including:

- *Scientific change* – promoting interdisciplinary approaches to addressing complex environmental problems and improving predictive techniques.
- *Design and engineering change* – informing design decisions and directing project designs that account for environmental capacity.
- *Corporate change* – ensuring accountability and corporate social responsibility.
- *Stakeholder empowerment* – increasing awareness of development impacts, building knowledge and capacity, and providing an opportunity for a voice in decisions.
- *Institutional change* – influencing consent decisions about development that are informed by science, knowledge, and values and providing for an accountability framework for decisions taken.

Expectations

What EA can or should deliver often depends on the lens through which it is viewed. For example, members of a local community about to be adversely affected by a development may see EA as a public relations tool used by developers and politicians to justify decisions already made, or they may see it as a way to ensure the accountability of developers and provide a means of negotiating consensus solutions such that local concerns are considered in the development process. Non-government environmental organizations may view EA as a tool to improve stakeholder involvement in development decision-making, or they may see it as a means of preventing development from proceeding but as a "rubber stamp" when it is unsuccessful in doing so. A proponent may view EA as a time-consuming and expensive regulatory hurdle that must be overcome to receive development approval or as a means of improving project design, earning a social licence to operate, and contributing to the social and economic well-being of potentially affected communities.

Cashmore (2004) suggests a broad spectrum of philosophies and values about EA (Table 1.2). At one end of this spectrum is the perception that the scientific method provides the basis for EA theory and practice. To be credible, EA must be based on scientific objectives, modelling and experimentation, quantified impact predictions, and hypotheses-testing. The results must lead to new scientific understanding. At the other end of the spectrum is the belief that EA is a decision tool used to empower stakeholders and promote an egalitarian society, with a strong green interpretation of sustainability. In this regard, EA must be deliberative, promote social justice, and help to achieve community self-governance.

Understanding these different values and expectations is essential to ensuring a meaningful, credible, and efficient EA process. For example, if a special interest group is fundamentally opposed to a certain type of development (e.g., a bitumen pipeline), no amount of participation or collaboration in the EA process is likely to change that perspective. Any bitumen pipeline approved under the EA process will likely be seen as an EA failure. The scientific community may equally be dissatisfied

Table 1.2 Spectrum of EA Philosophies and Values

Applied science ↑↓	*Analytical science:* EA serves to inform decisions and enhance scientific understanding. Science in EA is applied, experimental, and naturalistic.
	Environmental design: EA serves to inform and influence design decisions. Science in EA is applied environmental science for design and engineering.
	Information provision: EA serves to inform decisions. Science in EA is largely based on the natural sciences, with limited role for the social sciences.
Civic science ↑↓	*Participation:* EA is about participatory decision-making. There is an extensive role for both the natural and the social sciences.
	Environmental governance: EA is about deliberative democracy. Science in EA is largely based on the social sciences, with limited role for the natural sciences.

Source: Based on Cashmore, 2004

with EA when it does not live up to the expectations of robust scientific design and advancing the state of science—even though under most regulatory systems the EA process is neither designed, nor intended to operate, as a tool for scientific research and discovery.

There is no "best" model that characterizes EA, and practitioners should treat these as reference points rather than recipes (Morgan, 2012). In practice, formal EA typically functions somewhere in between the information provision and participation models. EA is an information-gathering, assessment, and provision exercise to ensure that those responsible for development decisions can make informed decisions. Information is comprised of science and technology and the knowledge gained through the participation of affected communities, Indigenous peoples, interests, and other experts. EA functions as a mechanism to ensure the accountability of developers and regulators and provides a means to negotiate solutions such that the concerns of those potentially affected by development are accounted for in the planning, decision-making, development, and impact management process.

Who's Who in the EA Process

There are many different actors involved in EA, each bringing important knowledge and experience to the process. Although the specific actors vary from one context to the next depending on the jurisdiction, the nature and location of the proposed development, and the stakes involved, the main actors typically include:

Project proponents – The party that proposes the development and, ultimately, is responsible for its implementation, operations, and impact management. Proponents may be private corporations, government entities, or private–public partnerships.

Regulators or government authorities – The government agency or agencies responsible for the EA (and related regulatory processes) and ensuring due process. This usually includes setting the terms of reference for what the EA must consider, issuing approvals and authorizations, ensuring opportunities for public participation, fulfilling legal obligations for consultation with Indigenous peoples, and following up post–project approval to ensure compliance with the terms and conditions of project approval.

Decision-makers – Although various parties can make recommendations in the EA process, including government-appointed review panels, the ultimate decision-maker in most cases is the responsible elected minister, such as a minister of environment, who weighs the information provided by the proponent and other affected interests, considers government policy objectives or commitments, and determines whether it is in the public (and environmental) interest that a project proceed.

Affected interests – Anyone or any group potentially affected by a proposed development. This may include Indigenous governments or communities (i.e., often considered rights-holders), the public, (e.g., affected communities or interested individuals), private landowners (e.g., agricultural landowner, woodlot owner), private businesses or corporations, and special interest groups (e.g., environmental organizations, industry consortiums, lobby groups). Not all affected interests are involved in the EA process in the same capacity or share the same legal standing.

Practitioners – The consultants who carry out the technical analysis and compile the EIS on behalf of the proponent. Practitioners also work for communities, government agencies, and various review boards, panels, or commissions involved in the EA process. In some jurisdictions, EA is carried out by a branch of government, especially for more routine development proposals, in which case practitioners are government staff.

The Honest Broker

Meeting regulatory or industry standards is a necessary but minimum responsibility of the EA practitioner. A good practitioner will recognize that the minimum regulatory standard is sometimes an unacceptable standard for vulnerable ecosystems or for those directly affected by a project's activities. However, the EA process is adversarial. This is particularly the case for large development projects with the potential for significant adverse impacts. For example, it is often the role of practitioners working for a project proponent to present the strongest possible technical case for a project; it is the role of other practitioners, perhaps hired by those wishing to challenge the project, to identify the weaknesses in the proponent's technical position. Practitioners thus have choices in how they interact with project proponents, the public, decision-makers, and the EA process.

Pielke (2007) proposes four different roles for experts in decision-making, which are applicable to practitioners and other experts involved in EA. The first role is the *pure scientist* for whom the focus is only on the facts. Information is gathered, scientifically analyzed, and the results are presented in the absence of value-based interpretations to "speak for themselves." The second role is the *science arbiter*, or one who answers specific and factual questions that may be posed during an EA, but the information presented is limited to those facts that are relevant to the specific question at hand. Third is the *issue advocate*, the individual who seeks to reduce the scope of information available by presenting or interpreting information in a certain way—to try and influence the decision.

Finally, the *honest broker* is one who seeks to expand and clarify the nature and scope of information and options available such that a decision-maker, a proponent, and the public are more aware of the potential implications of different decision actions. As an arena for public debate about the acceptability of a proposed development, the EA process benefits from all four roles; however, practitioners must be honest brokers—upholding ethical conduct and ensuring the free flow of complete, unbiased, transparent, and accurate information to decision-makers, proponents, and affected interests.

Getting the Big Picture

The focus of project EA is on the impacts and regulatory and permitting processes associated with a single project proposal (WCEL & NWI, 2016). Many of the important decisions that determine the trajectory of development—such as whether mining is even appropriate for a particular region, what is an appropriate level of disturbance, or whether petroleum exploitation is consistent with Canada's climate commitments—are decisions that are (or at least should be) made long before development projects are proposed and EA processes triggered.

Project EA is intended to function in a much larger and integrated system of policy, planning, and decision-making (Figure 1.2). The Council of Canadian Academies (2019) expert panel on integrated natural resource management describes this system as underpinned by treaties and legislation that shape policies and regulations, which provide the context for land-use plans whereby specific land-use goals and longer-term priorities are identified, including land-use zoning or designations. Land-use plans inform regional and strategic EAs (see Chapters 11 and 12), which focus on the **cumulative effects** of alternative land uses, the spatial organization and pace of development programs, and setting environmental standards or targets. This provides direction for what types of projects are acceptable, when, and where, and shapes and simplifies project EA processes. Licensing and permitting follow, whereby compliance is assured. Monitoring, evaluation, participation, and learning occur across this continuum, ensuring that environmental objectives and targets are being achieved and developing new strategies as required.

This is an ideal model. In practice, there are significant disconnects between land-use planning, where it exists, environmental monitoring, and project EA (Council of Canadian Academies, 2019; Cronmiller & Noble, 2018). As a result, many issues emerge when projects are proposed that are well beyond the scope of a single project review. For example, in 2014 Spectra Energy (now Enbridge Inc.) submitted an EA application to the British Columbia Environmental Assessment Office to develop the Westcoast Connector Gas Transmission Project—an 850-kilometre natural gas pipeline from northeast British Columbia to Ridley Island. The project would traverse the traditional territory of the Blueberry River First Nation. In a letter to the province's minister of the environment, the First Nation asked the minister not to approve the pipeline (Yahey, 2014). The First Nation expressed concern about the project's impacts, but the greatest concerns were about the history of development in their traditional territory. Between 2012 and 2016, more than 2600 oil and gas wells, 1884 kilometres of petroleum access and permanent roads, 1500 kilometres of new pipelines, and 9400 kilometres of seismic lines had been authorized for development in the territory (Macdonald, 2016). Concerns about the project were tightly coupled with concerns about the prospects of future LNG activity, including **induced developments**, the upstream impacts of hydraulic fracturing and natural gas exploration, and the desire for a more comprehensive approach to direct land use and development in the traditional territory (Noble, 2017).

Project EA is not well equipped to deal with these bigger-picture issues, including sector-wide development and regional land-use planning. A major challenge to current EA practice in most jurisdictions, however, is that there is often no other

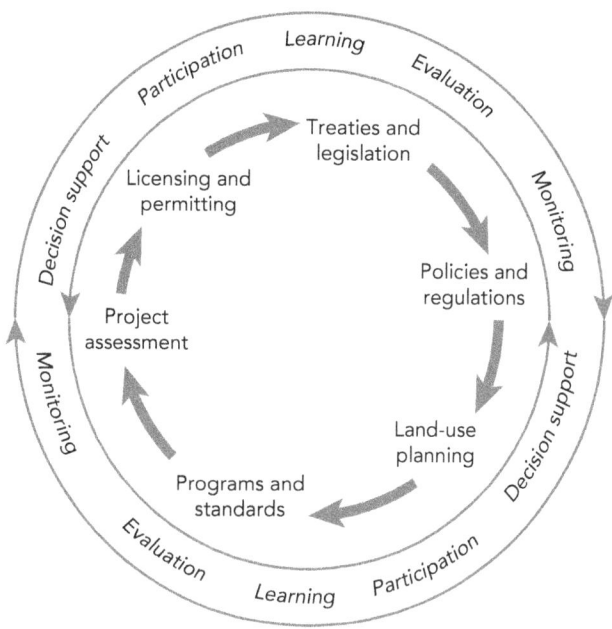

Figure 1.2 Project EA as part of an integrated system of policy, land-use planning, and natural resource management

outlet to address these issues before projects are proposed. Because of this, the EA process is left to grapple with issues and drivers that are far larger than any one project (Hegmann & Yarranton, 2011).

What to Expect

This book is about the foundational principles of EA and supporting good practices. It is written for the current and future practitioner, the regulator, the decision-maker, and the community member or interest group. The principles and practices addressed in this book are not grounded in any single regulatory system, since EA rules and regulations vary considerably across Canada and indeed globally. This book is about what is needed to do the right things in EA and how to do things right. The term EA is adopted throughout the book, reflecting the dominant terminology used internationally and across Canadian provincial and territorial jurisdictions. This includes a holistic understanding of environment that is inclusive of *both* physical and human systems and their interactions.

Chapter 2 provides an overview of EA systems in Canada—systems that are constantly evolving, especially at the federal level, and have likely changed even since the time this text was written. The chapters that follow explore the basic components of a typical EA process. Emphasis is on project-level assessment and good-practice principles. Attention then turns to some of the more complex issues in EA, including cumulative effects and strategic-level assessments. The final chapters step back from the components of EA to address ethical practice and to reflect on the future of EA.

There are four important points to remember about EA when using this book and when practising or seeking new ways to improve EA. *First*, for many people, their views about EA are shaped by what they have experienced (or read about on social media)—which, unfortunately, is heavily skewed by a few high-profile and conflict-ridden cases, such as the rejection of major energy projects or court challenges that garner significant political attention. For most projects, EA unfolds without incident and without much media interest. *Second*, EA is not a "magic bullet" for all environmental and social challenges—it is not a scientific research project, and it is not about policy formation. EA is a critically important environmental management and decision support tool, but it is designed to complement other environmental policy, planning, and management instruments—not to replace them (Fuggle, 2005). *Third*, sometimes a community or interest group is philosophically opposed to a project or an entire resource sector, no matter what the distribution of benefits and impacts. Any EA, no matter how good the process, that results in project approval will thus be considered a "bad" EA. *Fourth*, sometimes decision-makers just make the *wrong* choices about a proposed development, regardless of the quality and comprehensiveness of the evidence provided by the EA process. Poor or unpopular decisions can result despite good EA.

Key Terms

cumulative effects
environmental assessment (EA)
environmental impact statement (EIS)
health impact assessment (HIA)
human access
induced development
industrial concessions
National Environmental Policy Act (NEPA)
social impact assessment (SIA)
strategic environmental assessment (SEA)
sustainability assessment (SA)

Review Questions and Exercises

1. What are the potential benefits of conducting an EA?
   a. For communities affected by development?
   b. For proponents undertaking development?
2. How can EA contribute to sustainable development?
3. Obtain a small sample of EISs from your local library or government registry, or access them online. Examine the table of contents of each, and create a general list of common elements or issues addressed in each assessment. Are the contents similar in terms of the types of issues and topics addressed?
4. What requirements currently exist in your country or province for conducting EA? When were these requirements implemented, and how have they evolved over time? Is the environment defined to include both physical and human systems? Who is responsible for conducting EA? How many assessments have been completed to date? Is there a sector or type of EA that dominates? Why might this be so?

5. EA has evolved to become one of the most widely applied environmental management tools in the world. Are there certain ethical, cultural, or other elements that should be considered when applying EA in developed versus developing nations? What elements of the process, if any, might vary from one socio-political or cultural context to the next?
6. What land-use and natural resource issues have dominated the media in your country or region in recent months? What has been said about EA? Based on your understanding of the issue and knowledge of EA, are the expectations about EA aligned with the role or purpose of EA?

References

Bringezu, S., Ramaswami, A., Schandl, H., O'Brien, M., Pelton, R., Acquatella, J., . . . & Zivy, R. (2017). Assessing global resource use: a systems approach to resource efficiency and pollution reduction. Nairobi, Kenya: International Resource Panel, United Nations Environment Programme.
Carson, R. (1962). *Silent spring*. Boston: Houghton Mifflin.
Cashmore, M. (2004). The role of science in environmental impact assessment: process and procedure versus purpose in the development of theory. *Environmental Impact Assessment Review* 24, 403–26.
Cashmore, C., Gwilliam, R., Morgan, R., Cobb, D., & Bond, A. (2004). The interminable issue of effectiveness: substantive purposes, outcomes and research challenges in the advancement of environmental impact assessment theory. *Impact Assessment and Project Appraisal* 22(4), 295–310.
CEAA (Canadian Environmental Assessment Agency). (2013). Basics of environmental assessment. www.ceaa.gc.ca.
Cheng, R., & Lee, P. (2014). Canada's industrial concessions: a spatial analysis. GFWC Bulletin. Ottawa, ON: Global Forest Watch Canada.
Council of Canadian Academies. (2019). *Greater than the sum of its parts: toward integrated natural resource management in Canada*. Ottawa, ON: The Expert Panel on the State of Knowledge and Practice of Integrated Approaches to Natural Resource Management in Canada.
Cronmiller, J., and Noble, B.F. (2018). Integrating environmental monitoring with cumulative effects management and decision making. *Integrated Environmental Assessment and Management* 14(3), 407–17.
Fuggle, R. (2005). Have impact assessments passed their "sell by" date? *Newsletter of the International Association for Impact Assessment* 16(3), 1, 6.
Gilpin, A. (1995). *Environmental impact assessment: cutting edge for the 21st century*. Cambridge, UK: Cambridge University Press.
Hegmann, G., & Yarranton, G.A. (2011). Alchemy to reason: effective use of cumulative effects assessment in resource management. *Environmental Impact Assessment Review* 31(5), 484–90.
IAAC (Impact Assessment Agency of Canada). (2019). Impact assessment process overview. https://www.canada.ca/en/impact-assessment-agency/services/policy-guidance/impact-assessment-process-overview.html.
IAIA (International Association for Impact Assessment). (2012). Impact assessment. *Fastips* 1. Prepared by M. Partidário, L. den Broeder, P. Croal, R. Fuggle, and W. Ross. Fargo, ND: IAIA.
IAIA & IEA (International Association for Impact Assessment & UK Institute of Environmental Assessment). (1999). *Principles of environmental impact assessment best practice*. Fargo, ND: IAIA.

Jay, S., et al. (2007). Environmental impact assessment: retrospect and prospect. *Environmental Impact Assessment Review* 27, 287–300.

JRP (Joint Review Panel Established by the Federal Minister of Environment and Climate Change and Alberta Energy Regulator). (2019). *Report of the Joint Review Panel Established by the Federal Minister of Environment and Climate Change and the Alberta Energy Regulator Decision 2019 ABAER 008: Teck Resources Limited, Frontier Oil Sands Mine Project, Fort McMurray Area.*

Leadem, T. (2013). Environmental assessment in Canada and Aboriginal law: some practical consideration for navigating through a changing landscape. Aboriginal Law Conference, Paper 1.2. Vancouver, BC.

Macdonald, E. (2016). Atlas of cumulative landscape disturbance in the traditional territory of the Blueberry River First Nations. Victoria, BC: Ecotrust Canada.

Morgan, R. (2012). Environmental impact assessment: the state of the art. *Impact Assessment and Project Appraisal* 30(1), 5–14.

Morrison-Saunders, A. (2018). *Advanced introduction to environmental impact assessment.* Cheltenham, UK: Edward Elgar.

Morrison-Saunders, A., & Bailey, J. (1999). Exploring the EIA/environmental management relationship. *Environmental Management* 24(3), 281–95.

Morrison-Saunders, A., Pope, J., Gunn, J., Bond, A., & Retief, F. (2014). Strengthening impact assessment: a call for integration and focus. *Impact Assessment and Project Appraisal* 32(1), 2–8.

Munn, R.E. (1979). *Environmental impact assessment: principles and procedures.* New York, NY: Wiley.

Natural Resources Canada. (2018). 10 key facts on Canada's natural resources. Ottawa, ON: Natural Resources Canada.

Noble, B.F. (2011). Environmental impact assessment. In *Encyclopedia of life sciences.* Chichester, UK: John Wiley & Sons. doi: 10.1002/9780470015902.a0003253.pub2.

Noble, B. (2017). *Getting the big picture: how regional assessment can pave the way for more inclusive and effective environmental assessments.* Ottawa, ON: Macdonald-Laurier Institute.

Noble, B.F., & Udofia, A. (2015). *Protectors of the land: toward an environmental assessment process that works for Aboriginal communities and developers.* Aboriginal People and Environmental Stewardship Series. Ottawa, ON: Macdonald-Laurier Institute.

Orenstein, M. (2018). What now? The fate of projects: a review of outcomes from the federal EA approval process. Policy Brief. Calgary, AB: Canada West Foundation.

Pielke, R.A. (2007). *The honest broker: making sense of science in policy and politics.* Cambridge, UK: Cambridge University Press.

Riley, S.J. (2019). Why the proposed Frontier oilsands mine is a political hot potato. *The Narwhal* 12(12). https://thenarwhal.ca.

Rossini, F.A., & Porter, A.L. (1983). *Integrated impact assessment.* Social Impact Assessment Series. New York, NY: Perseus Books.

Serran, J.N., & Creed, I. (2016). New mapping techniques to estimate the preferential loss of small wetlands on prairie landscapes. *Hydrological Processes* 30(3), 396–409.

Smith, W., & Cheng, R. (2016). *Canada's intact forest landscapes updated to 2013.* Ottawa, ON: Global Forest Watch Canada.

WCEL & NWI (West Coast Environmental Law & Northwest Institute). (2016). *Regional strategic environmental assessment for northern British Columbia: the case and the opportunity.* Vancouver, BC: West Coast Environmental Law and Northwest Institute.

Yahey, M. (2014). Letter from Chief Marvin Yahey, Blueberry River First Nations, to Mary Polak, Minister of Environment, Government of British Columbia, regarding the Spectra Energy's Westcoast Connector gas transmission project environmental assessment. 28 October. https://projects.eao.gov.bc.ca/p/westcoast-connector-gas-transmission/docs.

2 Environmental Assessment in Canada

Overview of Environmental Assessment in Canada

The Canadian Constitution assigns certain responsibilities to each level of government, but environment is not explicitly noted in the Constitution. Provinces own the majority of resource rights in Canada and have power to make laws dealing with the development, conservation, and management of nonrenewable resources and forestry resources (Thompson, 2013). Cotton and Zimmer (1992) explain that the federal government derives its authority to legislate with respect to environmental matters from several sections of the Constitution Act—namely, those that deal with criminal law, navigation and shipping, federal lands, fisheries, taxation, and Indigenous reserve lands. Of equal importance is the power of the federal government to enact laws for the "peace, order, and good government of Canada," which allows federal legislation on several matters of national concern—including environmental matters (Cotton & Zimmer, 1992).

EA in Canada is enshrined in the law of the provinces, territories, Indigenous governments, and the federal government (Figure 2.1). Canada is recognized internationally as a nation that has contributed significantly to the development and advancement of EA law, policy, and practice. Since its inception in the early 1970s, thousands of EAs have been completed in Canada, most of them for relatively small-scale, routine development projects and undertakings. As of mid-2020, there were more than 170 active EAs at the federal level for various resource projects across Canada and hundreds more under provincial and territorial EA systems. This chapter presents a brief overview of Canadian EA and snapshots of jurisdictional EA systems.

Provincial EA Systems

Ontario was the first province in Canada to legislate EA, in 1975. All provinces and territories now have some form of EA process, whether under broad environmental or resource management law, such as New Brunswick's Clean Environment Act and Newfoundland and Labrador's Environmental Protection Act, or under EA-specific law, such as Saskatchewan's Environmental Assessment Act. In 1998, to facilitate coordination of EA applications and responsibilities between the government of Canada and the provinces and territories, the Canadian Council of Ministers of the Environment signed an accord to improve cooperation on EA. The purpose of the accord is to streamline the EA process and eliminate duplication between federal

18 Introduction to Environmental Assessment

Figure 2.1 Provincial, territorial, and land-claims–based EA systems in Canada
Source: Map produced by J. Martin, University of Saskatchewan

and provincial or federal and territorial processes in cases where a proposed project could potentially trigger dual EA systems—referred to as a "one-window approach."

Harmonization does not transfer jurisdictional authority from the federal government to the provinces or territories but rather ensures that federal and provincial or territorial approval processes are coordinated such that a project proponent does not need to present the same impact statement twice or in two different formats (Meredith, 2004). Notwithstanding harmonization provisions, however, the coordination of EA regulatory requirements among jurisdictions remains a challenge (Fitzpatrick & Sinclair, 2009). Further, it is possible, and it has happened, that conflicting decisions can be reached for projects subject to EA review under both federal and provincial processes. In 2010, for example, Taseko Mines' proposed Prosperity gold-copper mine in British Columbia was issued an environmental certificate of approval by the province but rejected under the federal EA process because of concerns about significant adverse effects to fish and fish habitat that could not be mitigated.

British Columbia

EA in British Columbia was first introduced by way of policy guidelines, with assessment processes developed separately for different types of projects—such as coal mines and energy projects (Rutherford, 2016). The province's first comprehensive EA legislation was introduced in 1995 under the Environmental Assessment Act, coming into force in 1996. The act was repealed and replaced in 2002 with a new Environmental Assessment Act. Projects subject to EA in British Columbia are defined in the province's Reviewable Projects Regulation, which lists projects and project thresholds (e.g., project size or production capacity) that will trigger assessment. The EA process applies only to "major projects," meaning that smaller projects that fall under the size threshold, such as clusters of individual run-of-river hydroelectric projects, do not trigger EA review. The Government of British Columbia recently announced a province-wide initiative to better capture the effects of these types of developments under a province-wide cumulative effects framework (see Chapter 11). The province is also undergoing an EA revitalization strategy to enhance public confidence and transparency in EA and to advance reconciliation with First Nations. Other proposed improvements include funding for Indigenous peoples to participate in EA, strengthening the independence of the EA process through expert and peer reviews, and enabling regional assessment in legislation. Bill 51—the Environmental Assessment Act—recently passed in the legislature to implement these and other changes.

Alberta

Unlike British Columbia's EA legislation, Alberta's Environmental Protection and Enhancement Act, introduced in 1993, consolidates environmental legislation in the province, meaning that EA is only one part of the act. The consolidation was, in part, an effort to streamline regulatory processes so as not to hinder development. Hanna and Parkins (2016) suggest that whether this approach has resulted in better environmental protection is debatable. Whether a project will undergo EA is determined in part by the complexity of the project, the potential for significant impacts, and the scale of the undertaking. Regulations under the act identify projects that are always subject to EA, such as pulp and paper mills, sour gas processing plants, oil sands mines, and coal mines of a certain production capacity and projects that are exempt from EA, including small-scale operations such as drilling an oil well. Inclusions and exemptions are usually determined based on the size of the project operations, such as production capacity or emissions, or technology. A long-standing issue in Alberta is the effectiveness of EA application in the oil sands industry. The Alberta oil sands region is one of the most heavily monitored regions in Canada. Alberta does have a comprehensive land-use planning system, but the connections between land-use planning, long-term environmental monitoring, and EA decisions remain fragmented.

Saskatchewan

Saskatchewan first introduced EA in 1971, with a formal EA policy established in 1976. Between 1976 and 1980, more than 150 project proposals were reviewed under the EA policy and more than 50 EAs conducted (Westbrook & Noble, 2016). It was also during this time that the province, in collaboration with the federal government, initiated the Bayda Commission Cluff Lake Board of Inquiry—a landmark assessment of the environmental, social, and cultural impacts of its uranium mining industry and exploration of whether the province should even pursue uranium mining in the North. The first EA legislation in Saskatchewan was passed in 1980, the Environmental Assessment Act. Saskatchewan does not have a list of projects that require EA; rather, all project proposals are considered on a case-by-case basis for their potential impacts and based on public concern. This means that projects of any size or characteristic may be subject to assessment—it also means that they can be exempt. On average, about 30 per cent of projects submitted to the province undergo a full EA review—most of these are mining-related, followed by transportation. The only activity in Saskatchewan for which EA is always required is 20-year forest management planning.

Manitoba

The Environmental Assessment and Review Process was adopted as Manitoba provincial policy in 1975. It was not until 1988 that EA in Manitoba was entrenched in legislation, under the Manitoba Environment Act. The intent of the act was to develop and maintain an environmental management system in Manitoba. EA is one part of this environmental management system, but it is linked directly to the project licensing process (Lobe & Sinclair, 2016). Like Saskatchewan, the triggers for EA in Manitoba are broad and include such factors as whether a proposed activity will affect rare or endangered features of the environment, substantially use or alter a resource, or have a significant effect. Activities that meet these criteria are termed "developments," which fall into different classes that then determine the nature and scope of EA required. Classes are defined based on the nature of the potential effects, including their magnitude and public concern. The act also established the Clean Environment Commission—an arm's-length provincial agency responsible for conducting investigations and holding public hearings for development proposals. Recent reports by the Clean Environment Commission related to hydroelectric development in northern Manitoba have expressed concerns about the ability of the EA process to capture the legacy effects of development and emphasize the need for complementary regional assessment (Noble, Liu, & Hackett, 2017).

Ontario

The Ontario Environmental Assessment Act came into force in 1975. The act was restrictive in its application, applying only to Ontario government department and agency undertakings. In 1977, the act was expanded to include other public sector activities, such as municipal infrastructure and conservation authorities but was rarely applied to private sector activities. Of concern during the early years of Ontario's EA process was its inefficiency and high cost. Hostovsky and Graci (2016) report that EAs were widely criticized by proponents as being too time-consuming and costly, often millions of dollars over several years. After much criticism and the rejection of several major high-profile infrastructure projects, EA underwent a major reform in 1996 to reduce the "red tape" associated with development. Included among the reforms was the introduction of a class EA process. Only a small number of projects in Ontario now undergo an individual EA. A class EA document sets out a standardized planning process that covers the routine activities, including mitigations, associated with certain types or categories of developments. A proponent who receives approval for an undertaking within a class does not need to undergo a separate EA approval. Included among the class EAs are provincial transportation facilities (e.g., new highways), flood and flood erosion control projects (e.g., water diversion), waterpower projects (e.g., dams, powerhouses), and certain activities under the province's Mining Act (e.g., rehabilitation). Other types of projects, such as renewable energy projects, previously subject to class EA, are no longer included; they are assessed and regulated under the Renewable Energy Approvals regulation.

Quebec

The Quebec EA context is different from that of other provinces, given the role of the **James Bay and Northern Quebec Agreement**. The Environmental Quality Act (1978) is the legislative basis for EA in Quebec. Part 1 of the act deals with EA in Quebec in general, and part 2 deals with EA provisions for the James Bay and northern Quebec territory. The act has undergone several major amendments since first introduced, including provisions for the government to be able to treat several projects with a common objective as a single project (Bardati, 2016). EA regulations under the act apply to activities that are considered a "major development project," meaning that they are likely to trigger significant public concern and have the potential to generate significant adverse impacts. Such projects include electrical transmission lines, major roads, river diversions, oil pipelines, and toxic waste disposal facilities, to name a few. Activities exempt from EA are also included in the regulations, such as fossil-fueled power stations less than 3000 kW, transmission lines less than 75 kV, and all forestry developments included in forest plans approved under the Forest Act. In northern Quebec, EA operates under the James Bay and Northern Quebec Agreement, signed in 1975 between the Cree and Inuit, the federal and provincial governments, and Hydro-Québec. EA in northern Quebec is characterized by much stronger Indigenous representation and engagement when compared to the rest of Quebec; however, there are no formal provisions under the northern EA regime for monitoring and follow-up of effects after projects are approved (Bardati, 2016).

New Brunswick

The Clean Environment Act of 1987 and the Environmental Impact Assessment Regulations formally established EA in New Brunswick. However, EA was practised long before this by way of an EA policy that set out expectations for major project reviews that were funded by the province. Wilgenburg (2016) reports that the beginnings of EA in New Brunswick were as early as 1972, with the Lorneville study of the impacts of a proposed supertanker port and thermal energy generating station. The current EA system in New Brunswick now captures all types of projects, including those submitted by private developers, municipalities, and provincial government departments and agencies. The types of projects subject to an initial EA review are defined in regulations and include such undertakings as the commercial extraction or processing of minerals, transmission lines exceeding 69,000 volts or five kilometres in length, and all projects involving the transfer of water between drainage basins. For oil and gas developments, a phased EA process has been introduced. The phased EA process allows the project review to start much earlier during the project planning stage, when detailed project designs may still be pending, and for certain types of exploration activities to be undertaken alongside the EA process. Since 2010, about 250 projects have been registered for decision under New Brunswick's EA system (Environment and Local Government, 2019).

Nova Scotia

In Nova Scotia, EA was initially administered under the Water Act and the Environmental Protection Act. It was in 1989 that the Environmental Assessment Act and supporting regulations were formally established. The act and regulations were amended in 1995 and again in 2008. Individuals, private companies, and government agencies are required to apply to the responsible minister for a determination as to whether an EA is required for their proposed undertakings. As in New Brunswick, proponents are provided with a specific list under the regulations that describes the types of activities that must be registered and thus screened for potential EA. Unlike New Brunswick, Nova Scotia has two lists: one for minor undertakings, which are reviewed and often screened out of EA, and one for major undertakings, for which proponents are required to carry out an EA. Major undertakings include such activities as nuclear facilities, highway construction when the proposed highway is longer than 10 kilometres and of four or more lanes, and hydroelectric generating facilities with a production capacity greater than 10 MW. Nova Scotia also differs from the other Atlantic provinces in that it has an EA board, like Manitoba, that has the authority to conduct public hearings, investigations, and other studies (Wilgenburg, 2016). Between 2000 and 2019, approximately 170 EA reviews were completed in Nova Scotia (see https://novascotia.ca/nse/ea/projects.asp). In Nova Scotia's offshore environment, EA for energy projects is administered separately by the Canada–Nova Scotia Offshore Petroleum Board, pursuant to federal EA acts and regulations.

Prince Edward Island

Project proponents in Prince Edward Island were expected to prepare an EA for their activities as early as 1973, under the Executive Council Minutes. The Environmental Protection Act of 1988 established the first statutory provisions for EA. The requirement for EA under the act is at the discretion of the responsible minister—in this case, the minister of fisheries, aquaculture, and the environment. Proponents of a construction, industry, operation, or other project or any alteration of any existing project which will or may cause the emission or discharge of any contaminant into the environment are required to register their project with the province. This includes projects that may have an effect on rare and endangered species or generate significant public concern about potential effects. Projects are screened by an interdepartmental committee, and only those deemed to have the potential for a significant impact are subject to EA. In certain cases, should the responsible minister deem that public concerns about the project are significant, an independent assessment review panel may be appointed to conduct public hearings and further investigations into the project's potential impacts. Since 2012, 63 projects were reviewed: eight were screened out, with no EA required; one project was withdrawn; and 54 were approved (see https://www.princeedwardisland.ca/en/feature/projects-under-environmental-review-undertakings#).

Newfoundland and Labrador

Newfoundland and Labrador was the first jurisdiction in Atlantic Canada to legislate EA, under the Environmental Assessment Act of 1980. The act was replaced in 2002 by a more encompassing act, the Environmental Protection Act, which set out the EA process under specific regulations. Under current EA regulations, the government can outright reject any project without full EA review if it believes that the project contradicts a current law or policy or if the project is deemed to be not in the public interest (Wilgenburg, 2016). A regulations list designates projects that must be registered, for which an EA may be required. Listed projects range from large undertakings, such as waterway modifications and transmission lines, to smaller undertakings, such as scrapyards, quarries, or ATV trails. The registration describes the proposed project and outlines how it will affect the biophysical and socio-economic environments. The responsible minister may then determine that the project may be released from EA and proceed, subject to any applicable terms of regulations that additional information may be required in the form of an environmental preview report, that an EA may be required, or that the undertaking is rejected. Since 2000, nearly 1200 projects have been subject to some form of EA review and decision. In Newfoundland's offshore environment, EA for energy projects is administered separately by the Canada–Newfoundland and Labrador Offshore Petroleum Board, pursuant to federal EA acts and regulations.

Northern EA

North of 60°, EA is a mixed system of federal jurisdiction, federal–territorial agreements, and regulation under numerous land-claims and co-management boards. From the initial Mackenzie Valley pipeline inquiry of 1974–7 to the more recent Mackenzie gas project, EA in Canada's North has undergone several significant regulatory and legislative changes. In 1973, the Government of Canada announced a policy that would permit northern Indigenous groups to seek compensation in the form of a land-claims agreement and to have more control over development activities on their traditional lands. The very first land-claims-based EA process was initiated shortly thereafter, in 1975, when the governments of Canada and Quebec and the Cree and Inuit of northern Quebec signed the James Bay and Northern Quebec Agreement. In addition to the James Bay agreement, four different EA processes now apply in northern Canada, established under the **Nunavut Land Claims Agreement**, the **Mackenzie Valley Resource Management Act**, the **Inuvialuit Final Agreement**, and the Yukon Environmental and Socioeconomic Assessment Act.

Nunavut

The 1993 Nunavut Land Claims Agreement represents by far Canada's largest land-claims settlement and land-claims-based EA process. It provided the Inuit with self-government title to approximately 350,000 square kilometres of land in what had been the eastern half of the Northwest Territories. The negotiated settlement area included mineral rights for approximately 35,000 square kilometres. The agreement also included an undertaking for the implementation of a new territory, Nunavut, which was formally established in 1999. The agreement provides for the establishment of an EA review board, the Nunavut Impact Review Board (NIRB), which is the primary authority responsible for EA activities in the land-claims area. The overall purpose of the NIRB is to protect and promote the future well-being of residents and communities of the Nunavut settlement area, including its ecosystem integrity.

Mackenzie Valley

EA in the Mackenzie Valley of Canada's Northwest Territories is established under the Mackenzie Valley Resource Management Act (MVRMA) (1998). For the purposes of the MVRMA, the Mackenzie Valley includes all of the Northwest Territories except for Wood Buffalo National Park and the Inuvialuit Settlement Region. The Northwest Territories Devolution Act (2004) places decision-making powers into the hands of the Government of the Northwest Territories, including the regulatory regimes that guide the licensing and permitting processes for major projects. A co-management board was also established for the Sahtu and Gwich'in settlement areas of the Northwest Territories to delegate responsibilities for land-use planning and for issuing land-use permits and water-use licences. As part of this establishment, a valley-wide public board, the Mackenzie Valley Environmental Impact Review Board (MVEIRB), was created to undertake EAs in the jurisdiction of the MVRMA. There are three potential stages of the assessment process: preliminary screening, EA, and the environmental impact review when a panel, joint panel, or public hearings are necessary. Federal assessment law no longer applies in the Mackenzie Valley except for very specific situations involving issues that are transboundary in nature or for which it is deemed necessary by the board and the federal minister of environment. In such cases, a joint review panel is established.

Inuvialuit Settlement Region

The Inuvialuit Settlement Region encompasses approximately 1,000,000 square kilometres and includes six communities. The settlement region is an outcome of the Inuvialuit Final Agreement, signed between Canada and the Inuit of the western Arctic. One of the outcomes of the agreement was the establishment of two co-management agencies, the Environmental Impact Screening Committee and the Environmental Impact Review Board, to deal with EA screenings and reviews in the Inuvialuit Settlement Region. During the agencies' first 10 years, more than 150 development proposals were screened and two public reviews undertaken for offshore oil and gas exploration programs. In 2000, Canada and the Inuvialuit Environmental Impact Review Board signed an EA agreement for the Inuvialuit Settlement Region. The agreement concerns how the EA process under the Environmental Impact Review Board can be substituted for a panel review under federal assessment legislation and details the process and the steps involved should the Environmental Impact Review Board request such a substitution.

Yukon

In 2003, federal responsibility for land and resource management in Yukon territory was devolved to the Yukon government. As part of the devolution, Yukon was required to pass legislation that mirrored the federal impact assessment process (Slocombe, Hartley, & Noonan, 2016). The Yukon Environmental and Socio-economic Assessment Act was enacted in 2003, setting out regulations and an EA process to assess the environmental and socio-economic effects of projects and other activities. The act also established the Yukon Environmental Socio-economic Assessment Board (YESAB) and six regional designated offices to manage the EA process. Three types of assessment are provided under the act: an evaluation by a designated office, which applies to relatively routine and small-scale projects; an executive committee screening, by YESAB, for projects referred to it by designated offices or for larger projects; and a panel assessment, reserved for projects with the potential for significant adverse effects. Regulations under the act set out those activities that may require an evaluation, but generally the act applies when the Yukon government is the proponent of a project, funds a project, licenses or authorizes a project, or disposes of lands for a project. Between 2008 and 2017, more than 1900 projects were reviewed under the Yukon EA process—the majority of which were recommended to proceed (see http://www.yesab.ca/about-yesab/assessment-statistics). On a per capita basis, more projects are reviewed in the Yukon than perhaps in any other jurisdiction in Canada. This is not a reflection of the level of resource development activity per se but of the encompassing nature of the EA requirements.

Origins and Development of EA in Canada

The current system of EA in Canada is the product of nearly 50 years of development and reform. EA in Canada is currently required at the federal level by means of EA-specific law. When EA was first formally introduced in Canada in the early 1970s, however, it was never the intent of the federal government that EA would someday have a legal basis.

Table 2.1 A Brief History of Federal EA in Canada

Period	Development
1973–4	Cabinet makes policy commitment to review the environmental effects of federal decisions. The Environmental Assessment Review Process (EARP) is established to screen projects to "ensure that they do the least possible damage to our natural environment."
1984	EARP is reinforced and codified by the *Environmental Assessment Review Process Guidelines Order*. The Federal Environmental Assessment Review Office (FEARO) is established to administer EARP.
1990	The minister of environment announces a reform package for EA. The package includes new EA legislation, an assessment process for policy and program proposals, and a participant funding program to support public participation in EA.
1992	Consultations on and a parliamentary review of Bill C-13, the Canadian Environmental Assessment Act, take place. The Canadian Environmental Assessment Act receives royal assent.
1995	The Canadian Environmental Assessment Act comes into force, providing a legal basis for EA in Canada. The Canadian Environmental Assessment Agency is established to administer the act, replacing FEARO.
1999	The minister of environment launches a five-year review of the Canadian Environmental Assessment Act. The review is a requirement of the provisions of the act.

continued

Table 2.1 Continued

Period	Development
2001	The minister of environment introduces Bill C-19 to amend the Canadian Environmental Assessment Act.
2003	Bill C-9, formerly Bill C-19, receives royal assent, and the amended Canadian Environmental Assessment Act comes into force.
2010	The Canadian Environmental Assessment Act is amended to strengthen the role of the Canadian Environmental Assessment Agency and improve the timeliness of EA.
2012	The 2012 Economic Action Plan on Jobs, Growth and Long-term Prosperity is presented in the House of Commons and includes a commitment to reforming the regulatory system in Canada's natural resource sector by shortening timelines for EA, reducing duplication and regulatory burden on development, and enhancing Aboriginal consultation. The Jobs, Growth and Long-term Prosperity Act introduces a new Canadian Environmental Assessment Act, 2012 and repeals the former act.
2016	Minister of environment and climate change establishes an Expert Panel to review EA processes in Canada, engaging the public and interest groups in a national discussion about the state of EA in Canada and needed reforms.
2017	The Expert Panel completes its work and releases its report *Building common ground: a new vision for impact assessment in Canada*. Several recommendations call for substantive change, including strengthening the role of regional and strategic assessment, an early planning and engagement phase, and rebranding EA to "Impact Assessment" (IA) to reflect the broader scope of factors to be considered.
2018	Bill C-69 is introduced, An Act to enact the Impact Assessment Act and the Canadian Energy Regulator Act, to amend the Navigation Protection Act, and to make consequential amendments to other Acts.
2019	The Impact Assessment Act (2019) receives royal assent, and the Impact Assessment Agency is formed to replace the Canadian Environmental Assessment Agency.

Based on the Canadian Environmental Assessment Agency's (2012) *Introduction to the Canadian Environmental Assessment Act, 2012* training manual.

Early Beginnings

EA in Canada originated with the establishment of a task force in 1970 to examine a policy and procedure for assessing the impacts of proposed developments. In 1973, the Cabinet Committee on Science, Culture and Information agreed on the need for a formal process to assess the potential impacts associated with project development. At that point, the first Canadian EA process, the federal **Environmental Assessment Review Process (EARP)**, was formed, and in 1984 the Federal Environmental Assessment Review Office (FEARO) was created to administer its implementation. The scope of EARP was broad and captured any initiative for which there was a federal decision-making authority—from individual resource development projects to regional resource development strategies.

At the time, EARP was the only formal, federal process that provided a window for public debate about the potential impacts that accompanied major resource development proposals (Noble & Udofia, 2015). In practice, participation under EARP typically involved only oral or written presentations to a formal review panel. There were no provisions for Indigenous engagement or for the specific consideration of the impacts on Indigenous peoples' traditional lands and culture. EARP provided that the public may be consulted during the development of guidelines for conducting an EA, but early and ongoing engagement was not mandatory (Dorcey, 1986). There were not even guarantees that input from affected communities would be sought until after a project assessment was submitted for regulatory review and decision (Wondolleck, 1985). Unlike NEPA in the United States, however, the Canadian EARP had no legislative basis and was therefore not legally enforceable; rather, project impact reviews were intended to be cooperative and voluntary. The result was that impact assessments under EARP were carried out inconsistently and, in some cases, not carried out at all.

Mackenzie Valley Pipeline Inquiry

In the formative years of EA, there also emerged, external to the EA system, several major initiatives that would shape expectations about what impact assessment and the engagement of communities and Indigenous peoples in resource development should look like (Noble & Udofia, 2015). Most notable was the Mackenzie Valley pipeline inquiry, commonly referred to as the **Berger Inquiry**, lasting three years and engaging dozens of Indigenous communities along the Mackenzie River, Northwest Territories, about a proposal in the early 1970s to develop an energy pipeline corridor from the Mackenzie River Delta in the Beaufort Sea through the Northwest Territories to tie into gas pipelines in northern Alberta. The proposed energy pipeline corridor would be the longest in the world. In 1974, a federal royal commission, led by Justice Thomas Berger, was appointed to consider the proposal and the environmental, cultural, social, and economic impacts on the North.

The commission's report, *Northern frontier, northern homeland*, was issued in 1977. Concerns identified in the report were not only about the potential impacts of a pipeline corridor on the environment, especially caribou, but also about the potential adverse cultural and social impacts of northern industrial development more broadly on Indigenous communities. Berger argued that the greatest need in the North was not accelerated resource development but opportunities for Indigenous people to determine their own future. The report recommended that pipeline development along the Mackenzie Valley be delayed for 10 years and that land-claims agreements be negotiated and settled prior to any pipeline development. Development was delayed for much longer than 10 years, since by the time Berger had completed his inquiry world energy prices had fallen, making the project uneconomical.

Noble and Udofia (2015) report that what was more impactful on the evolution of EA was not Justice Berger's conclusions but rather how he arrived at them. In his report, Berger writes:

> To hear what they [northerners] had to say, I took the Inquiry to 35 communities.... All those who had something to say—white or native— were given an opportunity to speak.... The impact of the industrial system upon the native people has been the special concern of the Inquiry, for one thing is certain: the impact of a pipeline will bear especially upon the native

people. That is why I have been concerned that the native people should have an opportunity to speak to the Inquiry in their own villages, in their own language, and in their own way.

Bocking (2007) explains that Berger's process "really shook conventional thinking.... Berger demonstrated that the best decision requires not just the right information, but the right process." Some scholars suggest that the Berger Inquiry set international expectations for the critical and cross-cultural public assessment of natural resource development undertakings (Gibson & Hanna, 2009).

Legislative Foundations

It was 30 years following EA's introduction to Canada before it would receive legislative backing, owing in part to the actions of environmental groups over the proposed **Oldman River dam** irrigation project, Alberta (Box 2.1). Introduced in Parliament in 1992 by the Conservative government as Bill C-78, the Canadian Environmental

Box 2.1 Oldman River Dam Project, Alberta

Throughout the 1960s and into the 1970s, agriculture was of growing importance in Alberta. Periodic droughts, characteristic of the region's semi-arid continental climate, meant considerable uncertainties in crop production and sustainability of local water supplies. Combined, agricultural growth and climatic unpredictability reinforced earlier initiatives for the development of large-scale irrigation systems. In 1976, plans were announced to construct an earth- and rock-filled dam at the Three Rivers site in Alberta (Oldman, Castle, and Crowsnest rivers). The proposed dam was to be approximately 75 metres in height and more than 3000 metres long, with a reservoir capable of storing nearly 500 million cubic metres of water (Oldman River Dam Project, 1992). Several sites for the dam were considered, but the Three Rivers site was favoured from the outset, notwithstanding studies that identified potentially significant adverse effects on fish and wildlife as well as impacts on ranchers whose properties would be partially divided and flooded.

The controversy over the proposed project occurred over two periods. First, between 1976 and 1980 landowners whose lands would be flooded opposed the dam, and the Peigan Indian Reserve argued that the project was intruding on lands within the Blackfoot Nation's territory. This first controversy in part ended in 1980 when the final project site was selected. The second, and more significant, controversy emerged in 1987 when an environmental group, Friends of the Oldman River, challenged the project through the provincial courts. The environmental group was initially unsuccessful, and based on assessments under Alberta's environmental assessment process and with the approval of the federal government under the Navigable Waters Protection Act, project construction commenced in 1988. No federal assessment under the EARP Guidelines Order was initiated, notwithstanding the involvement of a federal authority. The Oldman dam project was nearly 40 per cent complete when Friends of the Oldman River once again challenged the project in 1989 but this time in the federal court. The environmental group lost the court case, but in an interesting turn of events, the court of appeals in 1992 reversed the initial decision and quashed approval for the Oldman dam project under the Navigable Waters Protection Act. The court of appeals ordered that the EARP did in fact have the force of law, and the federal government was compelled to comply with the Guidelines Order. While the decision was too late to reverse any damage that had already been created by the dam project, it did pave the way for a legal requirement for EA in Canada, which would ensure that EA would be implemented before development took place.

Assessment Act replaced EARP, providing teeth to federal EA and introducing new requirements for public participation, including new requirements for issuing public notice of an EA application and making EA documentation available in a public registry (Noble & Udofia, 2015).

While the act was intended to make EA more rigorous and systematic, it also limited the reach of EA to include only project actions and not broader planning or policy issues. Under EARP, for example, the potential reach of EA was broad, as indicated by the Beaufort Sea hydrocarbon review (1982–4) and Atomic Energy of Canada Limited's nuclear fuel waste management concept (1988–94)—both were area-wide reviews and concept-based assessments. With the introduction of the Canadian Environmental Assessment Act, regional and concept-based reviews became divorced from the formal EA process. The act also set out responsibilities and procedures for the environmental assessment of projects involving federal authorities and established a process to ensure that impact assessment was applied early in the planning stages of proposed project developments. When the act applied to a proposed project, the relevant federal government department was designated as the "responsible authority" and was to ensure that EA unfolded according to the principles of the act.

In 1994, the Canadian Environmental Assessment Agency was created to oversee the act and replaced FEARO. The act itself was proclaimed and came into force in 1995.

Era of Reform

In 1999, Bill C-13, An Act to Amend the Canadian Environmental Assessment Act, was introduced, and the Standing Committee on the Environment and Sustainable Development organized public consultations on EA reform. Following consultation, the standing committee reintroduced the bill to Parliament as Bill C-9, which received royal assent on 11 June 2003, and the revised Canadian Environmental Assessment Act became law on 30 October 2003. Among the strengthened commitments under the new act were specific provisions to: request follow-up studies or additional information before making a project decision; improve the consideration of cumulative effects; exempt certain smaller projects from EA requirements provided that specified environmental conditions were met as a means to find efficiencies in approval processes; extend EA application to include Crown corporations and the assessment of transboundary effects; better coordinate EAs between federal and provincial jurisdictions; and expand opportunities for public input and ensure better incorporation of Indigenous perspectives and values into the decision-making process.

Canadian Environmental Assessment Act 2012: Focus on efficiency
In 2008, in the midst a global economic crisis, the Conservative government of Canada declared its top priority as being "to support jobs and growth and to sustain Canada's economy" (House of Commons Standing Committee on Finance, Sub-committee on Bill C-38, 2012). Perceiving inefficiencies in the current EA process as a hindrance to economic development, the government included provisions in its federal budget implementation bill (Bill C-38, the Jobs, Growth and Long-Term Prosperity Act) to replace the federal EA process (the Canadian Environmental Assessment Act) with the Canadian Environmental Assessment Act, 2012 (CEAA, 2012) (Becklumb & Williams, 2012). Ensuring an expedited review process and removing barriers to economically attractive resource development ventures, such as the highly contested Enbridge Northern Gateway project, were among the primary drivers for the new act.

Cited in the federal government's Economic Action Plan 2012 were several examples of EA delays, including delays caused by the existence of multiple federal approval processes for any single project application. In the case of the Enbridge pipeline project, for example, several federal departments and agencies did not issue their regulatory approval of the pipeline until almost two years after the National Energy Board had approved it. Under CEAA, 2012, the number of federal authorities responsible for EA was reduced from any federal department or agency potentially involved in issuing permits or authorizations for a project to only three organizations—the Canadian Environmental Assessment Agency, the National Energy Board, and the Canadian Nuclear Safety Commission.

An independent review by de Kerckhove, Minns, and Shuter (2013) of EA approval timelines criticized the government for using exceptions versus the norm to make their case for regulatory change. Based on a sample of 122 authorizations under the Fisheries Act, de Kerckhove et al. concluded that notwithstanding larger projects, reviews of projects over the previous decade had been within the government's stated preferred timelines. Further changes under the act also restricted the scope of federal EA to apply to fewer undertakings, meaning that fewer projects would potentially be subject to assessment. Kirchoff, Gardner, and Tsuji (2013) estimated that more than 90 per cent of projects that required EA under the former act were exempt under CEAA, 2012. The Council of Canadian Academies (2019) Expert Panel on Integrated Natural Resource Management in Canada cites a significant dissatisfaction about the state of natural resource management during this time period, illustrated by increasing litigation related to EA matters (Figure 2.2)—notwithstanding several thousand fewer EAs being conducted than under the previous federal EA process.

Impact Assessment Act, 2019: Broadening Scope
In 2016, following a federal election campaign promise to "make environmental assessment credible again," the minister of environment and climate change appointed an expert panel to consult with the public, Indigenous groups, industry, stakeholders,

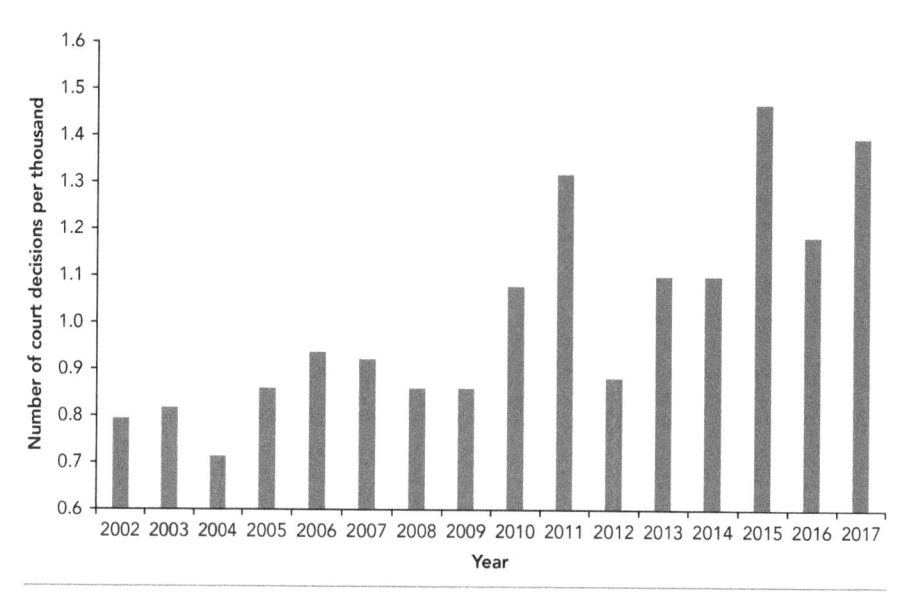

Figure 2.2 Proportion of Canadian court decisions that include the term "environmental assessment," 2002 to 2017

Source: Council of Canadian Academies, 2019, based on data compiled from CANLII

and jurisdictions across Canada and develop recommendations on how to improve federal EA processes. The expert panel released its report in 2017, *Building common ground: a new vision for impact assessment in Canada*, which included several recommendations for EA reform, including the establishment of a single impact assessment agency, the consideration of a broader range of potential project impacts, and earlier and enhanced engagement of the public and Indigenous groups during the planning stages for assessment. Bill C-69, an omnibus bill to enact the **Impact Assessment Act** and the Canadian Energy Regulator Act, to amend the Navigation Protection Act, and to make consequential amendments to other acts, was subsequently introduced in Parliament in early 2018. The bill received royal assent in June 2019, and the Impact Assessment Act is now the legislative basis for the federal assessment process.

The Impact Assessment Act sets out a new process for federal assessment (Figure 2.3) and establishes the Impact Assessment Agency of Canada—a single agency responsible for the management and coordination of EA—a change from the three agencies responsible for EA under CEAA, 2012. For a detailed overview of the federal EA process, see https://www.canada.ca/en/impact-assessment-agency/services/policy-guidance/impact-assessment-process-overview.html. The Impact Assessment Act introduces several changes to federal EA—some fundamental and others perhaps superficial. A major critique of CEAA, 2012, for example, was the reduction in the number of projects subject to federal review under the act. As noted above, CEAA, 2012 reduced the number of projects that would be subject to EA by more than 90 per cent, effectively limiting most small projects from federal EA. Under the Impact Assessment Act, the introduction of new regulations

❶ Planning
The need for EA is determined. The public and Indigenous peoples are invited to contribute to planning the assessment. A determination is made as to whether the project will be assessed by the Impact Assessment Agency or Review Panel. Guidelines for the impact statement are issued.

❷ Impact Statement
The proponent conducts studies and engages the public and Indigenous peoples to develop a detailed impact statement, which provides a technical analysis of the project and its impacts.

❸ Impact Assessment
The Impact Assessment Agency assesses the proponent's impact statement and submits an impact assessment report and report on consultations to the minister of environment and climate change.
The Review Panel assesses the proponent's impact statement, holds public hearings, and submits an impact assessment report to the minister of environment and climate change.

❹ Decision-Making
The minister determines if adverse effects are within federal jurisdiction and if the effects are in the public interest or refers the determination to the Governor in Council. A decision statement is issued that includes the reasons for the determination and conditions.

❺ Post-decision
The proponent is responsible for meeting any conditions and for carrying out any follow-up or monitoring programs. The Impact Assessment Agency is responsible for verifying that the project is in compliance with the minister's decision statement.

Figure 2.3 Simplified overview of the basic steps involved in the federal impact assessment process under the Impact Assessment Act

defining those projects subject to review did not reinstate the broad reach of federal EA that existed prior to CEAA, 2012. For some projects, the proposed thresholds to be met for triggering an assessment are higher than under the CEAA, 2012, such as those established for certain mining operations and pipeline and highway rights-of-way, indicating that there may even be fewer EAs for certain types of projects.

The most obvious change is the use of "impact assessment" instead of "environmental impact assessment." The intent is to signal a broader consideration of potential project impacts beyond the biophysical environment, including, for example, impacts to health and social aspects, gender-based impacts, impacts on the economy, and impacts on Indigenous peoples. In practice, however, in most jurisdictions and applications in Canada and internationally, "environment" in EA is already broadly defined to include social, cultural, and economic factors in addition to biophysical ones.

Under the Impact Assessment Act, decision-makers must determine whether a project is "in the public interest," having given consideration to five key factors: i) the extent to which the project contributes to sustainability; ii) the extent to which adverse effects are significant; iii) the implementation of mitigation measures; iv) the extent to which the effects of the project hinder or contribute to Canada's climate commitments; and v) the impacts on Indigenous peoples and Section 35 rights. In principle, the "public interest" determination differs from that in CEAA, 2012 in which the key determination was whether a project was likely to result in "significant adverse environmental effects" and, if so, whether those effects were justified under the circumstances. The Impact Assessment Act presents a different flavour than CEAA, 2012, but how this informs a different practice remains to be seen. Significance determination remains a key criterion of the public interest test, and other factors such as mitigation measures have always been part of significance determinations. A project with the potential for significant adverse effects may still be justified as being in the public interest.

The sustainability criterion, climate test, opportunities for public participation, and impacts on Indigenous peoples and Section 35 rights indicate forward thinking when compared to CEAA, 2012. Gibson (2019) suggests that clauses 63(a) and 63(e) of the act, which require decision-makers to consider the extent to which an assessed project would contribute to sustainability and hinder or contribute to meeting Canada's obligations and climate change commitments, set the bar far higher than previous federal assessment laws. That said, there is limited guidance as to what a sustainability test involves, and thus far guidance for climate commitments seems to be based largely on GHG accounting of individual projects.

Under CEAA, 2012, public participation in EA was limited to those deemed to be "directly affected" by the project, restricting the role of EA in supporting public debate about development proposals. The Impact Assessment Act does away with this restriction and acknowledges the importance of public participation early in the assessment process, including during the planning stages of assessment. However, notwithstanding strengthened provisions that the public be "provided with an opportunity to participate," such opportunities are not provided at all stages, and there is limited instruction on how to ensure participation that is meaningful. The Impact Assessment Act does, however, include enhanced recognition of Indigenous rights, interests, and knowledge, but it also stops short of referencing the **United Nations Declaration on the Rights of Indigenous Peoples**.

A critical lens might indicate that the new Impact Assessment Act does not differ much from the foundational principles and good practices of EA that informed the initial Canadian Environmental Assessment Act of 1992. It does, however, present a shift from—and is less based on the principles endorsed by—the Canadian Environmental Assessment Act, 2012.

Continuous Learning Process

Initiated as a policy with little grip, EA has since evolved to become a rigorous regulatory and legal framework. How this evolution has unfolded over the past 50 years, however, might be described either as the dislocation of an established order to improve regulatory requirements, as a natural transition from one system to another, or as a series of unnatural breaks triggered by forces of social and legal change. However one chooses to characterize the history of EA in Canada, Gibson (2002, p. 156) captures it best in suggesting that "Canadian environmental assessment policies and laws have evolved slowly . . . this evolution has been hesitant and uneven, though overall it has been positive." At the time this book was written, Canada had recently enacted a new federal assessment act and other jurisdictions, including British Columbia, had introduced EA reform. The effectiveness of these new EA processes and reforms remains to be seen. What is certain, however, is that EA is an evolving process and expectations about EA are constantly changing.

Key Terms

Berger Inquiry
Environmental Assessment Review Process (EARP)
Impact Assessment Act
Inuvialuit Final Agreement
James Bay and Northern Quebec Agreement
Mackenzie Valley Resource Management Act
Nunavut Land Claims Agreement
Oldman River dam
United Nations Declaration on the Rights of Indigenous Peoples

Review Questions and Exercises

1. What are some of the critical events that have shaped the course of Canadian EA at the federal, provincial, and territorial levels?
2. When was EA enacted in your province (or territory)? What is the name of the legislation that requires EA? Describe some of the main features of your province or territory's EA legislation, such as its purpose, how it defines "environment," and when an EA is required. Compare this to the EA legislation of another Canadian province or territory.
3. It has been suggested that all of Canada should be governed by a single EA regulatory framework and assessment process. Do you agree? What might be the advantages and challenges to removing EA responsibility and authority from provinces and territories and transferring it to a single, composite federal EA system?
4. What are the major differences between the Impact Assessment Act, the Canadian Environmental Assessment Act, and the Canadian Environmental Assessment Act, 2012? Identify a theme or topic, such as public participation, traditional knowledge, or cumulative effects, to frame your answer. Do you think these changes are positive or negative for strengthening EA in Canada?

References

Bardati, D.R. (2016). Environmental assessment in Quebec. In K. Hanna (ed.), *Environmental impact assessment: practice and participation*. Don Mills, ON: Oxford University Press.

Becklumb, P., & Williams, T. (2012). Canada's new federal environmental assessment process: background paper. Publication no. 2012-36-E. Ottawa, ON: Library of Parliament.

Bocking, S. (2007). Thomas Berger's unfinished revolution. *Alternatives Journal*. http://www.alternativesjournal.ca/energy-and-resources/thomas-bergers-unfinished-revolution.

Cotton, R., & Zimmer, J.S. (1992). Canadian environmental law: an overview. *Canada-United States Law Journal* 18(10), 63–84.

Council of Canadian Academies. (2019). *Greater than the sum of its parts: toward integrated natural resource management in Canada*. Ottawa, ON: The Expert Panel on the State of Knowledge and Practice of Integrated Approaches to Natural Resource Management in Canada.

de Kerckhove, D.T., Minns, C.K., & Shuter, B. (2013). The length of environmental review in Canada under the Fisheries Act. *Canadian Journal of Aquatic Sciences* 70(4), 517–21.

Dorcey, A. (1986). Techniques for joint management of natural resources: getting to yes. In J.O. Saunders (ed.), *Managing natural resources in a federal state*. Toronto, ON: Caswell.

Environment and Local Government. (2019). Registrations and determinations. Government of New Brunswick. https://www2.gnb.ca/content/gnb/en/departments/elg/environment/content/environmental_impactassessment/registrations.html.

Fitzpatrick, P., & Sinclair, A.J. (2009). Multi-jurisdictional environmental impact assessment: Canadian experiences. *Environmental Impact Assessment Review* 29, 252–60.

Gibson, R.B. (2002). From Wreck Cove to Voisey's Bay: the evolution of federal environmental assessment in Canada. *Impact Assessment and Project Appraisal* 20(3), 151–9.

Gibson, R.B. (2019). Assessment law is still too vague to achieve lasting green goals. *Policy Options*. https://policyoptions.irpp.org/magazines/october-2019/assessment-law-is-still-too-vague-to-achieve-lasting-green-goals.

Gibson, R., & Hanna, K. (2009). Progress and uncertainty: the evolution of federal environmental assessment in Canada. In K. Hanna (ed.), *Environmental impact assessment: practice and participation*, 2nd edn. Toronto, ON: Oxford University Press.

Hanna, K., & Parkins, J. (2016). Alberta's environmental assessment system. In K. Hanna (ed.), *Environmental impact assessment: practice and participation*. Don Mills, ON: Oxford University Press.

Hostovsky, C., & Graci, S. (2016). Environmental assessment in Ontario: moving from comprehensive planning and decision-making to streamlined approvals. In K. Hanna (ed.), *Environmental impact assessment: practice and participation*. Don Mills, ON: Oxford University Press.

House of Commons Standing Committee on Finance, Sub-committee on Bill C-38. (2012). *Bill C-38, Part 3. Responsible Resource Development*. Ottawa, ON: House of Commons.

Kirchhoff, D., Gardner, H., & Tsuji, L. (2013). The Canadian Environmental Assessment Act, 2012 and associated policy: implications for Aboriginal peoples. *The International Indigenous Policy Journal* 4(3), 1–14.

Lobe, K., & Sinclair, J. (2016). Environmental assessment: Manitoba approaches. In K. Hanna (ed.), *Environmental impact assessment: practice and participation*. Don Mills, ON: Oxford University Press.

Meredith, T. (2004). Assessing environmental impacts in Canada. In B. Mitchell (ed.), *Resource and environmental management in Canada*, 3rd edn, pp. 467–96. Toronto, ON: Oxford University Press.

Noble, B.F., Liu, G., & Hackett, P. (2017). The contribution of project environmental assessment to assessing and managing cumulative effects: individually and collectively insignificant? *Environmental Management* 59(4), 531–45.

Noble, B.F., & Udofia, A. (2015). *Protectors of the land: toward an EA process that works for Aboriginal communities and developers*. Ottawa, ON: Macdonald-Laurier Institute.

Oldman River Dam Project. (1992). *Report of the Environmental Assessment Panel*. Report no. 42. Hull, QC: FEARO.

Rutherford, M. (2016). Impact assessment in British Columbia. In K. Hanna (ed.), *Environmental impact assessment: practice and participation*. Don Mills, ON: Oxford University Press.

Slocombe, S., Hartley, L., & Noonan, M. (2016). Environmental assessment and land claims, devolution, and co-management: evolving challenges and opportunities in Yukon. In K. Hanna (ed.), *Environmental impact assessment: practice and participation*. Don Mills, ON: Oxford University Press.

Thompson, A.R. (2013). Resource rights. *The Canadian Encyclopedia*. https://thecanadianencyclopedia.ca/en/article/resource-rights.

Westbrook, C., & Noble, B. (2016). Environmental assessment in Saskatchewan. In K. Hanna (ed.), *Environmental impact assessment: practice and participation*. Don Mills, ON: Oxford University Press.

Wilgenburg, H. (2016). Atlantic Canada: a story of EIA adaptation. In K. Hanna (ed.), *Environmental impact assessment: practice and participation*. Don Mills, ON: Oxford University Press.

Wondolleck, J.M. (1985). The importance of process in resolving environmental disputes. *Environmental Impact Assessment Review* 5, 341–56.

3

Pre-project Planning and Public Engagement

Introduction

Demonstrating the need for and purpose of a proposed project, exploring viable alternatives to meet the project's objectives, and engaging the public in doing so are important requisites to good EA. Through pre-project planning, options can be considered to improve the project's design or minimize or avoid potentially adverse impacts from the outset and to build important relationships between the project proponent, interested or affected communities, and governments. Potentially contentious issues can be resolved prior to an application for development being submitted for project screening and determining the need for EA, resulting in a more meaningful and efficient assessment process.

Roles and Responsibilities

The project proponent is primarily responsible for pre-project planning and engagement. The success of an EA, and project, often depend on whether and how the proponent engages early in the pre-project planning and design stages. It is thus in a proponent's best interest to engage early with potentially affected communities to identify key concerns, build relationships, and manage expectations and to consult with the responsible government agencies to understand what will be required of them should the project proceed to an EA. Pre-project planning is an opportunity for the proponent to test the public and political feasibility of their project and address any potential conflicts early in the process. However, other parties and interests also play important roles in pre-project planning (Table 3.1). It is important for potentially affected publics to identify key issues of concern and indicate how they would like to be engaged. It is important for Indigenous groups to similarly identify key issues and concerns, but they also may negotiate certain protocols for engagement, negotiate agreements with the proponent about project benefits, and advise the proponent on proper engagement strategies. Government agencies and regulators are also important players in pre-project planning. They must be responsive to questions of proponents, communities, and Indigenous groups, facilitate public and Indigenous participation, and ensure that regulatory rules and obligations are being met early in the project's design.

Project Need and Consideration of Alternatives

Projects are proposed to meet a defined need or to realize an opportunity—be it social, economic, or environmental. This is often referred to as the "need for" and "purpose of" a proposed project. The need for a proposed project is the demand or opportunity that the project is intended to address or satisfy. The purpose of a

Table 3.1 Key Roles and Responsibilities in Pre-project Planning

Project proponent	Public
Engage with the public and potentially affected Indigenous groups	Identify key issues and concerns
Prepare a detailed description of the project, explore viable alternatives, and identify key issues and concerns	Participate in engagement opportunities hosted by the proponent or government agencies to learn about the project
Develop an engagement strategy for involving the public	Provide input to the proponent and government agencies on project issues and the scope of issues to be considered in an EA
Negotiate agreements with Indigenous groups, as needed	Apply for available funding to support participation and engagement
File a project application with the responsible authorities	Share knowledge with other publics and community members

Indigenous groups	Government agencies
Identify key issues and concerns, including those pertaining to Indigenous rights	Consult with potentially affected Indigenous groups about impacts on Indigenous rights
Participate in engagement opportunities hosted by the proponent or government agencies to learn about the project	Provide financial resources to facilitate Indigenous and public participation
Provide input to the proponent and government agencies on project issues and the scope of issues to be considered in an EA	Ensure that project information is made available to the public
Apply for available funding to support participation and engagement	Identify key issues and concerns, including those raised by the public and Indigenous groups, and make them known to the proponent
Share knowledge with community members	Provide guidance to project proponents on what is required of them for EA or regulatory compliance
Negotiate agreements for engagement or project benefits sharing, where applicable	Provide data to support the work of the proponent, communities, and Indigenous groups

proposed project is what is intended to be achieved by carrying out the project. For example, the need for a proposed wind energy project may be defined based on the growing demand for electricity supply and the inability of the current system to meet that demand. The purpose of the project might be to supply cost-efficient, carbon-neutral, and reliable electricity. The need for and purpose of a project provide context for the identification and evaluation of project alternatives.

The identification and consideration of alternatives early in the pre-project planning stage is central to ensuring that the identified need or opportunity is met with only the "best" possible project. Positive benefits can be discovered, and many adverse impacts can be avoided before irrevocable project choices, such as project type and location, and design decisions are made by considering alternatives to those proposed (Morrison-Saunders, 2018). There are two categories of alternatives that should be considered in EA—alternatives to the identified project and alternative means of carrying out the project (Figure 3.1).

```
"Alternatives to"         No action (A)
                          Proposed project (B)
                          Demand reduction program (C)
                          Another project (D)

"Alternative means"       Proposed location (A)
 –location                Alternative location (B)
                          Alternative location (C)
                          Alternative location (D)

"Alternative means"       Proposed design (A)
 –design                  Alternative design (B)
                          Alternative design (C)
                          Alternative design (D)
```

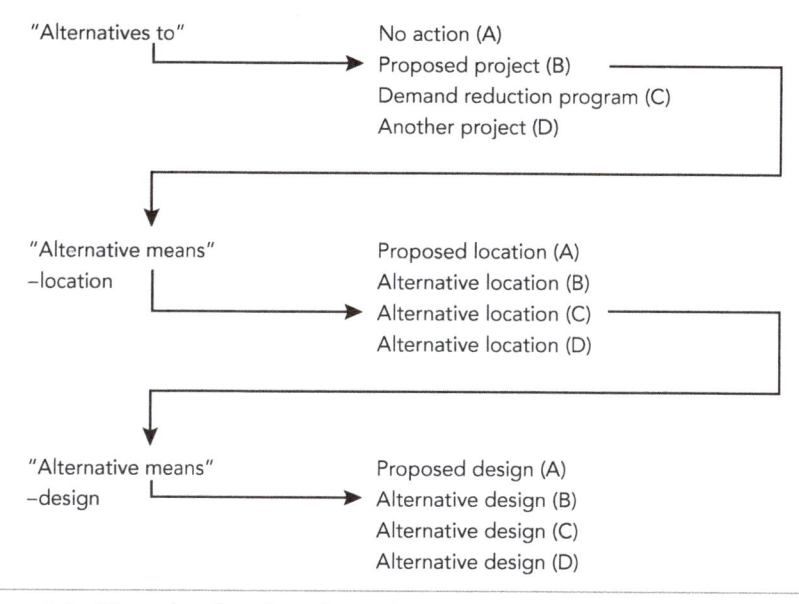

Figure 3.1   Hierarchy of project alternatives

Alternatives to a project are functionally different ways of meeting the need and purpose of the described project, including the option of no project. For example, a hydroelectric project may be proposed to meet growing energy demands in a region. Alternatives to the project may include the consideration of different ways to meet the demand (e.g., small modular nuclear reactors, wind energy) or not meeting the demand (e.g., the project is replaced by energy conservation programs).

The consideration of "alternatives to" is a requirement under some EA systems such as in Europe under the EIA Directive 2014/52/EU and under the Canadian federal Impact Assessment Act. Section 22(1) of the act specifies that when a project is being assessed by the Impact Assessment Agency or a review panel, consideration must be given to any alternatives to the project that are technically and economically feasible and are directly related to the project. In principle, the consideration of "alternatives to" ensures that EA is not reduced to the defence of a single project proposal but rather ensures a proactive consideration of viable options to determine an overall positive course of action (DEAT, 2004). In practice, however, "alternatives to" are rarely considered comprehensively in project EA. When considered, they often reflect narrow agency or proponent goals or are biased toward the proposed project. Alternatives to a proposed project are typically limited to the "no action" alternative, which is interpreted as either "no change" from the current or ongoing activity or "no activity," which refers to not proceeding with the proposed development—typically accompanied by a carefully crafted argument that no project, or a substantially different project, would result in an important social or economic need or opportunity not being met.

To a certain extent, this bias is understandable in EA in that a project proponent—for example, a mining company—has a vested interest in developing mineral resources and by the time a project is conceived the proponent has already dedicated considerable resources to it, such as feasibility studies and market analyses. There are few "alternatives to" available to a mining company, whose objective is to develop and bring to market mineral resources, other than not to mine. Many alternatives that may be more environmentally sound or socially beneficial have thus already been foreclosed by the time early planning commences for a single project development. "Alternatives to" are best addressed much earlier in the policy and planning cycle during the stages of strategic environmental assessment (Chapter 12), during which objectives can be thoroughly examined to determine the best path forward without the pressures, expectations, and financial risks of a project on the table. However, a mining company may consider different alternatives to how electricity might be supplied to the mine site, such as coal-fired generation from the regional grid or the development of local hydroelectric resources and off-grid wind power dedicated to supplying the mine with low-emissions energy options.

Alternative means can effectively be addressed at the pre-project planning stage. "Alternative means" refers to different options for carrying out a project when it has been accepted that the proposed project is the most suitable alternative to meet the need or opportunity at hand. The consideration of alternative means is a common requirement under most EA systems. Alternative means typically include alternative ways that a project can be implemented, such as technical or engineering design, or timing of project operations, or alternative locations (e.g., alternative routings of pipelines, transmission lines, or roads). Candidate alternatives should be systematically compared to identify the preferred option(s) among those that remain for detailed impact assessment.

All alternatives considered in EA must be technologically feasible, relevant to the need or objectives at hand, and economically viable (DEAT, 2004). This means that it may not always be possible to consider alternative means for all aspects of all developments (Box 3.1). There are few possible alternative locations for a proposed mining operation—siting is clearly dependent on the location of the mineral reserve. That said, alternative means of extracting the resource to minimize surface disturbances, as well as alternative means of and locations for tailings management, or alternative locations for access roads to avoid culturally important places or ecologically significant habitats can be considered.

Box 3.1 Alternative Means and the James Bay Lithium Mine, Quebec

In 2019, Galaxy Lithium submitted an EIS to develop the James Bay Lithium Mine project. Lithium is an alkali metal with several industrial and commercial applications, including pharmaceutical polymers, portable electronic devices, manufacture of glass and ceramics, and lithium ion batteries. Global demand for lithium is increasing, and lithium use in renewable energy storage, including electric vehicles, has increased significantly in recent years. The proposed mine site is in the Nord-du-Québec administrative region on the territory of the Eeyou Istchee James Bay Regional Government, about 100 kilometres east of James Bay. The project includes an open-pit mine and concentrator facility, tailings, waste rock, ore and overburden storage areas, as well as related infrastructure. The project would have a mine life of 15 to 20 years and produce on average 5480 tonnes of ore per day. The mining property is an approximate 2160-hectare area.

The proposed mining operation was subject to EA under both the Quebec Environmental Quality Act and the Canadian Environmental Assessment Act, 2012. As per the requirements of both acts, the proponent was required to consider *alternative means* for the project. However, the proponent noted in its EIS that given the location of the mineral resource, the location and accessibility of existing roads and power line infrastructure, and the need to maintain safe distances around the open-pit operation, it was not possible to consider viable alternatives for certain aspects of the project, namely: the mining and material extraction method, given the shallow depth of the resource and the economic viability of an open-pit operation; location of the concentrator for material processing, ensuring proximity to the pit and considering the distribution of wetlands in the area; worker campsite location, given the importance of proximity to the mine site and of reducing on-site vehicle use; and water supply, given the remote location of the site and the only feasible option being to construct water supply wells.

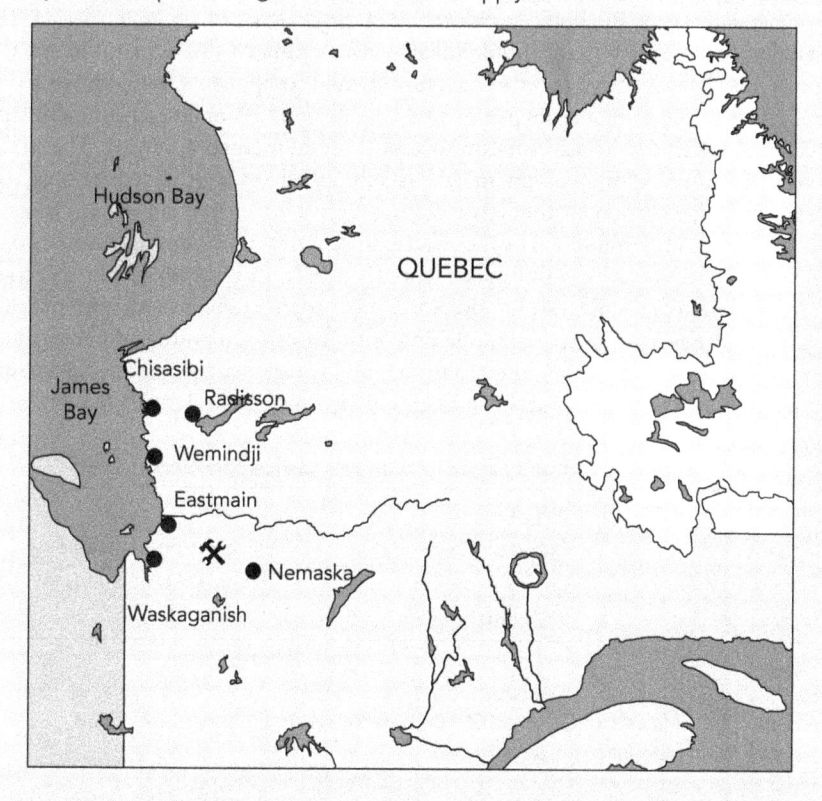

For other aspects of the project, several alternative means were considered by the proponent, including those associated with alternative project technology and location. For example, the EIS identified five options for mine tailings disposition, four alternative locations for waste rock and tailings stockpiles, two locations for

continued

overburden stockpiles, four technologies for domestic wastewater treatment at the work camp, and two options for power supply at the mine site. For each set of alternatives, the proponent described the options, including a rationale for those options, and the factors that led to the preferred options that would constitute the project's design. The final design options were carried forward for detailed assessment.

Source: WSP Canada Inc., 2019

Methods to Support Alternatives Assessment

Several methods are available for comparing or evaluating alternatives (Table 3.2), but whatever the method, it is important to compare alternatives using similar criteria or objectives. Alternatives can be evaluated using a simple rating or ranking procedure based on potential economic, social, or environmental impacts or contributions. Ideally, the consideration and evaluation of alternatives is a participatory process conducted early in the project planning and design stage, with opportunity for input from those most affected by the project's actions.

One approach is the **Peterson matrix** (Peterson, Gemmel, & Shofer, 1974). The Peterson matrix consists of three individual matrices: a matrix that depicts the impacts of project actions or causal factors on environmental components; a matrix depicting the impacts of the resultant environmental change on the human environment; and a vector of weights or relative importance of those human components. The initial project–environment interaction matrix is multiplied by a matrix depicting secondary human-component impacts resulting from project-induced environmental change. The result is multiplied by the relative importance of each of the human components to generate an overall impact score (Box 3.2). In cases where the selection among alternatives is not so straightforward, more sophisticated evaluation methods may be required—such as spatial modelling or **multi-criteria evaluation** procedures (Nielsen, Noble, & Hill, 2012) (see Environmental Assessment in Action: Alternative Means for Managing the Impacts of a Highway Twinning Project on Wetlands).

Table 3.2 Example Methods for Comparing or Evaluating Alternatives

Ranking	Weighting	Benefit-cost ratio
Decision trees	Paired comparison	Multi-criteria evaluation
Contingent valuation	Choice experiments	Life-cycle evaluations
Matrices/checklists	GIS/spatial analysis	Q-method

Box 3.2 Peterson Matrix for Comparing Project Route Alternatives

The following Peterson matrix presents a simple example of alternative routing options for a proposed high-voltage transmission line across the traditional territory of an Indigenous community. Scores denoting the potential impacts of each routing option on the biophysical environment [I] were derived using the expert judgment of the proponent's consultants and based on experiences with similar transmission projects through northern boreal environments. Scores denoting how changes in biophysical conditions may impact local community values [II] (e.g., access to country foods, disruption of traditional hunting patterns, etc.), including the relative importance of those values [III], were derived through community forums and focus groups in the potentially affected community.

In the example, impact scores are scaled in such a way that a potentially minor adverse impact is indexed as 1 and a potentially major adverse impact is indexed as 4. A minor impact is defined simply as one that can be fully mitigated with standard, known impact management actions and, based on previous experiences, is unlikely to cause long-term irreversible effects. A major impact is defined as an impact likely to have a long-term effect that cannot be fully reversed within the time frame of the project's life cycle.

[I] Project routing options and impacts on biophysical components

Project routing options	Potentially affected biophysical (B) components				
	B1	B2	B3	B4	B5
(i)	3	2	4	1	4
(ii)	1	3	2	4	1
(iii)	3	2	4	3	2
(iv)	1	1	3	2	1
(v)	4	2	1	1	1

[II] Impacts on community values resulting from a change in the quality of biophysical components

Affected biophysical components	Potentially affected community (C) values				
	C1	C2	C3	C4	C5
B1	2	2	1	1	3
B2	1	1	2	2	3
B3	2	2	1	3	2
B4	1	2	4	3	2
B5	2	2	3	2	3

[III] Relative importance (weighting) of potentially affected community values, with higher weights indicating greater importance

Community values	Importance (weight)
C1	0.80
C2	0.25
C3	0.60
C4	0.40
C5	0.20

continued

To calculate the Peterson matrix, multiply the primary biophysical effects matrix [I] by the secondary community effects matrix [II], and multiply the resulting matrix by the vector of community values [III]. The final result is an index for each project routing option as follows: (i) 62, (ii) 50, (iii) 62, (iv) 36, (v) 37. The higher index is indicative of a higher overall effect and therefore a less preferred routing option to be considered in the project's design. The options can also be scaled or standardized to generate a relative ranking of routing options by using the following equation:

$$\frac{i_{max} - i}{i_{max} - i_{min}}$$

Where: i_{max} is the value of the routing option with the highest index; i_{min} is the value of the routing option with the lowest index; and i is the value for each option (i) through (v). This result will yield a preferred option that is always 1, a least preferred option that is always 0, with all other scores falling somewhere in between. In this example, the scaled results are as follows: (i) 0.00, (ii) 0.46, (iii) 0.00, (iv) 1.00, (v) 0.96. Routing options (iv) and (v) are clearly the preferred options and more than twice as preferred as the next competing option (ii). Options (i) and (iii) are clearly the least preferred routing options.

The advantage of the Peterson matrix lies in its simplicity and in the multiplicative properties of matrices. That said, the Peterson matrix can oversimplify the complex relationships and interactions between projects, environmental change, and impacts on local community values. Such matrices should be used as supporting tools to summarize and communicate information or as a rapid appraisal process rather than for detailed and quantified assessments of project alternatives.

Environmental Assessment in Action

Evaluation of Alternatives for Managing Impacts to Wetlands for the Louis Riel Trail Highway Twinning Project, Saskatchewan

Wetlands provide important ecological services, including habitat provision, nutrient cycling, carbon sequestration, and flood control. However, wetlands are under considerable threat due to human-induced surface disturbances. Next to agriculture, road development is among the most significant sources of wetland degradation in Prairie Canada. In 2007, the Saskatchewan Ministry of Highways and Infrastructure proposed to twin an approximately 110-kilometre highway between Saskatoon and Prince Albert, the Louis Riel Trail. The highway is one of the busiest highways in the province. The project application was subject to the conditions of an aquatic habitat protection permit and standard upland wetland mitigation strategies. The project would not require an EA under the Saskatchewan Environmental Assessment Act, based on the notion that a properly designed wetland mitigation strategy for managing impacts to wetlands within the provincially regulated 31-metre highway

3 | Pre-project Planning and Public Engagement   45

right-of-way would result in no other significant adverse effects to wetlands. The 31-metre right-of-way is measured outward from the centreline of a highway and includes the highway itself and the roadside ditches. Because of concerns about the project routing and design, including the potential for cumulative wetland loss, an independent assessment of mitigation alternatives was carried out by researchers at the University of Saskatchewan in partnership with Ducks Unlimited Canada.

The assessment was based on a one-kilometre-wide corridor, centred on the proposed highway centreline. Previous science has shown that adverse effects to wetland functions can occur up to distances of 500 metres from highways

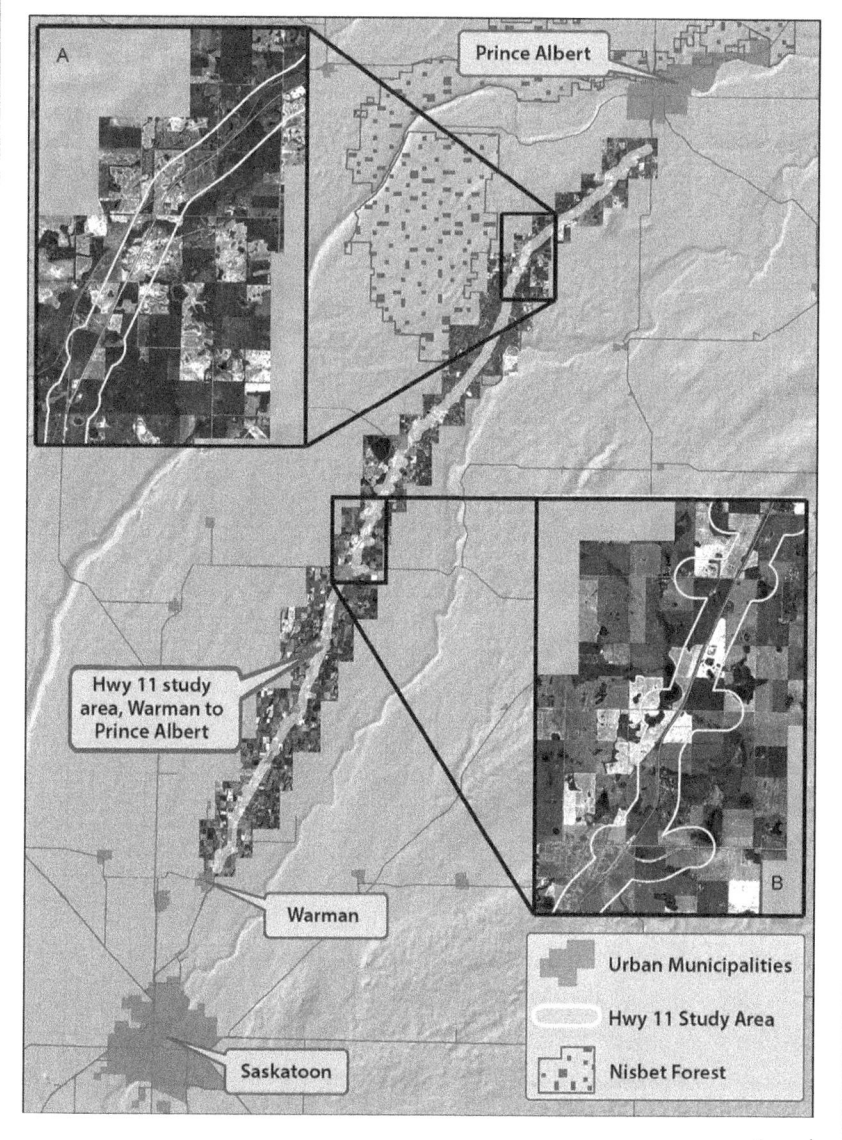

continued

(Houlahan, Keddy, Makkay, & Findlay, 2006). A field-based analysis was completed to identify and map the total number and area of wetlands within a 500-metre buffer zone on either side of the centreline of the proposed highway. Project disturbance, wetland numbers, and wetland area were inventoried using aerial photos and panchromatic satellite imagery, ground-truthed for accuracy using a Trimble GeoXM field computer and mapped using Geographic Information Systems (GIS). Wetland area was used as a proxy to assess potential effects on both wetland habitat and function (Dahl & Watmough, 2007).

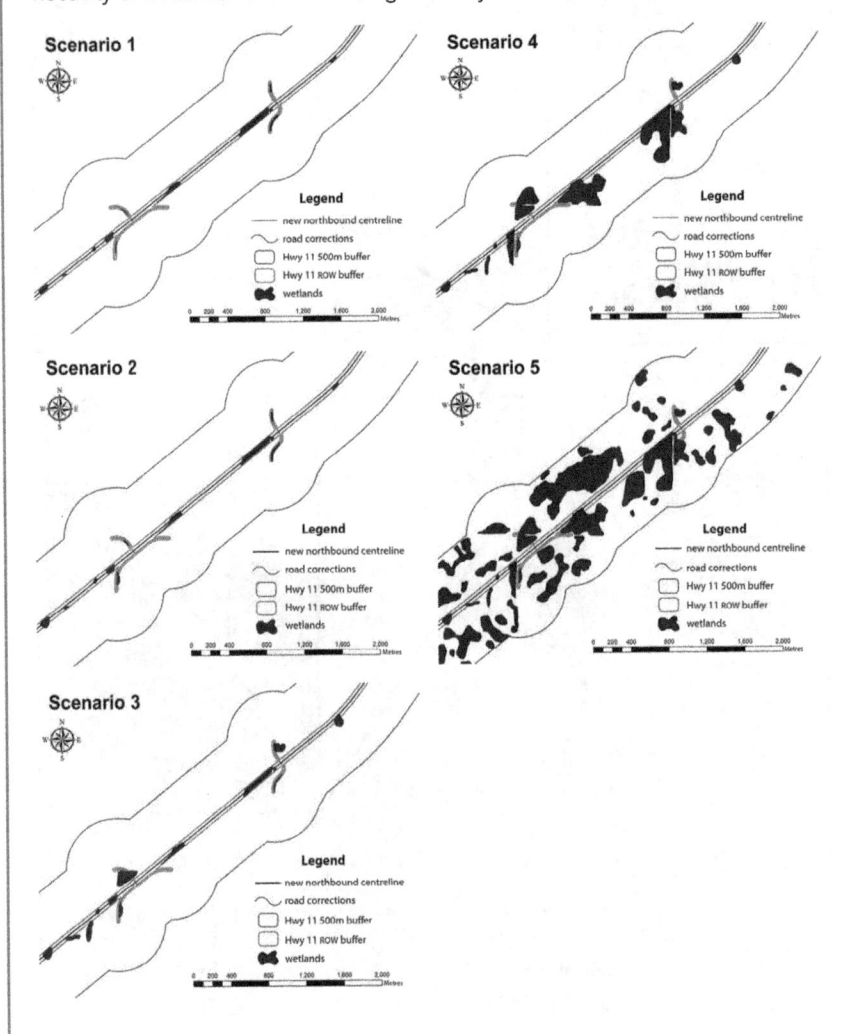

Using GIS, mapped wetlands within the 500-metre zone were classified based on their potential to be affected by project activities, either directly or indirectly. Potentially affected wetlands were determined by delineating "impact zones" of increasing distance from the proposed highway and based on visual surface connectivity of wetlands. The 31-metre right-of-way was considered the "zone of direct impact," and wetlands contained in this area were considered at high risk of a complete loss of both habitat and function. Wetlands with more than 50 per cent of their area overlapping the zone of direct impact were also considered at high risk. The distance from the edge of the 31-metre right-of-way up to 500 metres was classified as the "zone of indirect impact." Wetlands in this zone were considered at risk because of functional degradation as a result of surface connectivity to directly affected wetlands and because of potentially induced impacts. Five alternatives for managing the impacts to wetlands were then considered, each more or less spatially ambitious. A section of the highway depicting each alternative is shown on page 46.

Multi-criteria analysis, a structured approach to evaluating options, was used to compare alternative management strategies based on the input of a panel of government and non-government experts in wetland conservation, hydrology, and transportation engineering. Each alternative was evaluated based on the financial cost of implementation; compliance with a federal policy of no-net-loss of wetland functions; technical feasibility with regard to implementation; the potential to capture cumulative effects; administrative requirements in terms of permitting needs and land access for implementation; and public acceptability.

A total of 458 wetlands (1115 ha) were identified as at risk within a 500-metre buffer on either side of the proposed highway. Approximately 50 ha of wetland habitat was located within the zone of direct impact, denoted by the 31-metre highway right-of-way and at high risk of complete loss. An additional 1065 ha of wetlands were determined to be at risk of some level of functional degradation. Under the permitted approach to wetland management for highway developments, only 50 ha of wetlands would be subject to management—the wetland area completely inside the regulatory 31-metre highway right-of-way. Of the 458 wetlands identified inside the 500-metre buffer, more than half were wetlands less than 1.0 ha in size, including many seasonal wetlands that are often not captured in wetland management plans.

Results showed the need to consider a much more ambitious approach to wetland management in the project design than was included in the conditions attached to the project permit. The preferred alternative was the most ambitious in comparison to current practice, demonstrating a decision outcome based on the importance of wetland conservation and dissatisfaction with the proposed project design. The distribution of wetland size in the study area is typical of a prairie landscape and indicative of the need for an approach to project design and impact management that does not overlook the cumulative loss of many small and seasonal wetlands.

Source: Based on Nielsen et al., 2012

Public Engagement

Public participation or engagement refers to the involvement of individuals and groups that are positively or negatively affected by a proposed intervention subject to a decision-making process or are interested in it (André et al., 2006). The provision of opportunities for some form of public participation or engagement is required in most EA systems around the world. Under most Canadian provincial and territorial EA systems, there is a requirement that the public have an opportunity to provide input on the scope of an EA or the terms of reference for a project's assessment. At the federal level, the Impact Assessment Act establishes that the public and potentially affected Indigenous peoples must be involved in pre-project planning, which includes a requirement for public participation for certain designated projects that may require an assessment.

The practice of public participation in EA has been highly criticized, with engagement processes rarely consisting of highly participatory approaches in which proponents are willing to significantly alter project design or implementation plans. Arnstein (1969) described different levels of public participation, ranging from manipulation of the public to citizen control (Figure 3.2). At the bottom is "non-participation"—involvement of the publics in a way that does not include direct participation. This approach consists of what Arnstein labels rubberstamp committees and efforts to inform or educate the public rather than genuinely seeking their involvement. Public education about a proposed project is important to engagement, but it should not be the only engagement strategy. At the opposite end of the spectrum is citizen power in which the affected public is granted full control and authority in development decision-making—this is rarely, if ever, the case in EA. In practice, public participation in EA has focused primarily on *consulting* the public and sometimes negotiating trade-offs, with few options for greater participation. In other words, public participation in EA has typically involved *providing information* to the public about the proposed project.

Meaningful Engagement

Engagement of those communities and interests potentially affected by a project must occur early in the pre-project planning stages, it must occur often throughout the EA process and the project's life cycle, and it must be meaningful.

Nature of involvement	Participation	Degree of power
Citizens have complete control	Citizen control	Degrees of citizen power
Citizens are delegated certain powers	Delegated power	
Trade-offs are negotiated	Partnership	
Advice is received but not acted on	Placation	Degrees of tokenism
Citizens are heard but not always heeded	Consultation	
Citizens' rights and options are identified	Informing	Non-participation
Powerholders educate citizens	Therapy	
Rubberstamp committees	Manipulation	

Figure 3.2 Arnstein's ladder of participation

Source: Arnstein, 1969

Meaningful engagement in EA is when those potentially affected by development, or who have a vested interest in development, are enlisted into the planning, assessment, and decision process to contribute to it, thus providing opportunities for the exchange of information, opinions, interests, and values. It also means that those initiating the process of engagement (e.g., proponents, governments) are open to the potential need for change in a proposed development and are prepared to work with different interests to alter plans or to amend or even drop existing proposals. Meaningful engagement in EA extends beyond issuing public notice that an undertaking is about to occur or making project information available and soliciting public feedback (Table 3.3). Meaningful engagement helps to promote the legitimacy of the project, the proponent, and the EA process (Prno & Slocombe, 2012; Nakamura, 2008). It also helps to ensure that EA addresses both the needs and values of the proponent and of the communities or interests affected (Voutier, Dixit, Millman, Reid, & Sparkes, 2008; Booth & Skelton, 2011).

Sometimes proponents or governments will argue that it can be more efficient to exclude the public from the early stages of EA, given that the public often lacks the project-specific expertise necessary to contribute to such matters as project design and impact management strategies, arguing instead for "educating the public" about the project. However, early and ongoing public engagement is beneficial at all stages of the EA process, from initial project planning and design to post-decision analysis and monitoring (Table 3.4). By involving the public, proponents and governments can access a wider range of information about a project's potential impacts; improve project design and impact management strategies; identify socially acceptable solutions; ensure more balanced decision-making; and build capacity in local communities. While engaging the public early can extend the time needed during the initial project-planning phases, this initial investment is usually returned later in the

Table 3.3 Basic Characteristics of Meaningful Engagement in EA

- Early notice to those potentially affected about the prospects of a development proposal and opportunities for engagement.
- Access to complete and accurate information about a proposed development.
- A working relationship with potentially affected communities to identify potential problems and concerns and to work together on developing solutions.
- Transparency, whereby development plans, decisions, and decision-making processes are publicly accessible.
- Ensuring that affected communities have the necessary resources (financial, technical, human) to engage in the EA process and remain engaged post-EA approval.
- Affected communities are willing to engage for the purpose of improving project design, managing impacts, and providing information of relevance to the decision-making process.
- Opportunity for formal, legal challenge or intervention should community concerns not be adequately addressed or due process for engagement not followed.
- Proponents and communities have a genuine interest in working together to understand the issues and concerns of both parties and to resolve them.
- Opportunity to influence a project's design and the outcomes of the EA process.

Source: Noble & Udofia, 2015

Table 3.4 Objectives of Public Engagement throughout the EA Process

EA stage	Public involvement objectives
Pre-project planning Project description Screening	Early identification of affected interests and values to minimize conflict. Build public understanding about the project and its need and purpose. Input on project design features, including alternatives and impact minimization. Establishing an engagement plan for participation in the EA process. Opportunity for public review and challenge of the EA screening decision.
Scoping and baseline assessment	Identification of potentially affected values and desired outcomes or conditions. Elicitation of local knowledge for baseline assessment and understanding change. Identification of other areas of public concern and suggestions for management.
Impact assessment	Elicitation of local knowledge to assist impact predictions and characterization of potential impacts.
Impact management	Identification of opportunities for managing impacts, from impact avoidance to compensation strategies.
Significance determination	Identification of publicly acceptable limits, targets, or thresholds. Elicitation of public input about the significance of effects to values.
Submission and review of EIS	Informing public of impact assessment results. Identification of errors or omissions in the EIS. Providing public an opportunity to challenge EIS assumptions and predictions.
Recommendations and decision	Resolution of potential conflicts. Final integration of responses from EIS review in project approvals or conditions.
Implementation and follow-up	Maintenance of trust and credibility. Identification of management effectiveness. Elicitation of local knowledge in data collection and ongoing monitoring of environmental change.

Source: Based on Petts, 1999

process because it minimizes or avoids conflict and facilitates project approval and implementation (Noble, 2004). From a corporate perspective, Eckel, Fisher, and Russell (1992) suggest that consultation with different groups—regulators, shareholders, governments, and communities—can also help to clarify expectations about environmental performance and boost a firm's reputation with the public.

Participant Funding

In some jurisdictions, public engagement in EA processes is supported by **participant funding programs**. Federally, for example, when an EA is carried out by the federal Impact Assessment Agency, the agency must also establish a participant funding program to provide financial resources to facilitate public participation in:

a. the agency's preparations for a possible impact assessment of—or the impact assessment of and the design or implementation of follow-up programs in relation to—designated projects that include physical activities that are designated by regulations under the act;
b. the impact assessment of, and the design or implementation of follow-up programs in relation to, designated projects that are referred to a review panel; and
c. regional assessments and strategic assessments.

Noble and Udofia (2015), however, report that the level of funding made available through participant funding programs is often inadequate to ensure meaningful engagement. To address this challenge, especially for large or controversial projects that impact Indigenous lands and resources, project proponents may also provide potentially affected communities with financial resources to support their engagement in EA processes. One such example, described by Noble (2016), is the Sisson tungsten-molybdenum project, New Brunswick (Box 3.3).

Box 3.3 Participant Funding for the Sisson Project, New Brunswick

The Sisson tungsten-molybdenum project, proposed in 2008 by Northcliff Resources Ltd, involved the development of an open-pit mining operation, mineral processing facility, and water treatment plant, located near Napadogan, approximately 60 kilometres northwest of Fredericton. The mining operation would extract an average of 30,000 tonnes per day of tungsten- and molybdenum-containing ore for on-site processing over its 27-year operating lifespan (Sisson Partnership, 2015). The project was subject to review under New Brunswick's Environmental Impact Assessment Regulations and the Canadian Environmental Assessment Act.

The proposed mine was on the traditional territory of the Maliseet First Nations—in a region of historical significance to the Maliseet and used for hunting, fishing, and gathering of medicinal and ceremonial plants. The closest First Nations communities, the Woodstock, St Mary's, and Kingsclear First Nations, also expressed concerns about potential impacts on land and resources used for traditional purposes and impacts on archaeological resources. To support their participation in the EA process, Northcliff signed a capacity funding agreement with the St Mary's First Nation, the Woodstock First Nation, and the Assembly of First Nations Chiefs of New Brunswick, which at the time collectively represented all First Nations in New Brunswick. The funding assisted the First Nations in retaining their own technical experts to assist them in reviewing the project and assessing potential impacts on traditional lands and resources. Northcliff's support was in

continued

addition to funding the First Nations received under the Canadian Environmental Assessment Agency's participant funding program.

The capacity funding was an important step in creating an environment to support meaningful participation in the Sisson EA, but the provision of funding does not guarantee the success of a project or lasting support. In the spring of 2016, the Mi'kmaq Chiefs of New Brunswick declared their opposition to the Sisson mine, and five Maliseet First Nations, including St Mary's, called on the federal government to reject the project, following a federal study that concluded the mine would have a significant impact on some communities.

Source: Noble, 2016

Identifying the Publics

The Council of Canadian Academies (2019) explains that many potential actors or interests can be engaged in the natural resource management process, but not every decision requires the same level of engagement or collaboration of every actor. There is no such thing as "the public" in EA; rather, there are many "publics"—some of whom may emerge at different times during the EA process depending on their concerns and the issues involved. Mitchell (2002) makes a distinction between **active publics** and **inactive publics**. The active publics are those who affect decisions, such as industry associations, environmental organizations, quasi-statutory bodies, and other organized interest groups. Inactive publics are those who do not typically become involved in environmental planning, decisions, or issues and may include the "average" town citizen (Diduck, 2004). When involving the publics in EA, it is important not to overrepresent the active publics and to ensure adequate representation of the inactive publics. Particular attention should thus be given to those who reside in the area where the project will be implemented and who may be directly affected by the project.

One way to approach identification of the publics for involvement in EA is to consider the "influence" of the public group versus the group's "stake in the outcome" (Figure 3.3). Different publics may need to be involved in different capacities and at different stages of the EA process. For example, publics with little stake in the outcome (i.e., they will not be directly affected) and with limited power and influence over the project decision and EA process may be considered spectators and involved only indirectly through public communications, news releases, and education about the project. Those with a high stake in the outcome, such as an affected local population, but with limited power and influence should be intimately involved throughout the EA process to ensure that their concerns are addressed, even though they may have limited authority to influence decisions. This second group is often referred to as the "victims" of project development: they may experience some benefits from development, but they also have the most to lose if project impacts are not properly managed. It is for this group that funding to participate, such as through participant funding programs, is important to ensure that the public has the necessary resources and capacity to become involved.

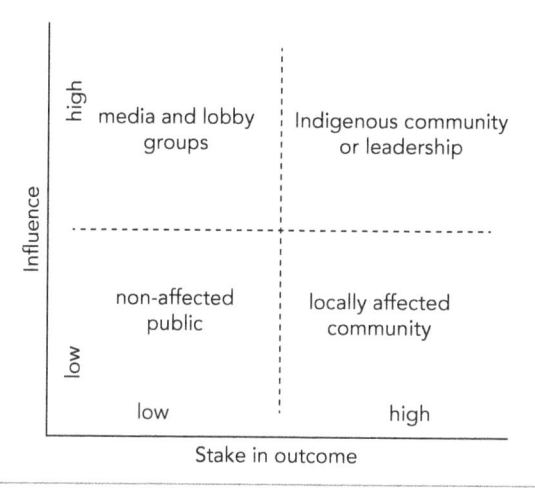

Figure 3.3 Influence versus stake in outcome when identifying publics

At the other end of the spectrum are publics who have a relatively limited stake in the outcome but are highly influential. This group might include quasi-regulatory bodies, the media, and other special interest groups. Caution must be taken to ensure that the voice of this highly influential group is not overrepresented relative to that of the "victims." The final group of publics is those with both high stakes in the outcome and a high degree of influence over the process. For regulators and proponents, this is the most complex group of publics, since they have a potential for considerable gains and losses from project development and at the same time are highly influential over the EA process and project success. In the Canadian context, influential Indigenous groups and quasi-government bodies could be classified as high-stake and high-influence interests.

Supporting Tools

There are many tools for public engagement and communication, both in the pre-project planning stages and throughout the EA process. The capability and capacity of these tools vary considerably (Table 3.5). For example, a proponent may hold information seminars early in the project design stage for the communities or interests most likely affected by a project, whereas other communities or publics outside of the direct project impact area may be informed at this point only through the mass media. In other instances, especially when Indigenous communities are involved, a project proponent may engage in more collaborative processes and development partnerships, such as the negotiation of benefits-sharing arrangements (Chapter 7). The specific tools selected for public involvement in any EA depend on such factors as the proponent's objectives and commitment to engagement; legal requirements for participation; level of public interest; sensitivity of the biophysical or human environments potentially affected by the project; available time and resources of both the proponent and the affected publics of interest to engage; and the level of conflict or controversy surrounding the development—which can be high in the case of novel technologies, highly uncertain impact management strategies, large-scale resource extractive projects, or projects where land rights are contested.

Table 3.5 Selected Techniques for Public Involvement and Communication

Capability meets the criterion = ✓ Capacity high = ◉ medium = ◉ low = ○	Capability				Capacity		
	Provide information	Obtain feedback	Resolve conflict	Identify problems and values	Two-way communication	Caters to special interests	Number of people involved
Public meetings	✓	✓		✓	◉	○	◉
Public displays	✓	✓			◉	○	◉
Presentations to small groups	✓	✓		✓	◉	◉	○
Workshops		✓	✓	✓	◉	◉	○
Advisory committees		✓		✓	◉	◉	○
Access to project documents	✓	✓		✓	○	◉	◉
Information brochures	✓				○	◉	◉
Press release inviting comments	✓	✓			○	○	◉
Public hearings		✓		✓	○	○	◉
Task forces				✓	◉	◉	○
Site visits	✓			✓	◉	◉	○
Information seminars	✓	✓			◉	◉	○
Mass media (television, radio)	✓				○	○	◉
Design charrettes	✓	✓	✓	✓	◉	◉	○
Social media	✓	✓		✓	○	○	◉

Sources: Updated based on original tools and evaluation by Sadar, 1996; Westman, 1985

Project Description

Once it is determined that the proposed project is the "best" option and is necessary for meeting the identified need or opportunity, a detailed project description must be prepared. The project description plays an important role in the determination of the need for an EA and, if needed, the scope of the assessment. Many governments provide project proponents with guidance on the types of information that must be included in their project description—referred to in some jurisdictions as the project application. The level of detail requested can vary by jurisdiction and based on project scale. When preparing a project description, it is important to include at least the following:

Project purpose: A description of the rationale for the project, including its need, purpose, and consideration of viable alternatives to the project.

Project information: A description of the project's location and design, including viable alternatives considered; activities and activity scheduling associated with construction, operation, and decommissioning phases; design specifications; resources and material requirements; waste production, discharges, and management.

Policy or regulatory context: Identification of relevant legislative, regulatory, and policy requirements and their interpretation in relation to the project; licences and applicable mitigation standards.

Engagement strategy: Identification of results of early public engagement, including key issues and concerns, affected interests, and the plan for meaningful engagement throughout the EA process.

Assessment information: Preliminary description of baseline environment, land uses, key issues, and potentially affected components in a manner that will allow those responsible to determine the need for assessment and establish the initial scope for assessment.

Key Terms

active publics	meaningful engagement
alternative means	multi-criteria evaluation
alternatives to	participant funding program
inactive publics	Peterson matrix
ladder of participation	public participation

Review Questions and Exercises

1. What are some of the key responsibilities of a project proponent during the pre-project planning phase?
2. Using Box 3.1 as an example, work in small teams to calculate a Peterson matrix to evaluate alternatives for supplying electricity to a mining project being proposed in the northern, remote parts of Canada. Discuss the advantages and limitations of the Peterson approach.
3. What are the requirements for alternatives consideration under Canada's federal impact assessment system? Are there requirements under your provincial or territorial EA system?
4. Why is public engagement important in the pre-project planning stages?
5. Assume that you are the responsible authority for project submissions and have received an application from a remote northern community to construct a diesel-fired electric generation project to meet growing energy demands. What reasonable alternatives to the project would you expect the proponent to consider? Discuss whether in all situations it is appropriate to require that a proponent consider alternatives to their project proposal, other than the "no project" option.
6. A proponent recently submitted an EA application for a large-scale wind energy facility to be located on the boundaries of your current municipality/city.

As a consultant for the proponent, you are responsible for early pre-project engagement with potentially interested and affected interests. Using Figure 3.3, identify and classify potential interests based on their influence and stake in the outcome. For each group of participants, identify a possible engagement technique from Table 3.5.

References

André, P., et al. (2006). Public participation: international best practice principles. Special Publication Series no. 4. Fargo, ND: IAIA.

Arnstein, S. (1969). A ladder of citizen participation. *Journal of the American Institute of Planners* 35, 216–24.

Booth, A.L., & Skelton, N.W. (2011). Industry and government perspectives on First Nations' participation in the British Columbia environmental assessment process. *Environmental Impact Assessment Review* 31, 216–25.

Council of Canadian Academies. (2019). *Greater than the sum of its parts: toward integrated natural resource management in Canada*. Ottawa, ON: The Expert Panel on the State of Knowledge and Practice of Integrated Approaches to Natural Resource Management in Canada.

Dahl, T.E., & Watmough, M.D. (2007). Current approaches to wetland status and trends monitoring in Prairie Canada and the continental United States of America. *Canadian Journal of Remote Sensing* 33(1), 17–27.

DEAT (Department of Environmental Affairs and Tourism). (2004). Criteria for determining Alternatives in EIA. Integrated Environmental Management, Information Series 11, Department of Environmental Affairs and Tourism (DEAT), Pretoria, South Africa.

Diduck, A. (2004). Incorporating participatory approaches and social learning. In B. Mitchell (ed.), *Resource and environmental management in Canada*, 3rd edn, pp. 497–527. Toronto, ON: Oxford University Press.

Eckel, L., Fisher, K., & Russell, G. (1992). Environmental performance measurement. *CMA Magazine* (March), 16–23.

Houlahan, J., Keddy, P.A, Makkay, K., & Findlay, C.S. (2006). The effects of adjacent land use on wetland species richness and community composition. *Wetlands* 26(1), 79–96.

Mitchell, B. (2002). *Resource and environmental management*. 2nd edn. New York, NY: Prentice Hall.

Morrison-Saunders, A. (2018). *Advanced introduction to environmental impact assessment*. Cheltenham, UK: Edward Elgar.

Nakamura, N. (2008). An "effective" involvement of Indigenous people in environmental impact assessment: the cultural impact assessment of the Saru River region, Japan. *Australian Geographer* 39(4), 427–44.

Nielsen, J., Noble, B.F., & Hill, M. (2012). Wetland assessment and impact mitigation decision support framework for linear development projects: The Louis Riel Trail, Highway 11 North project, Saskatchewan, Canada. *The Canadian Geographer* 56(1), 117–39.

Noble, B.F. (2004). Integrating strategic environmental assessment with industry planning: a case study of the Pasquai-Porcupine forest management plan, Saskatchewan, Canada. *Environmental Management* 33(3), 401–11.

Noble, B.F. (2016). Learning to listen: snapshots of Aboriginal participation in environmental assessment. Ottawa, ON: Macdonald-Laurier Institute.

Noble, B.F., & Udofia, A. (2015). *Protectors of the land: toward an environmental assessment process that works for Aboriginal communities and developers*. Ottawa ON: Macdonald-Laurier Institute.

Peterson, G.L., Gemmel, R.S., & Shofer, J.L. (1974). Assessment of environmental impacts: multiple disciplinary judgments of large-scale projects. *Ekistics* 218, 23–30.

Petts, J. (1999). Public participation and environmental impact assessment. In J. Petts (ed.), *Handbook of Environmental Impact Assessment*. London, UK: Blackwell Science.

Prno, J., & Slocombe, D.S. (2012). Exploring the origins of "social license to operate" in the mining sector: perspectives from governance and sustainability theories. *Resources Policy* 37(3), 346–57.

Sadar, H. (1996). *Environmental impact assessment*. 2nd edn. Ottawa, ON: Carleton University Press.

Sisson Partnership. (2015). *Sisson project: final environmental impact assessment report* (February). Vancouver: Sisson Mines Ltd. http://www2.gnb.ca/content/gnb/en/departments/elg/environment/content/environmental_impactassessment/sisson.html.

Voutier, K., Dixit, B., Millman, P., Reid, J., & Sparkes, A. (2008). Sustainable energy development in Canada's Mackenzie Delta–Beaufort Sea coastal region. *Arctic* 61(5), 103–10.

Westman, W. (1985). *Ecology, impact assessment, and environmental planning*. New York, NY: John Wiley.

WSP Canada Inc. (2019). James Bay Lithium Mine. Environmental impact statement. Trois-Rivières, QC.

4 Determining the Need for Assessment

Screening

The number of projects that could potentially be subject to EA is quite large. In some jurisdictions, EA is triggered only for *major* projects such as nuclear power plants, major energy pipelines, or large-scale hydroelectric projects. In other jurisdictions, even small or routine developments, such as road culverts, are subject to assessment. Depending on the jurisdiction, it may be the responsibility of the project proponent to determine whether their development is subject to EA laws and regulations and to inform the responsible government agency or department—usually an impact assessment branch or agency or an environment ministry. Typically, a project proponent provides the responsible government agency or department with a detailed project description or project development application, and the government agency or department makes a screening determination.

Screening is the narrowing of the application of EA to projects that require assessment because of the potential for adverse effects or because EA is required by way of certain regulations. The screening process asks: "Is an EA required?" The question will normally result in one of the following answers:

- no, an EA is not required;
- yes, an EA is required;
- a limited EA is required, consisting only of a preliminary assessment or mitigation plan; or
- further study is necessary, such as a preliminary environmental report or consultation study, to determine whether an EA is required.

The purpose of screening is to ensure that no unnecessary assessments are carried out but that developments warranting assessment are not overlooked.

Screening Approaches

There are ways to systematize the screening process, thus improving accountability and decision transparency in determining whether a proposal requires an EA. The most straightforward way is to compare the anticipated impacts of a project with the quality parameters of the environment specified in relevant legislation. However, listing quality parameters of the environment is not a common practice in EA legislation. Requirements for screening are thus highly variable from one EA system to another. Generally speaking, there are three main approaches to screening: list-based, case-by-case, and hybrid.

List-Based Screening

List-based screening, also referred to as prescriptive screening, involves a list of projects for which an EA is (or is not) required based on the potential of that project to generate significant effects or based on regulatory requirements and responsibilities. In California, for example, the California Environmental Quality Act lists projects for which EA must always be completed, based on project characteristics, thresholds, and geographic location. As well, a negative list outlines projects for which an EA is not required. Such lists are typically referred to as **project lists** and include projects that have either mandatory or discretionary requirements for EA. Under the Canadian Impact Assessment Act, for example, the Physical Activities Regulations defines those projects for which an EA is required, referred to as **designated activities** (Box 4.1).

Box 4.1 Designated Activities Requiring Assessment under the Impact Assessment Act

Impact assessment under the Impact Assessment Act applies only to designated activities. A designated activity means one or more physical activities that are carried out in Canada or on federal lands and are designated by the Physical Activities Regulations under the act or designated in an order made by the minister. It can also include any physical activity incidental to those activities. The Physical Activities Regulations include a schedule of those activities that may require assessment. The list includes, but is not limited to, certain projects in national parks, mines and metal mines of a specified capacity, nuclear facilities (including storage and disposal), certain oil and gas operations, renewable energy projects of a specified capacity, and transmission lines of a certain length or capacity. Activities on the list are largely defined based on project type and the physical characteristics of the undertaking versus the local context or nature of the receiving environment. For example:

- A new coal mine provided the coal mine has a production capacity of 5000 t/day or more.
- An *in situ* oil sands extraction facility if the production capacity is 2000 m^3/day or more and the project is not in a province in which there is existing legislation to regulate greenhouse gas emissions from *in situ* sites or is in a province in which provincial legislation is in force to limit greenhouse gas emissions produced by oil sands and that limit has been reached.
- A renewable energy project if the project is a hydroelectric facility with a production capacity of 200 MW or more, in-stream tidal with a production capacity of 15 MW or more, or new tidal power generation that is not an in-stream facility.
- A new all-season public highway that requires a total of 75 kilometres or more of new right-of-way.
- Expansion of an existing dam on a natural water body if the expansion would result in an increase in the surface area of the existing reservoir of 50 per cent or more and an increase of 1500 ha or more in the annual mean surface area of that reservoir.

Source: Physical Activities Regulations: SOR/2019-285

Project lists often include specified thresholds so that not all *types* of projects captured by a project list require assessment. Thresholds are often based on different types or classes of development or project size or magnitude (e.g., total reservoir size for a hydroelectric facility) or on environmental thresholds as established by regulations (e.g., total emission levels or concentrations). For example, it may be the case that not all wind turbine facilities require assessment, but turbine facilities that include more than a certain number of turbines, generate more than a specified level of energy output, are located within a certain distance of a migratory bird sanctuary, or occupy more than a specified land area require an assessment—whereas those that do not meet the threshold(s) do not require assessment. The province of Nova Scotia Environmental Assessment Regulations Schedule A—Class I and Class II Undertakings under Section 49 of the Nova Scotia Environment Act (1994–5, c. 1) lists specific projects and identifies thresholds for which an EA is required. For example, an EA is required for a facility for the incineration of municipal solid waste. An EA is also required for the construction of a water reservoir, provided the reservoir has a planned storage capacity that exceeds the mean volume of the natural water body by 10 million cubic metres or more.

Case-by-Case Screening

Case-by-case screening, also referred to as discretionary or criterion-based screening, involves evaluating project characteristics against a checklist of regulations, criteria, or general guidelines as projects are submitted. There is no prescribed list of projects that are subject to EA—two similar projects, but in different locations, may receive different screening determinations. In Saskatchewan, for example, the need for an EA is largely determined on the basis of screening criteria identified in the Saskatchewan Environmental Assessment Act. The act indicates that any project, operation, or activity, or any alteration or expansion of such, is subject to an assessment if it is likely to:

- have an effect on any unique, rare, or endangered feature of the environment;
- substantially utilize a provincial resource and pre-empt the use, or potential use, of that resource for other purposes;
- cause the emission of pollutants or create by-products or residual or waste products that require handling or disposal in a manner that is not regulated by any other act or regulation;
- cause widespread public concern because of potential environmental changes;
- involve a new technology that is concerned with resource utilization and that may induce significant environmental change; or
- have a significant impact on the environment or necessitate a further development that is likely to have a significant impact on the environment.

Case-by-case screening allows for maximum flexibility for EA application—it is sensitive to context, is dynamic, and provides for better consideration of the local and regional environment in which development is proposed. At the same time, however, case-by-case screening can be time-consuming, inconsistent, and sometimes difficult to defend if the screening criteria are too vague.

Both list-based and case-by-case screenings are vulnerable to *screening out* projects that should be assessed. For list-based screening, problems arise when projects fall just below a prescribed threshold. For example, if the EA threshold for an electrical transmission line is set at 50 kilometres in length, transmission lines that are only 40 kilometres in length are not subject to EA even though they are likely to generate very similar environmental effects. In an international study of screening mechanisms, Mayer et al. (2006) identified several additional challenges to list-based screening, including:

- listed project types, characteristics, or thresholds may be outdated and may not conform with the current state of the art;
- thresholds may be identified even though there is no sound reason for them;
- regional differences in the sensitivity of the environment are not considered;
- specified thresholds may not be appropriate in all situations and contexts;
- thresholds are set either too high or too low to properly capture significant impacts.

Such "small" projects noted above *could* be captured under case-by-case screening, but the decision is discretionary; there must be a political will to invest time and resources in the assessment of seemingly small development actions—which is the exception rather than the norm (Westbrook & Noble, 2016). The determination of the need for assessment depends heavily on the relative weight given to each criterion considered and on the context within which significant decisions about the project's potential impacts are being made (Snell & Cowell, 2006).

Environmental Assessment in Action

Screening Out Small Projects in a Grassland Ecosystem

Approximately 35 per cent of operating petroleum and natural gas (PNG) wells in the prairie ecozone are in the province of Saskatchewan. Approximately 80 per cent of PNG production over the next 10 years will come from currently undrilled wells. This is of major concern to grassland conservation efforts, given that nearly all of Saskatchewan's unconventional gas deposits are located underneath grassland ecosystems. A typical PNG site contains one to five wellheads and supporting infrastructure, which can include pumps, separators, and solution tanks for conventional reserves and water injection and disposal facilities for non-conventional reserves. The effects of PNG development on grasslands relate predominantly to surface disturbance associated with well sites and access roads and the disruption of nutrient and water exchange through the soil. An application for development of an individual PNG site is rarely considered significant enough to trigger an EA—under either the Saskatchewan provincial or federal EA acts and regulations (Noble, 2008; Westbrook & Noble, 2016).

continued

Typical PNG lease site in southwest Saskatchewan for conventional oil.

An independent assessment was conducted to examine the impacts of nearly 50 years of PNG development on southern Saskatchewan grasslands (see Nasen, Noble, & Johnstone, 2011). The assessment examined the effects of PNG activity on Prairie Farm Rehabilitation Administration lands, Agriculture and Agri-Food Canada, in the Swift Current–Webb Community Pasture. The pasture is approximately 9882 hectares, and 170 wells have been drilled within the pasture boundaries. The physical footprint of PNG infrastructure was examined using aerial imagery. Land-use and fragmentation metrics were calculated for five-year intervals, including patch density and edge density and the percentage of pasture occupied by roads, trails, and PNG lease sites. Field data were collected from a sample of 31 of the total 170 PNG sites. At each lease site, measurements were made along four transects that extended outwards from the lease-site centre. Environmental and biotic data were collected at sampling locations along the transect, and for every PNG site surveyed, a reference sample of non-PNG pastureland was also sampled. Data collected included percent herbaceous, percent bare ground, percent litter, percent aggregates, percent club moss, percent oil spill/contaminated, soil compaction, pH, conductivity, percent silt, percent sand, percent clay, and range health.

Between 1979 and 2005, the physical footprint of PNG infrastructure was found to have increased from 0.20 per cent to 1.00 per cent of the landscape. During this same time period, lease access roads increased from occupying 0.13 per cent to 0.42 per cent of the landscape. Patch density values per 100 hectares increased from 0.11 in 1979 to 0.42 in 2005. On average, over the 50-year period of PNG development, three site access roads were associated with each new PNG site constructed. The edge density of the 11 access roads in the study area in 1979 was

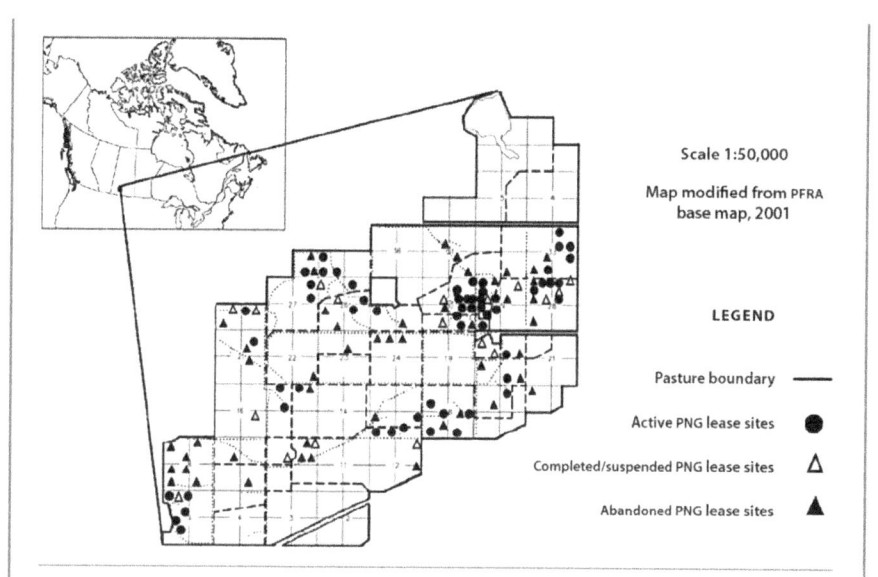

PNG activity on Swift Current–Webb Community Pasture, southwest Saskatchewan

3676 metres per hectare compared to 1954 metres per hectare for 41 access roads in 2005. Edge density for the 14 wells in the study area in 1979 was 419 metres per hectare, while edge density for 108 wells in 2005 was 595 metres per hectare. In 2006, 33 per cent of the total surface disturbance in the study area was attributed to PNG site access roads, 49 per cent to PNG lease sites, and 18 per cent to other roads and trails.

The 31 PNG lease sites surveyed were found to have lower percent cover of herbaceous plants, club moss, and litter and greater cover of bare ground than reference pasture sites. The upper 20 centimetres of soil at PNG sites had, on average, greater pH, electro-conductivity, and percent clay values than the reference pasture sites. The analysis of plant community data indicated that PNG sites had a greater undesirable species abundance and diversity than reference pastures. Active and suspended PNG sites also had range health values significantly lower than reference pastures.

The effects of PNG activity on grasslands were found to extend well beyond the direct disturbance of the physical infrastructure itself. On average, the impacts of PNG sites extended 25 metres from the PNG wellhead. In 1979, the percentage of the study area occupied by PNG sites and PNG access roads was 0.07 per cent and 0.13 per cent, respectively. When the spatial extent of the ecological effects of PNG activities is considered, the total footprint of PNG sites (25 metres) and access roads (20 metres) increased to 0.31 per cent and 0.67 per cent, respectively. The total footprint of PNG development pre-1979 accounted for 25 per cent of the total disturbed area in the pasture. By 2005, when the ecological effects of PNG sites and PNG access roads are considered, the spatial extent of PNG impacts

continued

had increased to 5.1 per cent, accounting for 75 per cent of the total disturbance in the area.

The effects of PNG development on grasslands in Saskatchewan is an example of how small disturbances on the landscape, based on their physical footprint, can be quite significant when assessed on the basis of their ecological footprint. Even more important, the assessment showed that a combination of seemingly insignificant projects can result in significant cumulative effects. In the case of the Swift Current–Webb Pasture, cumulatively, PNG development now accounts for 75 per cent of all disturbances to grasslands in the area. Moreover, PNG lease sites developed in 1955 have shown no significant improvements in terms of ground cover, species diversity, and range health compared to those developed more recently. Small projects do not always result in small or short-term impacts.

Source: Based on Nasen, Noble, & Johnstone, 2011

Hybrid Screening

Hybrid screening presents an alternative approach whereby list-based and case-by-case screening are used as complementary screening processes such that listed projects above specified thresholds or located in sensitive areas, for example, are subject to mandatory assessment, whereas projects that fall below the threshold or are not located in sensitive areas are screened for the need for EA on a case-by-case basis. Hybrid screening (Figure 4.1) is essentially a **threshold-based screening** system with allowances for case-by-case consideration. Under this approach, a prescriptive screening mechanism provides for a list-based screening approach to projects and thresholds for which assessment is always required, while a discretionary case-by-case screening mechanism is applied to projects that fall below mandatory thresholds, fall in between inclusion and exclusion thresholds or criteria, or are included in descriptive lists (DETR, 1998).

EA always required

....

[EA more likely to be required, but test remains likelihood of significant adverse effect]

Case-by-case consideration

[EA less likely to be required, but test remains likelihood of significant adverse effect]

....

EA not required

Inclusion threshold

Indicative threshold

Exclusion threshold (except projects in sensitive areas)

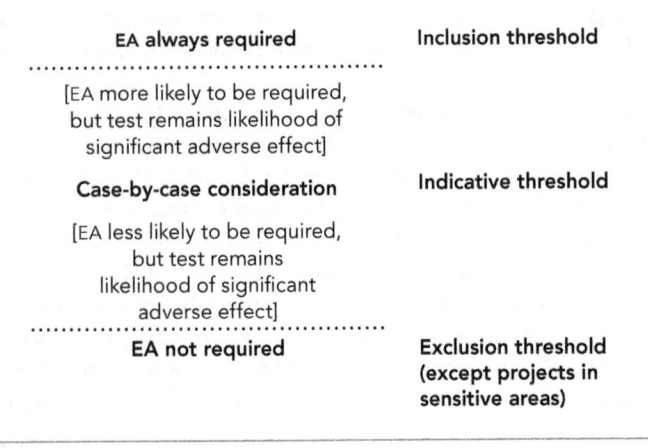

Figure 4.1 Hybrid screening model

Level of Assessment Required

If a determination is made that an EA is required, consideration must be given to the type of EA or level of effort necessary, given the potential for adverse impacts or public concern. Not all projects undergo the same level of assessment (Box 4.2). A proposed nuclear energy facility is likely to require a considerably larger and more

---

Box 4.2 EA Screenings and Reviews in Nunavut

The Nunavut Impact Review Board (NIRB), established under the Nunavut Land Claims Agreement, is responsible for EA in the Nunavut Settlement Area (NSA), which consists of the Kitikmeot, Kivalliq, and Qikiqtani regions. Project proponents must submit a project proposal to the Nunavut Planning Commission (NPC) to determine whether the proposed undertaking conforms to the relevant land-use plan under the Nunavut Project Planning and Assessment Act (NuPPAA). The NuPPAA provides a one-window approach to land and resource management in Nunavut, which integrates initial reviews for all matters related to wildlife management, land-use planning, EA, water licensing, and dispute resolution. If the project conforms, the NPC will determine whether the project is exempt from screening by the NIRB. Many projects considered "low-impact," such as individual ship movements, community resupply, and developments within municipalities, are exempt from screening and are only required to undergo assessment if there are concerns about cumulative effects.

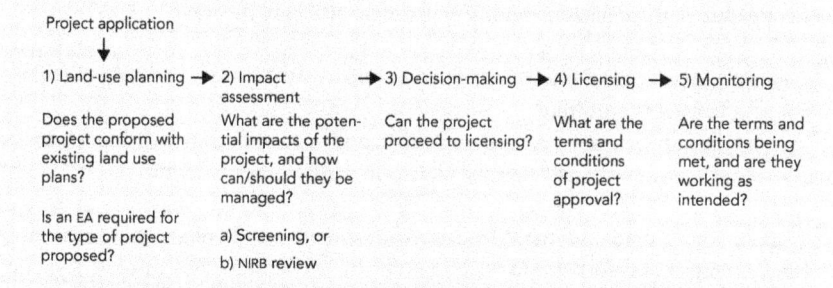

Project application
↓

1) Land-use planning →	2) Impact assessment →	3) Decision-making →	4) Licensing →	5) Monitoring
Does the proposed project conform with existing land use plans? Is an EA required for the type of project proposed?	What are the potential impacts of the project, and how can/should they be managed? a) Screening, or b) NIRB review	Can the project proceed to licensing?	What are the terms and conditions of project approval?	Are the terms and conditions being met, and are they working as intended?

If the project is not exempt, it is forwarded to the NIRB for a screening. Under the NIRB, a screening is a type of assessment to determine whether a full EA review is needed. The screening determines whether the project should proceed and receive the necessary permits or whether a more comprehensive review is required. The screening includes public consultations and an assessment of potential impacts, including information on the spatial extent of effects, ecosystem sensitivity, historical and cultural significance, probability of impact, reversibility, and the potential for cumulative impacts. Screening results in the project being either modified or abandoned, approved subject to terms and conditions, or referred for a full EA review.

A full EA review by the NIRB is required only for projects where significant public concern exists, where projects may be using untested technologies, or where there is

continued

> potential for a significant impact. An option also exists to refer a project to a federal review panel, though no such referrals have been made to date. The NIRB review process is comprised of a scoping phase, during which input is received from communities on the scope of the assessment and key issues, and the development of guidelines for the proponent to undertake the assessment. The proponent completes and submits to the NIRB a draft EIS, which undergoes both a public and a technical review, with feedback and requests for additional information and analysis provided to the proponent. The final EIS is then submitted, which may undergo further public and technical review, and a public hearing is held to determine whether the needs and concerns of the public have been sufficiently addressed. A recommendation is then made on whether the project should proceed or be modified or abandoned. Most project proposals in Nunavut are subject to screenings. Only 11 projects have proceeded to full EA review.
>
> Source: Nunavut Impact Review Board, 2013a; 2013b; 2013c

complex assessment than a forest access road extension project. Under Western Australia's EA system, for example, a notice of intent is submitted to the Australian Environmental Protection Agency describing the nature of the project, potential environmental effects, alternatives to the project, and proposed impact management measures. This is referred to as an **initial environmental examination** (IEE) or **environmental preview report** (EPR). This information is then used to determine whether an EA is needed and what level of assessment is required. Based on Australia's procedures, the decision stemming from an IEE may be that:

- no further assessment is required;
- a full EA is required;
- an examination by an independent commission of inquiry is necessary; or
- a less comprehensive public environmental report is required.

Under the Canadian federal Impact Assessment Act, there are three main types of assessment: a standard impact assessment by the Impact Assessment Agency; a panel review assessment; and a panel review assessment when there is a life-cycle regulator. Most designated projects are subject to assessment by the agency and follow a standard assessment process whereby the proponent prepares the EIS and the agency conducts the assessment and issues its recommendation(s). However, a designated project may be referred to a **review panel** should concerns emerge over the potential adverse impacts of the project, when there is due public concern about the impacts of the project, when there is an opportunity for EA cooperation with other jurisdictions, or when there are adverse impacts on the rights of Indigenous peoples. Review panels are responsible for holding public hearings and preparing the impact assessment report. For designated projects for which there is a life-cycle regulation, such as projects regulated under the Nuclear Safety and Control Act (e.g., nuclear reactors, nuclear fuel management), should an assessment be required it is automatically referred to an assessment by review panel. The assessment must meet the requirements of both the Impact Assessment Act and the respective regulatory act. After the project is developed, the life-cycle regulator is then responsible for monitoring project compliance.

Terms of Reference

If an EA is required, the government agency or EA office responsible for overseeing the assessment would normally prepare detailed **terms of reference** for the EIS. The terms of reference are informed by regulatory requirements, consider key issues raised during pre-project planning, and account for the detailed project description provided by the project proponent. The terms of reference provide a roadmap for project proponents, outlining the minimum requirements that must be met when preparing the EIS. The terms of reference also serve as a checklist for government agencies and the public once an EIS is submitted, ensuring that the proponent has provided the information required to proceed to the review stage. If the proponent has not met the requirements, the responsible government agency or EA office will normally return the EIS to the project proponent to provide the information.

The terminology used to denote the terms of reference for EA can vary by jurisdiction. In British Columbia, for example, the terms of reference are referred to as "application information requirements." In Newfoundland, the terms of reference are referred to as the "EIS guidelines" and are developed by the government for each project in consultation with the project proponent. Federally, under the Impact Assessment Act the terms of reference for EA are referred to as "tailored impact statement guidelines." The Impact Assessment Agency has developed a template for impact statements for designated projects (Table 4.1). Alberta has standardized terms of reference for EAs for certain types of projects, including coal mines, industrial plants, *in situ* oil sands projects, and oil sand mines. The intent is that the proponent tailor the standardized terms of reference to fit the specific context of their project. In most jurisdictions, there is an opportunity for the public to review and comment on the terms of reference before they are finalized.

Table 4.1 Generic Content of an EIS Terms of Reference

Introduction	Methodology used for effects assessment
Overview: proponent and project location	Predicted changes to the physical environment
Detailed project description, including components, activities, and workforce requirements	Effects on the environment
	Effects on human health
Project purpose, need, and alternatives considered	Effects on social and economic conditions
	Effects on Indigenous peoples
Description of public participation and views	Impact mitigation and enhancement measures
Description of engagement with Indigenous groups	Residual effects
Methodology, information sources, boundaries, and valued components for baseline assessment	Cumulative effects assessment
	Accidents or malfunctions
Baseline assessment—biophysical	Effects of the environment on the project
Baseline assessment—human health	Contributions of the project to sustainability
Baseline assessment—social and economic	Follow-up and monitoring programs
Baseline assessment—Indigenous peoples and land uses	Assessment summary

Source: Based on the Impact Assessment Agency of Canada's Tailored Impact Statement Guidelines Template

Screening and the Precautionary Principle

EA can foster precaution insofar as it provides a mechanism for anticipating adverse environmental impacts associated with a proposed development (Glasson, Therivel, & Chadwick, 1999). Within the context of screening, precaution focuses on determining what is a *sufficiently* sound or credible basis for requiring an EA. The screening stage is pivotal in this regard in that it ensures that activities that are likely to cause adverse impacts are given proper treatment and assessment. The problem is that it is often not possible at the screening stage for the competent authority to prove, with certainty, the significance or insignificance of the impacts of a project. The suggested approach is to err on the side of caution; however, as Snell and Cowell (2006) explain, avoiding harm in the absence of evidence does not always sit well with the ethos of EA in that decisions should be based on sound information.

The **precautionary principle** suggests that when scientific information is incomplete but there is a threat of adverse impacts, the lack of full certainty should not be used as a reason to preclude or to postpone actions to prevent harm. In other words, when considerable uncertainty exists as to whether a proposed activity is likely to cause adverse effects, the lack of certainty about adverse effects should not be a reason for approving the proposed activity, for not requiring an EA, or for not requiring rigorous mitigation and monitoring measures (IAIA, 2003). Erring on the side of caution, the precautionary principle as a screening guideline places the burden of proof on the proponent, making the proponent responsible for demonstrating "insignificance" (Lawrence, 2005).

However, the precautionary principle is not without problems, and the principle itself has been misused to justify everything from minimal change to rejecting any project proposal. Moreover, the interpretation of precaution within EA is troubled by concerns over the efficiency of the process as well as by the need to ensure that precaution about likely, significant impacts does not unduly constrain development (Snell & Cowell, 2006). That said, the underlying objective of adopting a precautionary approach to screening is to provide better assurance that potentially significant effects are captured during the screening process and that an EA is carried out when needed.

Key Terms

case-by-case screening
designated activities
environmental preview report
hybrid screening
initial environmental examination
list-based screening

precautionary principle
project lists
review panel
screening
terms of reference
threshold-based screening

Review Questions and Exercises

1. What is the purpose of screening? Should *all* proposed developments be subject to EA? What would be the advantages and limitations?
2. What type of screening process is used in your EA jurisdiction? Discuss the relative advantages and disadvantages of list-based versus case-by-case approaches to screening.
3. Consult the EA regulations in your jurisdiction, or consult the federal regulations, and determine whether the following project proposals would likely be subject to an EA:
   a. The construction of a fossil fuel–fired electrical generating station with a production capacity of 250 MW.
   b. The construction of a 100-kilometre-long, two-lane public highway through a migratory bird sanctuary.
   c. The construction of a coal mine with a production capacity of 4800 tonnes per day.
   d. The expansion of an existing petroleum storage facility from its current storage capacity of 500,000 m^3 to 700,000 m^3.
   e. The reclamation of an abandoned gold mine project and processing facility.

References

DETR (UK Department of Environment, Transport, and the Regions). (1998). *Draft town and country planning (assessment of environmental effects) regulations.* London, UK: DETR EC Directive 85/337/EEC First Schedule Part II Regulations 1989; DETR 1998 (proposed amendment to Directive 97/11).

Glasson, J., Therivel, R., & Chadwick, A. (1999). *Introduction to environmental impact assessment.* London, UK: Spon Press.

IAIA (International Association for Impact Assessment). (2003). Social impact assessment international principles. Special Publication Series no. 2. Fargo, ND: IAIA.

Lawrence, D. (2005). *Significance criteria determination in sustainability-based environmental assessment.* Report to the Mackenzie Gas Project Joint Review Panel. Langley, BC: Lawrence Environmental.

Mayer, S., Correia, M.R., Pinho, P., Cruz, S.S., Hrncarova, M., Lieskovska, Z., & Paluchova, K. (2006). Projects subject to EIA. Vienna, Austria: Improving the Implementation of Environmental Impact Assessment (IMP)3 Project Team, Austrian Institute for Regional Studies and Spatial Planning.

Nasen, L., Noble, B.F., & Johnstone, J. (2011). Effects of oil and gas lease sites in a grassland ecosystem. *Journal of Environmental Management* 92, 195–204.

Noble, B.F. (2008). Strategic approaches to regional cumulative effects assessment: a case study of the Great Sand hills, Canada. *Impact Assessment and Project Appraisal* 26(2), 78–90.

Nunavut Impact Review Board. (2013a). *NIRB Public Guide Series: introduction.* Cambridge Bay, NU: NIRB.

Nunavut Impact Review Board. (2013b). *NIRB Public Guide Series: screening.* Cambridge Bay, NU: NIRB.

Nunavut Impact Review Board. (2013c). *NIRB Public Guide Series: review*. Cambridge Bay, NU: NIRB.

Snell, T., & Cowell, R. (2006). Scoping in environmental impact assessment: balancing precaution and efficiency? *Environmental Impact Assessment Review* 26, 359–76.

Westbrook, C.J., & Noble, B.F. (2016). Environmental assessment in Saskatchewan. In K. Hanna (ed.), *Environmental assessment: practice and participation* (pp. 340–53). Toronto, ON: Oxford University Press.

5 Scoping and Baseline Assessment

Scoping

If the quality of an EA were measured only by how much shelf space (or megabytes) the EIS required, then the state of practice must surely be improving. Complex projects do require a significant volume of information; however, the number of volumes and maps contained in an EIS and the total page count are not indicators of the quality of an assessment. A large volume of information *could* be collected about the biophysical and human environment when conducting an EA, and a long list of techniques for collecting those data could be employed that draw on existing data sets (e.g., monitoring programs, historical records, land-use plans) or that rely on applied field methods (e.g., **remote sensing**, interviews, field sampling). EA is not about undertaking data collection for the purpose of generating a descriptive history or environmental study of a region. More information does not necessarily translate into better assessment or a more informed decision. Good EA is focused on those issues most important to managing the impacts of the proposed project and to informing the decision(s) at hand. Too often, EAs are rich in description and weak in analysis of trends and associations between environmental change and key drivers of change.

A central component of EA and one that strongly affects EA quality is **scoping**—determining the issues and parameters that should be addressed in EA, establishing the spatial and temporal boundaries of the assessment, and focusing the assessment on the relevant issues and concerns (Kagstrom, 2016). To consider *all* issues, impacts, and environmental components in any single EA is neither feasible nor desirable. Scoping is about determining what elements of the project to assess, what environmental components are likely to be affected, how these environmental components have changed over time and what factors have driven such change, and how these components may be affected by other actions or disturbances in the project's local and regional environment. Scoping limits the amount of information to be gathered in EA to a manageable level and identifies specific objectives and indicators to guide the assessment. Scoping thus serves a number of important functions in the EA process:

- ensuring early input from those affected by development;
- identifying public and scientific concerns and values;
- focusing the assessment and providing a coherent view of the issues;
- ensuring that key issues are identified and given an appropriate degree of attention;
- reducing the volume of unnecessarily comprehensive data and information;
- defining the spatial, temporal, and other boundaries and limits of the assessment; and
- ensuring that the EA is designed to maximize information quality for decision-making purposes.

Baseline Assessment

An **environmental baseline** considers the past, present, and possible future state of the environment *without* the proposed project or activity (Figure 5.1). A **baseline study** consists of identification and analysis of conditions over space and time for the purpose of delineating change, trends, patterns, or limits to assist in impact assessment, impact evaluation, and impact monitoring activities.

Baseline assessments in EA are comprised of three main parts:

1. identifying and selecting the **valued components** (VCs) for inclusion in the assessment;
2. establishing the spatial and temporal boundaries for the assessment; and
3. assessing the condition of, and changes or trends in, VCs.

Identifying and Selecting Valued Components

Baseline studies must be undertaken within the context of clearly defined scope and objectives; otherwise, too much information is acquired that is often of too little use. While it is important to ensure that all potentially affected environmental components are given consideration, attention should focus on key VCs. Valued components are aspects of the environment, physical and human, that people value and are considered important, thus warranting detailed consideration in an impact assessment.

In most cases, VCs are components of the environment such as fish and wildlife species or species groups whose distribution and abundance are of management interest because they are harvested, at risk, or sensitive to disturbance or aspects of the human environment, such as social services and infrastructure, health, or traditional land uses. However, VCs can also be identified in relation to ecological processes, such as carbon sequestration or flood control, and even broad or holistic concepts such as ecological integrity or human health and well-being.

The rationale for selecting a VC for consideration in EA is normally based on its ecological importance, societal importance, or regulatory importance. In the Mackenzie Gas Project EA, a proposed 1200-kilometre-long natural gas pipeline from the Beaufort Sea in the Northwest Territories to northern Alberta, for example, wildlife VCs in the pipeline corridor were identified because of their regulatory status,

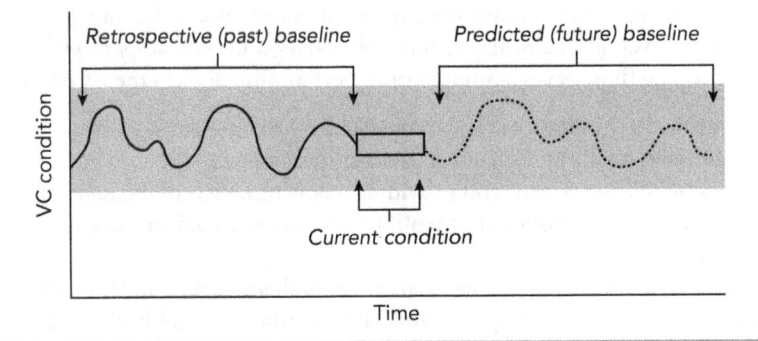

Figure 5.1 Past, present, and future baseline. Shaded region depicts the range of normal variation in VC condition.

ecological importance, socio-economic importance, and conservation concern. For the Voisey's Bay project, a nickel-copper-cobalt mining operation in Newfoundland and Labrador, 17 key VCs, including water, caribou, plant communities, Indigenous land use, and family and community, were identified on the basis of social, economic, regulatory, and technical values and concerns. In the case of the Elk Valley cumulative effects framework, VC selection was guided by local values, available knowledge, and also the linkages between the potential VC and regulatory decision-making (Box 5.1).

Olagunju and Gunn (2013) examined the rationale for VC selection across a sample of 11 Canadian road construction EAs and found that ecological importance and societal value were common rationales, but other rationales for VC selection included: fragility, scientific value for study or monitoring, importance to legal compliance, economic importance, professional judgment, biodiversity and conservation value, medicinal importance, recreational value, and spiritual importance. Ball, Noble, and Dubé (2012), however, in an analysis of VC selection across a sample of 35 federal and provincial EAs in the South Saskatchewan watershed, from multiple sectors, found that VC selection for the aquatic environment was based most often on the need for a proponent to address their exposure to liability under the federal Fisheries Act and the Species at Risk Act. Those VCs protected by punitive federal legislation were used more than twice as often as other aquatic VCs. Ball, Noble, and Dubé also observed that the number and diversity of VCs considered in any single EA was higher in those assessments that engaged some form of public consultation in the scoping process.

Determining VC interaction with project actions or activities
If there is no potential for interaction between the actions or activities of the proposed project and the VC, then the VC is less likely to be considered or assessed in detail in the project EA. To ensure that baseline studies are purposeful for managing project impacts and supporting decisions about project proposals and not simply a compilation of information about VCs, it is important to ask at least the following two questions:

Box 5.1 Principles for VC Selection in the Elk Valley, British Columbia

The Elk Valley is in the Rocky Mountains in the southeastern region of British Columbia, home to the communities of Elkford, Sparwood, Hosmer, Fernie, Morrissey, and Elko. A popular tourism destination and home to many important species such as grizzly bears, bighorn sheep, and American dipper elks, the Elk Valley also contains the largest producing coalfield in British Columbia. There are five surface metallurgical coal mines operating in the valley, all owned by Teck Coal Ltd. In 2012, in response to an approval condition for a coal mine expansion that required Teck Coal Ltd to engage in broader discussions about the impacts of development in the Elk Valley, a working group was formed by Teck and the Ktunaxa Nation Council, including various provincial government agencies, municipal and regional government representatives, environmental non-government organizations, and other resource companies operating in the region, to establish a cumulative effects management framework. The working group adopted a consensus-based

continued

approach to scoping the Elk Valley assessment, including VC selection. VC selection was based on the following principles:

- VCs must be based on the values identified by residents of the Elk Valley, and potential linkages between values will be considered;
- any selected VC must reflect several values (ecological, social, cultural, and economic);
- there must be reason to believe that the VC is or will be affected either directly or indirectly by current or future activities in the region;
- the VC must be sensitive to several important disturbing and supportive ecological and human processes, including interactions among these processes;
- there must be sufficient scientific knowledge about the VC to allow selection of indicators that will respond in a way that can be readily measured;
- any assessment of the VC must help to inform regulatory decisions likely to be included in permits and licences for development decisions;
- the VC must represent traditional knowledge and traditional uses of the land and water resources of the Elk Valley; and
- results of the assessment of the VC must be capable of contributing to more confident decision-making.

The Elk Valley initiative is an example of a collaborative and transparent approach to scoping and VC identification, guided by agreed-upon principles and grounded in the values of those most affected by development and with the most relevance to management decisions. The cumulative effects management framework was developed between 2012 and 2017 and is discussed in further detail in Chapter 11.

1. Is the VC likely to be affected, directly or indirectly, by project activities?
If there is no potential for interaction between the project and the VC, then there is no need to compile a comprehensive baseline about that VC. This is not to say that the VC should be disregarded, since it may still be of significant public concern and warrant some consideration in the impact assessment process. Or it may be determined important for understanding potential cumulative impacts and supporting regional monitoring programs even though it may not be affected by the project under consideration.

To identify the potential for interaction between VCs and project activities, practitioners often use simple **impact matrices**. While specific design and format vary from project to project, matrices are essentially two-dimensional checklists that consist of project activities on one axis and potentially affected VCs on the other (Figure 5.2). Matrices are commonly used for impact identification and communication in project baseline studies and are particularly useful for identifying first-order relationships between specific project activities and potential impacts and for providing a visual aid for impact summaries.

Perhaps the best-known and most comprehensive impact matrix is the traditional **Leopold matrix**, originally developed for the US Geological Survey by Leopold et al. (1971). The Leopold matrix consists of a grid of 100 possible project actions

5 | Scoping and Baseline Assessment 75

Forest management activity		Valued components						
		Moose	Woodland caribou	White-tailed deer	Fur-bearers	Bats	Rodents	Raptors
Infrastructure development	All-weather access roads	I	M	I	I	I	I	I
	Dry-weather access roads	I	M	I	I			
	Winter roads							
	Camps, timber and fuel storage sites		I		I			
Harvesting	Logging	M/*	M	M/*	I-M/*	M	I	I-M/*
	Slashing and woody debris management						I/*	I/*
	Timber storage							
Forest renewal	Site preparation		*		I		I	
	Tree establishment		*		*			
	Mechanical stand tending							
	Chemical stand tending	I	*				I	

Matrix legend: Blank cell = not applicable or no impact; "I" = insignificant / mitigable; "M" = significant / mitigable; "N" = significant / non-mitigable; * = positive

Figure 5.2 Section of an impact identification matrix from the Tembec, Pine Falls Operations, Manitoba, Forest Stewardship Plan 2010–2029, depicting potential interactions of forest management activities with valued wildlife components

Source: Select data from "Environmental Impact Statement of the Forest Management Licence 01, 2010-2029 Forest Stewardship Plan, Executive Summary," p. 3 https://www.gov.mb.ca/conservation/eal/registries/4572tembec/tembec_eis/_exe_sum.pdf

along a horizontal axis and 88 environmental considerations along a vertical axis, for a total of 8800 possible first-order project–component interactions. Each cell of the matrix consists of two values: a quantification of the magnitude of the impact and a measure of impact significance. Where an interaction is anticipated, the matrix cell is marked with a diagonal line in the appropriate row and column. The magnitude of any potential impact is then indicated within the top diagonal of each cell, typically on a scale from −10 to +10, and the importance of the impact or interaction is indicated in the bottom half of the diagonal (Figure 5.3). An advantage of the Leopold matrix is that it can easily be expanded or contracted based on the specific project and environmental context. The sheer magnitude of the Leopold matrix, however,

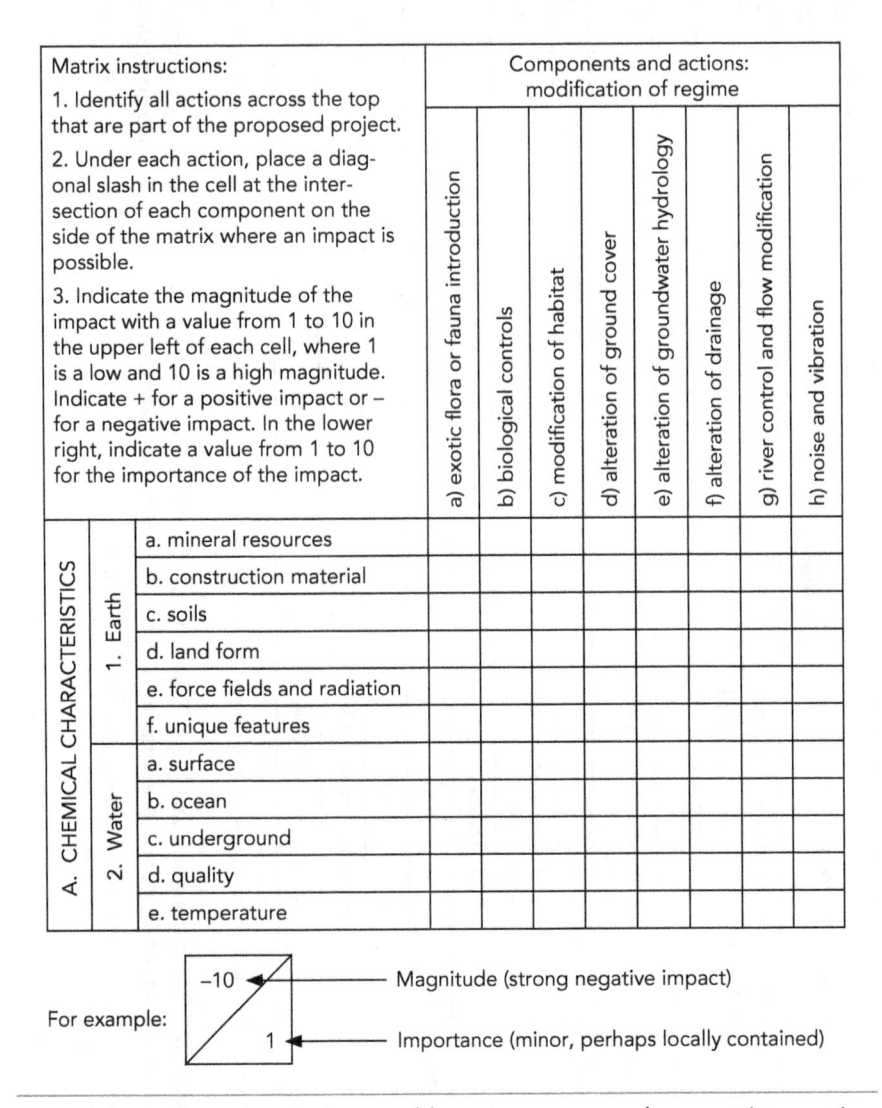

Matrix instructions:	Components and actions: modification of regime								
1. Identify all actions across the top that are part of the proposed project. 2. Under each action, place a diagonal slash in the cell at the intersection of each component on the side of the matrix where an impact is possible. 3. Indicate the magnitude of the impact with a value from 1 to 10 in the upper left of each cell, where 1 is a low and 10 is a high magnitude. Indicate + for a positive impact or − for a negative impact. In the lower right, indicate a value from 1 to 10 for the importance of the impact.		a) exotic flora or fauna introduction	b) biological controls	c) modification of habitat	d) alteration of ground cover	e) alteration of groundwater hydrology	f) alteration of drainage	g) river control and flow modification	h) noise and vibration
A. CHEMICAL CHARACTERISTICS / 1. Earth	a. mineral resources								
	b. construction material								
	c. soils								
	d. land form								
	e. force fields and radiation								
	f. unique features								
2. Water	a. surface								
	b. ocean								
	c. underground								
	d. quality								
	e. temperature								

For example: −10 ← Magnitude (strong negative impact); 1 ← Importance (minor, perhaps locally contained)

Figure 5.3 Section of a simple Leopold matrix to scope and communicate project impacts and characterize impact importance

Source: Based on Leopold et al., 1971

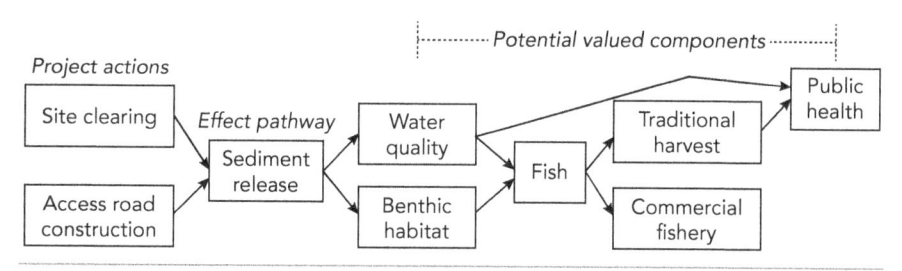

Figure 5.4 Example of a valued component effect pathway
Source: Based on British Columbia Environmental Assessment Office, 2013

can be a major drawback. The Leopold matrix was developed for use on many different types of projects. It thus tends to generate an unwieldy amount of information for any single project application (Glasson, Therivel, & Chadwick, 1999) unless the matrix is simplified and adapted to the specific context. Impact matrices are to be used to communicate interactions between project activities and VCs—they should be simple and communicate information that is useful for assessors and decision-makers.

2. Is information needed about the VC to support the assessment of project impacts on another VC?
Sometimes VCs are pathways for effects that might lead to effects or impact on other VCs. Guidance for VC selection by the British Columbia Environmental Assessment Office (2013) explains that a component may be one step in a pathway along which a project effect travels (Figure 5.4). Understanding the condition of, or change in, that component may be important to understanding the potential impact on a VC of concern. For example, water is often identified as a VC in baseline studies and also serves as an effects pathway that can potentially affect other VCs, such as freshwater fish habitat or waterfowl. In other cases, discharge from a mining operation, for example, may affect water quality and benthic habitat. Changes in these parameters may affect the health or survival of a valued fish species, which may disrupt recreational fishing activity or reduce a traditional food supply.

Selecting VC indicators

Identifying **VC indicators** is important to understanding and tracking actual change in VC conditions and, more important, to providing an early warning of potential adverse effects. VCs employ nomenclature used in EA, but they are not always measurable in their own right (Antoniuk et al., 2009). Often, indicators that are measurable need to be identified to assess or evaluate the status of, or threats to, VCs (Table 5.1).

Indicators provide a sign or signal that relays a complex message, potentially from numerous sources, in a simplified and useful manner (Jackson, Kurtz, & Fisher, 2000). In assessing the VC "surface water quality," for example, attention may focus on a specific indicator (e.g., phosphorus concentrations, benthic invertebrate abundance) that provides direct, measurable information about the condition or state of the VC. These are referred to as **condition-based indicators** (Table 5.2). In other cases, for VCs such as specific wildlife species or supporting habitat, attention may focus on measurable stress or disturbances that affect the VC (e.g., riparian habitat disturbance or stream crossing density for fish or the density of linear features for

Table 5.1 Example Environmental Components, Key Issues, VCs, and Indicators from the Cold Lake Oil Sands Project

Environmental System	Issue of Concern	Valued Component	Indicators
Air	Acid deposition, odour, GHG emissions	Air quality	Emitted gases transported over long distances (NO_x, SO_2)
Surface water	Lowering of lake levels, contamination	Water quantity and quality	Combined water volume withdrawals, water quality, drinking water standards
Aquatic resources	Contamination of fish, increased harvest pressure	Sport fish species	Northern pike
Wildlife	Loss, fragmentation of habitat, mortality due to hunting	Hunted and trapped species	Moose, black bear, lynx, fisher
Resource use	Decreased opportunities for resource harvesting—fish, traditional plants, trapping—increased road access	Timber harvest areas, furbearers, game species, new road access, recreational enjoyment	Aspen stands, beaver, moose, campsites

Source: Imperial Oil Resources, 2017

Table 5.2 Examples of Condition- and Stress-Based Indicators for Terrestrial and Aquatic VCs

Condition-based indicators for terrestrial VCs:	Condition-based indicators for aquatic VCs:
habitat effectiveness index; habitat suitability index; soil conductivity; range health; soil pH; soil erosion and compactness; presence/absence of invasive plant species; species richness; species population; species distribution; habitat corridor size; core habitat area; habitat patch size	total and dissolved organic carbon; pH; temperature; chloride; suspended sediments; total ammonia; metal indicators (various); nutrients; total dissolved phosphorous; benthic invertebrates (abundance, richness, evenness, index of biological integrity); fish population (relative abundance, catch per unit effort, condition factors, nutritional health, gonad size, fish tissue chemistry); riparian plant community; Hilsenhoff biotic index; ambient concentrations of regulated discharge parameters; quantity/flow
Stress- or disturbance-based indicators for terrestrial VCs:	**Stress- or disturbance-based indicators for aquatic VCs:**
industrial footprint; core habitat area loss; habitat patch size; habitat corridor size; linear features density; km roads/km²; total cleared area; percent area disturbed; total area burned; wildlife–vehicle collisions; number of hunting licences	stream access density; stream crossing density; number of hung culverts; km roads/km²; total cleared area; water abstraction rate; percent area disturbed by class of activity; disturbed riparian area; percent impervious surfaces; total area burned; species harvest rate; number of angling licences

Figure 5.5 **Right-of-way cleared for a liquefied natural gas pipeline development north of Kitimat, British Columbia**. The total length and density of linear features in a region, and the edge habitats created, are useful indicators for understanding habitat loss, risks to vulnerable species, and overall cumulative threat to ecosystem health and function.

terrestrial wildlife species) (Figure 5.5). These are referred to as **stress-based indicators** or **disturbance-based indicators**.

Indicators are the most basic tools for analyzing baseline conditions and changes in VCs—they allow practitioners and decision-makers to gauge environmental change efficiently by avoiding impact noise and focusing on parameters that are responsive to change, generate timely feedback, and can be traced effectively over space and time. The choice of indicators is often influenced by data availability (Cairns, McCormick, & Niederlehner, 1992). As such, the highest-priority indicators are often those for which there is already good scientific or local knowledge of how human activities and natural changes affect the indicator (Antoniuk et al., 2009). Good indicators must be indicative of the causes or sources of change in, or stress to, a VC and not only the existence of change. Guidance prepared by the British Columbia Environmental Assessment Office (2013) suggests that meaningful VC indicators for baseline assessments are:

- *Relevant*: The relationship between the indicator and the VC is based on scientific principles and knowledge and can be linked to the project's actions or activities.
- *Measurable:* Changes in the indicator reflect measured changes, or early warning of change, in the VC.

- *Predictable:* Indicators provide consistent and comparable assessments over time and, ideally, across projects and jurisdictions.
- *Understandable:* Indicators describe VCs that matter to people and can be communicated in a way that is understandable.
- *Responsive:* Indicators are likely to change, and are responsive to, project impacts.
- *Practical:* Data are readily available and/or data collection and assessment/analytical methods are practical and cost-effective.
- *Appropriate to scale:* Indicators are geographically and temporally relevant to the scale of the project.
- *Definable in a targeted condition:* A specific value for the indicators can be established *a priori* that will establish the predicted or desirable condition after mitigation measures are implemented.
- *Aligned with permitting or decision-making:* Indicators are relevant to regulatory actions or decisions about the project and its impacts.

Establishing VC objectives

Establishing objectives, usually based on benchmarks or thresholds, is important to understanding when a significant, acceptable, or unacceptable level of change in the indicator has occurred or is about to occur. A **benchmark** is a standard or point of reference against which change may be compared or assessed, such as the **range of natural variability** in an environmental phenomenon. A **threshold** is an established limit of change and may include:

- *acceptable limits of change:* what is acceptable from a broader societal perspective; and
- *desired VC conditions or objectives:* desired outcomes.

When environmental effects or conditions change beyond acceptable levels, or when VC objectives are not met, then project impacts are considered "significant" (see Chapter 8), and some form of impact management is necessary. Salmo Consulting Inc. (2004) suggest a tiered approach to setting thresholds or management targets, consisting of cautionary thresholds, target thresholds, and critical thresholds (Figure 5.6). At a **cautionary threshold**, monitoring efforts should be increased to more closely monitor VC or indicator conditions and the effectiveness of best-management practices verified to prevent any further adverse change. A **target threshold** is typically politically or socially defined—a margin of safety and a mandatory trigger for management action. A **critical threshold** defines maximum acceptable change, socially or ecologically, beyond which impacts may be long-term or irreparable. Such thresholds are sometimes referred to as **management targets** and represent the desired or target condition for VC indicators (Table 5.3)

Establish Assessment Boundaries

Scale matters in EA (Therivel & Ross, 2007). The choice of scale can have important implications for whether baseline assessments, and impact predictions, sufficiently capture the extent of the project's interactions in VCs—across space and over time. Scale also matters in terms of jurisdictional responsibility and whether the VCs of concern or the project's impacts on those VCs are transboundary or multi-jurisdictional

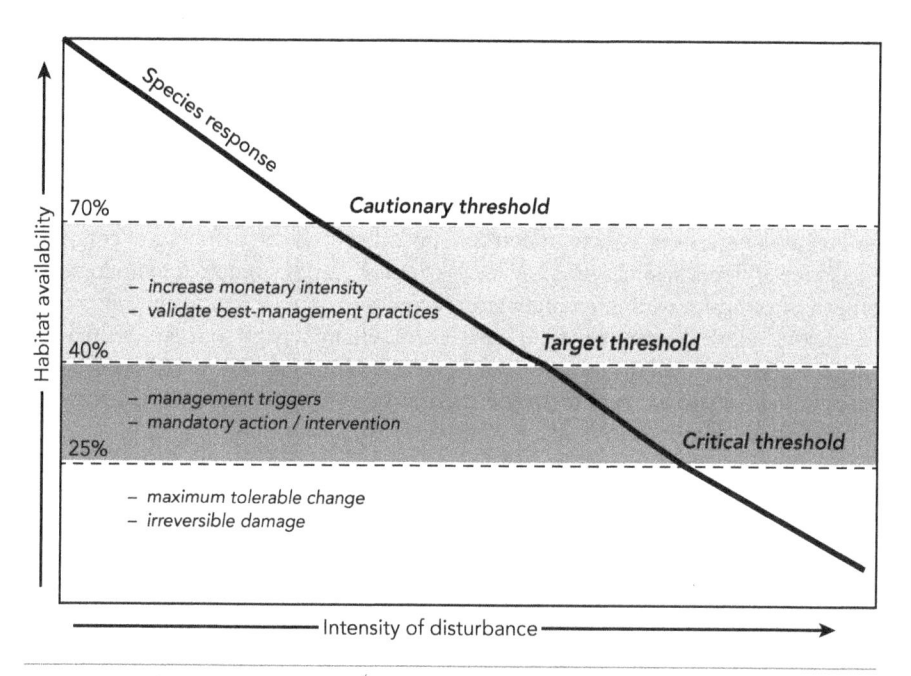

Figure 5.6 Cautionary, target, and critical thresholds
Source: Based on Salmo Consulting Inc, 2004

in nature. Three types of boundaries are thus considered when establishing baseline conditions: spatial boundaries, temporal boundaries, and jurisdictional boundaries.

Spatial bounding
What are the spatial limits of the assessment? In delineating the spatial limits of assessment, some consideration must be given to scale. João (2002) suggests that scale has two interrelated meanings relevant to EA: first, scale as the *spatial extent* of the assessment; second, scale as the amount of *geographic detail*. When the spatial scale of an EA covers a very large area, a high degree of geographic detail is less feasible.

Table 5.3 Example of Grizzly Bear as a Value and Its Assessment Components, Indicators, and Management Targets

Value	Assessment component	Indicator	Management target
Grizzly bear	Functional habitat condition	Road density	< 0.6 kilometres per square kilometre road density
	Habitat (food) supply	Percent structural stage distribution in forested land	< 30 per cent of forested land base in mid-seral age class
	Population size	# human-induced bear mortalities per year	< 6 per cent bear mortality due to human interaction in population unit

Source: British Columbia Ministry of Environment, 2014

In other words, as the geographic boundary of an assessment increases, the level of detail typically decreases. When considering the impacts of linear developments, such as a pipeline or transmission line project, on caribou, for example, a practitioner may adopt coarse stress- or disturbance-based indicators, such as the density of linear features or various landscape metrics (e.g., habitat intactness, industrial footprint). The spatial scale of an assessment is an important factor in the nature and types of indicators used to understand potential impacts to the VCs of concern.

Since different spatial scales and levels of detail are often required to understand project interactions with different environmental receptors, different environmental receptors should be examined at different scales, and for any given receptor there are various scales at which the receptor can be assessed (Box 5.2). For example, air quality can be measured across a metropolitan area or an airshed or within the confines of a particular administrative space. In other words, factors affecting air quality can be assessed across a variety of **functional scales**, ranging from point source to local, to regional (including regions of various sizes), to national (Chagnon, 1986).

It is also important to distinguish between two types of information: the activity information and the receptor information (Irwin & Rodes, 1992). **Activity information** characterizes the types of effects a project might generate, such as habitat fragmentation or increased use of non-renewable resources. **Receptor information** refers to the processes resulting from such effects. Both of these must be taken into consideration when delineating spatial boundaries. For example, the former Canadian Environmental Assessment Agency's operational policy for spatial bounding regarding offshore oil and gas exploratory drilling suggests that the spatial area of an assessment should be the "composite" of the spatial area of all affected environmental receptors and should consider uncertainties concerning the precise location of

Box 5.2 Basic Principles for Spatial Bounding

- Boundaries must be large enough to include relationships between the proposed project, other existing projects and activities, and the affected environmental components (Cooper, 2003).
- The scope of assessment should cross jurisdictions if necessary and allow for interconnections across systems (Shoemaker, 1994).
- Natural boundaries should be respected (Beanlands & Duinker, 1983).
- Different receptors will require assessment at different scales (Shoemaker, 1994).
- Boundaries should be set at the point where effects become insignificant by establishing a maximum detectable zone of influence (Scace, Grifone, & Usher, 2002).
- Both local and regional boundaries should be established (Canter, 1999).

Geographic boundaries for any particular assessment will vary depending on a number of factors, including the nature of the project itself, sensitivity of the receiving environment, nature of the impacts, extent of transboundary impacts, availability of baseline data, jurisdictional boundaries and cooperation, and natural physical boundaries.

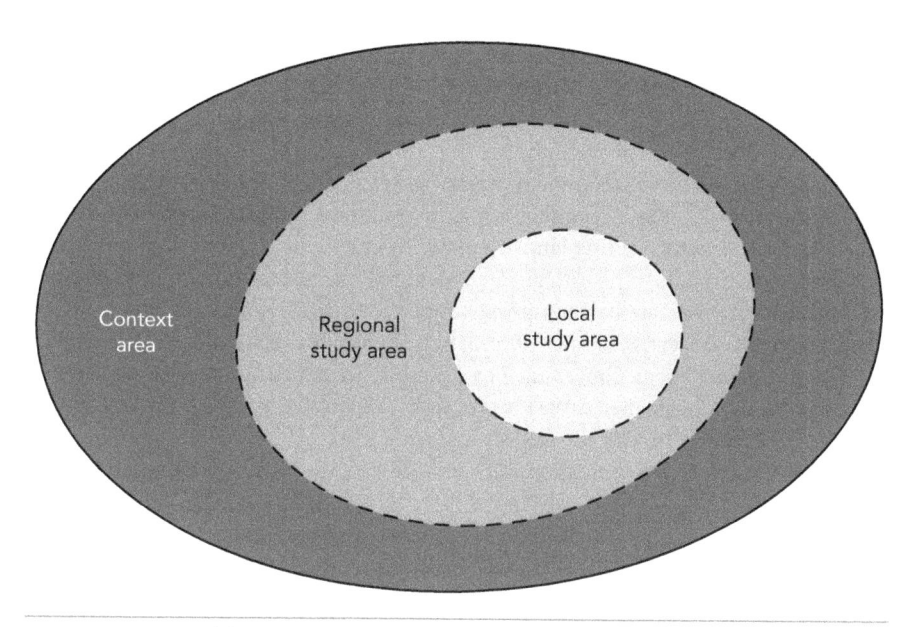

Figure 5.7 Spatial bounding based on local study area, regional study area, and broader context

proposed activities and effects, as well as the need to reconsider individual project bounding when multiple projects are being conducted in adjacent resource areas.

A common approach to spatial bounding is to adopt a set of nested study areas (Figure 5.7). Using this approach, a **local study area** is defined, delineating where the project's immediate or direct effects are likely to occur. This would include such things as site clearing or vegetation removal directly associated with project infrastructure. A **regional study area** is then defined for each VC that reflects the potential **zone of influence** of project effects and includes other activities within the region (e.g., other projects or human disturbances) that may also be affecting the VC of concern. The **context area** situates the VC of concern in a broader context, which may be required to understand background conditions, ecosystem functioning, or other factors that may help in identifying and understanding the nature and importance of effects caused by the project. In the absence of context, erroneous conclusions may be drawn about changes in the regional study area that may or may not have been caused by the project. Several methods can be used to assist in spatial bounding, including Geographic Information Systems, network and systems analysis, matrices, public consultation, quantitative and physical modelling, and expert opinion.

Temporal bounding
What are the temporal limits of the assessment? The temporal scale should include the past, present, and future. For baseline assessments, this involves consideration of previous and current activities in the project's region affecting environmental conditions. For impact predictions (Chapter 6), this also involves the consideration of reasonably foreseeable activities. For example, assessment of a proposed hydroelectric

transmission line project should consider previous linear disturbances that have affected the regional baseline environment, as well as current and proposed land-use activities that may contribute to additional habitat fragmentation.

How far into the past and how far into the future will depend on the quality and quantity of data available, the certainty associated with project environmental conditions, and what the assessment is trying to achieve. Examining past conditions may be as simple as examining land-use maps or census data, and it may be feasible to incorporate 50 years of historical data if deemed necessary for the proposed project. The most appropriate past temporal bounds for an assessment is a point in time when the VC of concern was most abundant or before the current drivers of change in the region were most dominant. For example, in the context of the Athabasca River system in Alberta, Seitz, Westbrook, and Noble (2011) suggest that an appropriate past temporal bound for impact assessment would be the late 1960s—prior to rapid agricultural intensification, oil sands development, and population growth in the watershed.

Establishing an appropriate future boundary for assessment may be based on the end of the operational life of a project, after project decommissioning and reclamation, or after affected environments recover. Future boundaries are much less certain, since data are often hypothetical or based on a range of future scenarios. Typically, future boundaries do not exceed a decade beyond project decommissioning; however, this may be too restrictive from a sustainability perspective or when dealing with legacy projects, such as mining operations and contaminated sites. Further attention is given to temporal boundaries in Chapter 11 in the context of assessing potential cumulative environmental effects.

Jurisdictional bounding and transboundary effects

In principle, EA should be spatially bounded based on ecological units, such as watersheds or eco-regions. In practice, administrative boundaries play an important role. Institutional boundaries and the administrative authority to implement impact management measures almost always temper ecological boundaries. The challenge is that certain, if not most, impacts spread across administrative boundaries and affect environments in other jurisdictions. In some cases, EA screening may indicate that the nature of the proposed project and its impacts requires the participation of multiple jurisdictions.

At the international level, the Convention on Environmental Impact Assessment in a Transboundary Context, commonly referred to as the Espoo Convention, establishes the responsibilities of nations assessing the transboundary impacts of projects. This convention outlines the obligations of the project's host country to notify, and include participation by, the affected neighbouring country. Members of the European Union and the East African Community have adopted these principles to varying degrees. The focus of attention is largely on consultation responsibilities concerning major developments or land-use changes in one jurisdiction that are likely to cause transboundary impacts in neighbouring countries.

Unfortunately, the majority of transboundary provisions for EA in the Canadian context are associated with physical infrastructure that is of a transboundary nature, such as pipelines or transmission lines that cross a provincial or territorial border. This is currently the major focus of transboundary screening and assessment in the Mackenzie

Valley of Canada's Northwest Territories, for example, where transboundary is primarily interpreted in relation to a development located partly in the Mackenzie Valley and partly in another area (see MVEIRB, 2004). The focus is on the physical activities associated with the project as opposed to the geographic extent of the effects of the project.

There are no specific rules for establishing jurisdictional boundaries in impact assessment or for determining when an adjacent jurisdiction should be involved in the assessment process. At a minimum, consideration should be given to:

- what jurisdictions are affected by the project;
- the extent to which adjacent jurisdictions are affected by the project;
- what jurisdiction or jurisdictions have decision-making authority;
- what jurisdiction or jurisdictions are responsible for managing project impacts and over what area; and
- the EA provisions and capacity of affected adjacent jurisdictions to manage potential impacts.

Condition and Change Analysis

The assessment of VC (or indicator) condition and trends over space and time is at the heart of baseline assessment. Three key questions should be considered in baseline change or trends assessment:

- What do we need to know about the baseline environment to make an informed decision about the acceptability of a project's impacts?
- What are the relevant background conditions that have influenced the current environment?
- What is the likely baseline condition in the future in the absence of the project?

A key component of establishing baseline conditions is identifying and establishing baseline trends and associations. There are a variety of tools available for identifying trends and associations, ranging from statistical correlations and regression trees to more complex Markov chain analysis. However, identifying trends and associations need not always adopt quantitative approaches. Sometimes simple, conceptual diagrams, such as network or system diagrams, are quite valuable for scoping potential relationships between project actions and potentially affected environment components or for simplifying complex system relationships.

Network or **system diagrams** are based on the notion that links and pathways of interaction exist between individual components of the environment such that when one component is affected there will be an effect on other components that interact with it (Cooper, 2012). The focus of network analysis is thus on identifying potential impact pathways through a series of linkages or network diagrams. Network diagrams are particularly useful for depicting sequential cause–effect linkages between project actions and multiple environmental components and can be used to describe how project activities could potentially lead to environmental changes that may affect certain components of the environment (Box 5.3). Network diagrams can be constructed based on mathematical representation of dynamic environmental processes or on more simplified and conceptual understandings of relationships.

Box 5.3 Network Diagrams for Mapping Impact Pathways

In June 2012, Manitoba Hydro and four Cree Nations together submitted an environmental impact statement (EIS) for the development of the Keeyask project—a 695-megawatt hydroelectric generating station at Gull Rapids on the lower Nelson River, northern Manitoba. The Keeyask project would consist of a powerhouse complex, a spillway, dams and dykes, and a reservoir. The project was subject to assessment under the Canadian Environmental Assessment Act and the Environment Act (Manitoba).

The EIS adopted an ecosystem approach to the identification of potential impacts to the terrestrial environment that would be triggered by project actions. The project's potential interactions with the terrestrial environment were examined by constructing several network or system diagrams (see below) to identify primary pathways of effects. This included, for example, identifying the project's primary impacts (e.g., project footprint, changes in water flow/levels) and the primary pathways associated with those impacts (e.g., altered hydrology, habitat edge, reservoir expansion) that linked to terrestrial environments and subsequent effects to terrestrial habitat and wetland function, wildlife movement, and terrestrial plants. Several approaches were used to develop and support the identification of potential pathways and linkages, including scientific knowledge concerning the causal relationships between ecosystem components (e.g., how soils are affected by flooding) and specific field-based data collection and models. Similar network or linkage diagrams were developed to identify aquatic ecosystem pathways of effects for the project. Details about the Keeyask project and the EIS, including technical and supporting analyses, can be found at www.keeyask.com.

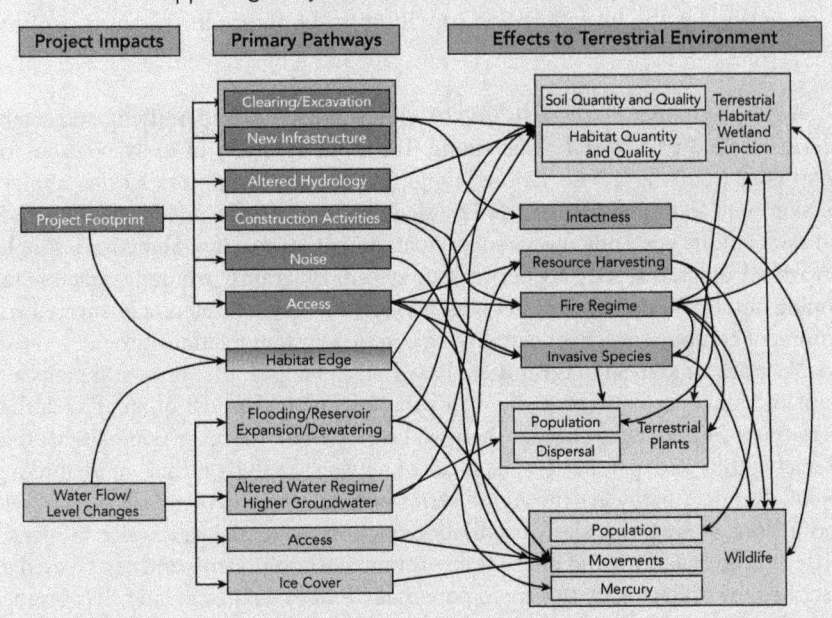

Source: Keeyask Hydropower Limited Partnership, 2012, Figure 6-16

Good baseline studies thus go beyond regional descriptions to provide an assessment of what is happening in the project's regional environment, specifically:

- What did VC or VC indicator conditions look like in the past?
- What do VC or VC indicator conditions look like at present?
- What are the trends, rates, patterns, and suspected drivers of change?
- What is the magnitude of change?
- Is change that has occurred in the VC or VC indicator to date within the range of normal variability or acceptability (e.g., benchmarks, limits, thresholds)?

Baseline assessments are also referred to as **retrospective assessments**, emphasizing that assessing baseline conditions is not simply focused on current conditions—but on how conditions have changed over time and the key drivers and interactions responsible for that change. Too often, the focus of baseline studies in EA is on a description of current conditions (Box 5.4) rather than on assessing how

Box 5.4 Shifting Baseline Syndrome

Too often, baseline studies are focused only on current conditions rather than also on changes in VC or indicator conditions from past to present and whether thresholds of concern are being approached or have already been exceeded. This approach defines only the current environment as the baseline for impact assessment. But because the existing environment is a result of the influence of past actions, this approach attributes the effects of past and present actions to the current condition rather than to contributions to cumulative change in VC or indicator conditions. The magnitude of the effects of past projects are discounted and treated as part of the current baseline condition.

Consider a scenario in which available core area habitat for caribou is already at a target threshold due to 25 years of linear disturbances caused by highways, pipelines, transmission lines, forestry operations, and mineral lease blocks. It is tempting for a project proponent to use the "current available habitat" as the baseline against which the project's potential environmental effects are assessed. In so doing, a proponent may be able to present a convincing argument that their project is contributing to only a minor loss of caribou habitat, perhaps less than 2 per cent of what is currently available. However, if past conditions are considered and it is observed that caribou core area habitat in the project's regional environment had declined by 60 per cent over the previous 25 years, the project's contribution of an additional 2 per cent loss now appears much more significant.

Current conditions are sometimes adopted as a "new normal" in baseline studies rather than considering current conditions *relative to past conditions* and evaluating the nature and significance of *change* in VCs in the study area. This was the approach adopted in the recent Manitoba Hydro Bipole III project, an approximately 1400-kilometre transmission line project from northern Manitoba, near Gillam, south to Winnipeg. The Manitoba Clean Environment Commission

continued

> report on public hearings for the project identified several concerns about limited baseline data against which to properly assess project impacts and ignoring the effects of past actions and changes in conditions over time (see Manitoba Clean Environment Commission, 2012). The proponent's EIS adopted past effects as the norm, thus precluding possible restoration or other mitigative actions that might be required to improve current conditions. This is poor practice.

conditions have changed over time and identifying the suspected drivers of change. Core to a good baseline study is an analysis of what VC indicators looked like in the past, how they have changed over time, and what types of stressors resulted in or triggered such changes. An attempt should be made to identify relationships between indicators of change in VC conditions (e.g., caribou population, water-quality indices) and measures of human or natural disturbance so as to determine trends and associations that can be used to predict VC conditions or responses to future change.

In cases where VCs or VC indicators are not amenable to interpolating trends, information on past conditions can be used alongside current conditions as benchmarks against which potential project-induced stress can be identified and assessed. If, for example, the trend in the project environment has been toward decreasing physician-to-patient ratios and the health care system is currently at or near capacity, then any induced stress caused by an increase in health care demand will be significant in terms of impacts on community health and well-being and thus warrant detailed consideration in impact assessment and mitigation.

Knowledge to Support Baseline Assessments

In a recent Canadian public survey by Nielsen, Delaney and Associates, and Publivate (2016) about the most important considerations when making environmental regulatory decisions, the top-ranked response was science, facts, and evidence. Scientific knowledge can come in many forms—it can be descriptive, qualitative, or quantitative and based on a variety of methodologies from multiple disciplines. It can be the result of experimental design, long-term monitoring, meta-analysis, or carefully designed social science data collection strategies, or it can be drawn from legal analysis. Baseline assessment in EA is not an exercise designed to advance scientific discovery, though it does require sound science. The state of science in EA, however, has been widely criticized—from poor understandings of causal relationships and improperly constructed predictive models to unverifiable predictions. That said, who should be responsible for ensuring sound science in the EA process remains unclear. Arguably, it is the shared responsibility of both those engaged in the science *inside* the EA process and those engaged in the science *outside* the EA process.

Greig and Duinker (2011) describe science outside EA as that conducted by researchers, research organizations, and monitoring agencies to study long-term processes, establish effect–cause relationships, and develop the models and identify the indicators for effectively predicting the impacts of environmental and social disturbance. Such scientific efforts are long-term and well beyond the scope of the science conducted during the narrow scope and short timelines of baseline assessments during the regulatory EA process; however, it is essential knowledge to inform the work of EA practitioners. Science *inside* the EA process is context-specific and focused on the practical application of the science from outside EA to address immediate and complex problems of specific relevance to the project at hand. Greig and Duinker argue that to enable uptake by the EA community, the science outside EA must pay attention to the needs of practitioners and regulatory decision-makers, but it is the responsibility of EA practitioners and decision-makers to adequately communicate their needs to those engaged in long-term research studies and monitoring programs (Wong, Noble, & Hanna, 2019).

Regardless of whether the information input to baseline studies is founded on science external or internal to the EA process, based on expert judgment or on the lived experience of local populations, good baseline assessment requires high-quality information. Guidance does exist on the basic principles of high-quality information for informed assessment, evaluation, and decision-making—such principles are not unique to EA. The *Science manual for Canadian judges* (National Judicial Institute, 2013), for example, defines several basic principles for the consideration and interpretation of expert-based information in the judicial system—namely, that such information is relevant to the issue, necessary in assisting the trier of fact, properly qualified, and verifiable. Guidance issued by the National Aeronautics and Space Administration (NASA) identifies many similar principles but specifically within the context of scientific and technical information. NASA's guidelines stipulate that information must be purposeful, accurate, clear, complete, unbiased, transparent, and reproducible. Poder and Luki's (2012) advice to EA practitioners is that all information, regardless of context, must be complete, correct, sufficiently detailed, and substantiated.

Guidance also exists on the hierarchy of evidence or rationales that should be considered, or provided, to substantiate the information provided as part of the assessment process. Page (2006), for example, suggests that the information provided in EA should satisfy a hierarchy of considerations commencing with legislative requirements (i.e., legal obligations, regulatory compliance, etc.), science and Indigenous knowledge, lessons from other cases or documented experience, and then the professional opinion or reasoned experience of the practitioner. Based on existing guidance from the scholarly and technical literature, information or input used to inform the EA process, especially baseline assessment, should reflect the following principles:

- *Compatible* – not contradictory to legal or regulatory obligations or commitments.
- *Relevant* – logically relevant to the project and its potential adverse effects, appropriate to the decision or assessment context.
- *Purposeful* – tailored toward an intended purpose, free of unnecessary information, and of value to the end-user.
- *Clear* – presented with clarity and coherence and free of vague language and imprecision.
- *Verifiable* – supported by data or information that are documented and available, including any assumptions, and transparent in methods and/or procedures.

- *Accurate* – reliable, correct, consistent with any relevant standards, and validated.
- *Complete* – includes, to the extent possible, the proper context to ensure completeness of the information presented, with full documentation and circumstances affecting quality identified and disclosed.
- *Unbiased* – considers all available evidence (both supporting and contradicting) and discloses agency objectives, priorities, or mandates that may relate to the issue(s) at hand.
- *Substantiated* – sufficiently supported or rationalized based on a hierarchy of evidentiary sources including legislative requirements (i.e., legal obligations, regulatory compliance), science and traditional knowledge, observation, documented experience, and professional opinion.

Key Terms

activity information
baseline study
benchmark
cautionary threshold
condition-based indicators
context area
critical threshold
disturbance-based indicators
environmental baseline
functional scales
impact matrices
Leopold matrix
local study area
management targets

network or system diagrams
range of natural variability
receptor information
regional study area
remote sensing
retrospective assessments
scoping
stress-based indicators
target threshold
threshold
valued components (VCs)
VC indicators
zone of influence

Review Questions and Exercises

1. What is the purpose of scoping in EA?
2. Select a small sample of EISs from your province or territory's online EA registry or from the Canadian impact assessment registry (https://www.canada.ca/en/impact-assessment-agency.html). Scan the EISs, and generate a list of the different types of VCs considered in the assessments.
    a. Develop a classification scheme (e.g., physical, ecological, social), and assign the VCs to categories. Explore whether certain types of VCs receive more attention in EA than others, and discuss the implications.
    b. How many VCs have clearly identified indicators?
    c. What problems emerge in EA when VCs do not have clearly identified indicators or limits/thresholds?
3. Obtain copies of topographic and political maps of your city or region. Suppose a new wind energy project is being proposed to meet a growing demand for electricity:
    a. Identify a project location on the map sheet.

b. Which VCs would you identify in the region as important to consider? Provide a statement as to why these components are identified as VCs. In other words, what makes them so important that you would include them in the assessment?
c. Given current VC conditions, what objectives would you attach to each VC?
d. Identify a list of indicators that you might use to assess and monitor the conditions of each VC—include both stressor and effect-based indicators.
e. What spatial boundary or boundaries would you establish for the assessment? Sketch these boundaries on a copy of the map. Identify the criteria you used to determine these boundaries.
a. What temporal boundaries might you consider? Explain your reasoning.
4. Defining temporal boundaries for baseline assessments can be challenging. Some argue that baselines should consider historical conditions, a time when a VC was undisturbed or "pristine," as the benchmark for assessing impacts. Others argue that attention should simply focus on a point in time when the VC was "less disturbed" than at present. In practice, a proponent will sometimes focus only on the "current" conditions of a VC as the baseline for assessing a project's potential impacts. What are the advantages and limitations of each of these three different reference points for baseline assessments in EA?

References

Antoniuk, T., et al. (2009). *Valued component thresholds (management objectives) project.* Report no. 172. Calgary, AB: Environmental Studies Research Funds.

Ball, M., Noble, B.F., & Dubé, M. (2012). Valued ecosystem components for watershed cumulative effects: an analysis of environmental impact assessments in the South Saskatchewan River watershed, Canada. *Integrated Environmental Assessment and Management,* doi: 10.1002/ieam.1333.

Beanlands, G., & Duinker, P. (1983). An ecological framework for environmental impact assessment. *Journal of Environmental Management* 18, 267–77.

British Columbia Environmental Assessment Office. (2013). https://www2.gov.bc.ca/assets/gov/environment/natural-resource-stewardship/environmental-assessments/guidance-documents/eao-guidance-selection-of-valued-components.pdf.

British Columbia Ministry of Environment. (2014). *Procedures for mitigating impacts on environmental values: environmental mitigation procedures.* Ecosystems Branch. Victoria, BC: Environmental Sustainability and Strategic Policy Division.

Cairns, J., Jr, McCormick, P.V., & Niederlehner, B.R. (1992). A proposed framework for developing indicators of ecosystem health. *Hydrobiologia* 263, 1–44.

Canter, L. (1999). Cumulative effects assessment. In J. Petts (ed.), *Handbook of environmental impact assessment,* vol. 1, *Environmental impact assessment: process, methods and potential.* London, UK: Blackwell Science.

Chagnon, S. (1986). Atmospheric systems: management perspective. In Canadian Environmental Assessment Research Council (CEARC) (ed.), *Cumulative environmental effects assessment in Canada: from concept to practice.* Calgary, AB: Alberta Society of Professional Biologists.

Cooper, L. (2003). *Draft guidance on cumulative effects assessment of plans.* EPMG Occasional Paper 03/LMC/CEA. London, UK: Imperial College.

Cooper, L. (2012). Network analysis in CEA, ecosystem services assessment and green space planning. *Impact Assessment and Project Appraisal* 28(4), 269–78.

Glasson, J., Therivel, R., & Chadwick, A. (1999). *Introduction to environmental impact assessment: principles and procedures, process, practice and prospects*. 2nd edn. London, UK: University College London Press.

Greig, L., & Duinker, P. (2011). A proposal for further strengthening science in environmental impact assessment in Canada. *Impact Assessment and Project Appraisal* 29(2), 159–65.

Imperial Oil Resources Limited. (2017). *Cold Lake Oil Sands expansion project environmental impact assessment and application for approval*. Calgary, AB: Imperial Oil.

Irwin, F., & Rodes, B. (1992). *Making decisions on cumulative environmental impacts: a conceptual framework*. Washington, DC: World Wildlife Fund.

Jackson, L.E., Kurtz, J.C, & Fisher, W.S. (2000). *Evaluation guidelines for ecological indicators*. Report no. EPA/620/R-99-005. Washington, DC: Environmental Protection Agency.

João, E. (2002). How scale affects environmental impact assessment. *Environmental Impact Assessment Review* 22, 289–310.

Kagstrom, M. (2016). Between "best" and "good enough": how consultants guide quality in environmental assessment. *Environmental Impact Assessment Review* 60, 169–75.

Keeyask Hydropower Limited Partnership. (2012). Keeyask generation project environmental impact statement. Winnipeg, MB: Manitoba Hydro.

Leopold, L.B., et al. (1971). *A procedure for evaluating environmental impact*. Geological Survey Circular no. 645. Washington, DC: United States Geological Survey.

Manitoba Clean Environment Commission. (2012). *Bipole III transmission project: report on public hearing*. Winnipeg, MB: Manitoba Clean Environment Commission.

MVEIRB (Mackenzie Valley Environmental Impact Review Board). (2004). Environmental impact assessment guidelines. Yellowknife, NT: MVEIRB.

National Judicial Institute. (2013). *Science manual for Canadian judges*. Ottawa, ON: National Judicial institute.

Nielsen, Delaney and Associates & Publivate. (2016). *Review of Canada's environmental and regulatory process—questionnaire report*. Prepared for the Government of Canada. Montreal, QC: Nielsen, Delaney and Associates.

Olagunju, A., & Gunn, J. (2013). Selection of valued ecosystem components in cumulative effects assessment: lessons from Canadian road construction projects. *Impact Assessment and Project Appraisal* 33(3), 207–19.

Page, J. (2006). Make it easy on your readers: ideas on environmental impact document focus, organization, and style. *Impact Assessment and Project Appraisal* 24(3), 235–45.

Poder, T., & Luki, L. (2012). A critical review of checklist-based evaluation of environmental impact statements. *Impact Assessme nt and Project Appraisal* 29(1), 27–36.

Salmo Consulting Inc. (2004). *Deh Cho cumulative effects study, phase 1: management indicators and thresholds*. Report prepared for the Deh Cho Land Use Planning Committee. Fort Providence, NT: Salmo Consulting Inc.

Scace, R., Grifone, E., & Usher, R. (2002). *Ecotourism in Canada*. Ottawa, ON: Canadian Environmental Advisory Council.

Seitz, N., Westbrook, C.J., & Noble, B.F. (2011). Bringing science into river systems cumulative effects assessment practice. *Environmental Impact Assessment Review* 31, 172–9.

Shoemaker, D. (1994). *Cumulative environmental assessment*. Waterloo, ON: University of Waterloo, Department of Geography.

Therivel, R., & Ross, W. (2007). Cumulative effects assessment: does scale matter? *Environmental Impact Assessment Review* 27(5), 365–85.

Wong, L., Noble, B.F., & Hanna, K. (2019). Water quality monitoring to support cumulative effects assessment and decision making in the Mackenzie Valley, Northwest Territories, Canada. *Integrated Environmental Assessment and Management* 15(6), 988–99.

6 Impact Prediction and Characterization

Impact Prediction

Environmental assessment requires both forethought and foresight about the potential implications of a proposed undertaking. Prediction in EA is about understanding, to the extent possible, the relationship between the environment(s) potentially affected, the source of the impact or activity that may be harmful or pose risk, and the pathway by which the activity is able to interact with the receptor (UK Environmental Agency, 2002). Prediction in EA can be highly technical and supported by complex modelling processes, but it can also be qualitative and based on learned outcomes of previous projects or the lived experiences and understanding of local communities and traditional resource users and knowledge holders. No matter how predictions are made, or the source of evidence used, there is always uncertainty. Predictions about future conditions based on a project that is yet to be developed, in an environment that is constantly changing, involve a host of possibilities and relationships that are not well understood.

Change and Project Effects

Important to prediction in EA is delineating the "effects" of the project from "change" in baseline conditions and from the effects of other activities and stressors (i.e., cumulative effects). Change is a temporal measure, and not all change is project-induced; an **environmental effect** is the difference between change conditions. A distinction can thus be made between **environmental change** and environmental effect (Figure 6.1). An environmental change is the difference in the condition of an environmental or socio-economic parameter, usually measurable, over a specified period. An environmental change is typically defined in terms of a process, such as soil erosion, that is set in motion by project actions, other actions, or natural processes. Actions such as road construction or dam construction, for example, contribute to environmental or condition change. An environmental effect is the change difference: the difference in the condition of an environmental parameter under project-induced change versus what that condition might be in the absence of project-induced change.

Understanding change in baseline conditions and the range of normal variability, as discussed in Chapter 5, is essential to predicting, characterizing, and ultimately measuring the impacts associated with a project. For example, if employment in the area for which the project is proposed has been steadily increasing over the previous five years, then this baseline trend should be predicted in the absence of the project to ensure a more informed decision about the significance of the predicted employment impact of the proposed project. Similarly, if a project is predicted to

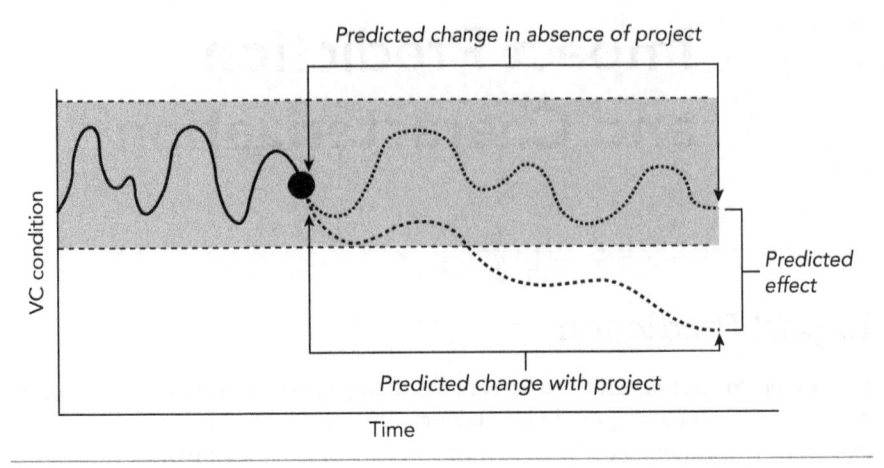

Figure 6.1 Change in VC condition and predicted effect, with shaded area depicting range of normal variability

have a negative effect on fish populations, then this information should be viewed considering past conditions and predicted trends in fish populations in the area in absence of the project.

The term "environmental effect" is often used interchangeably with **environmental impact**; however, some argue that impacts are estimates or judgments of the value that society places on certain environmental effects. Some consider this distinction little more than semantics. In this book, environmental "effect" and "impacts" are treated as synonymous and used interchangeably.

What to Predict

There are three main categories of impact predictions in EA: i) impacts of the project on the environment (biophysical and human); ii) impacts of the environment on the project; and iii) the cumulative impacts of the project.

Predicting the Impacts of the Project on the Environment

Predictions about the impacts of a project on VCs of the biophysical and human environment constitute the bulk of predictions in EA. Such predictions have been at the heart of EA since its inception under the US NEPA, but the scope and complexity of impact predictions has broadened considerably to include more holistic and integrated concepts, ranging from ecosystem services and climate change to impacts on well-being and Indigenous rights.

Valued components
At the most basic level, predictions in EA concern how a VC of the biophysical (e.g., water quality) or human (e.g., employment) environment might change under project conditions compared to future conditions in the absence of the project. These predictions are usually based on some indicator of change in the condition of a VC or indicator of the level of stress to a VC. For example, as discussed in Chapter 5,

predicted changes in *condition-based indicators* (e.g., phosphorous concentration, temperature, sediment load) may be used to identify potential impacts on water quality. In other instances, potential impacts on water quality may be predicted based on changes in *stress-based indicators* (e.g., stream crossing density, riparian area disturbance) under a project versus no-project future.

There is no single, comprehensive list of biophysical or human impacts to be predicted in all EAS. The specific impacts to be predicted depend largely on the nature of the project and the sensitivity of the local environmental and social setting and are defined by the EA scoping process and regulatory requirements. For biophysical impacts, attention is often focused on project-induced biological change, chemical change, physical change, and ecological change, from which specific effects or impacts emerge. For impacts on VCs of the human environment, in addition to those impacts attributed to biophysical change, predictions are often focused on various indicators of project-induced demographic change, cultural change, economic change, health and social change, infrastructure change, and institutional change (Table 6.1). These changes can lead to impacts that can be linked to project activities, such as local housing shortages or increased pressures on social services due to worker influx, the displacement of communities and psycho-social stress due to resettlement, or increased standard of living associated with new employment and income opportunities. Predicting the impacts of a project on VCs of the human environment is often much more complex and uncertain than predicting impacts on the biophysical environment. Attention is thus often limited to impacts over which the proponent has direct control, such as employment or infrastructure, whereas more complex impacts and interactions, such as impacts on culture or well-being or gender-based impacts, have traditionally received much less attention.

Table 6.1 Typical Biophysical and Human Environment Impacts of Projects and Examples of What to Predict

Air impacts: emission levels, rates, and types; pollutant concentration; dispersion; air temperature; wind speed; air quality/clarity

Terrestrial impacts: soil erosion; soil moisture content; soil compaction; vegetation loss/disturbed area; vegetation cover/composition; species population; species movement; species distribution; linear disturbance density; habitat fragmentation

Water impacts: surface and groundwater quality; surface and groundwater quantity; flow rate; extraction rate; turbidity; chemical change; biological change; riparian health; fish health; effluent discharge levels, rates, and types; heavy metals

Socio-economic impacts: employment and income; in-migration; population change; health risk; security (i.e., crime); community stress; mental and physical well-being; gender-based impacts

Infrastructure and service impacts: transportation; housing demand; housing prices; demand for health, education, and social services

Socio-cultural impacts: availability of traditional foods; food security; disruption to family structure; disruption to cultural practices; access to traditional lands and resources

Holistic concepts

Increasingly, EA is moving beyond predicting impacts on individual components of the biophysical and human environment and focusing also on more holistic concepts and understandings of impacts. These holistic concepts range from impacts on Indigenous rights (see Chapter 10), to **community health and well-being** (see Environmental Assessment in Action: Predicting the Impacts of the Lower Churchill Hydroelectric Generating Project on Community Health and Well-Being), to **ecosystem services**. Such holistic concepts are based on the notion that predicting individual VCs, though important, can sometimes oversimplify the complexity of biophysical and human systems.

Energy resource projects, for example, can create new employment and social development opportunities for communities, but they can also mean reduced access to traditional lands and disruption to cultural practices—either by way of physical restrictions on the land base or because of new employment opportunities disrupting traditional economies. Collectively, project-induced change can impact overall community health and well-being (Christensen & Krogman, 2012). Rather than focusing solely on a project's impacts on individual VCs, such as community services or access to traditional foods, predicting impacts on health and well-being is about identifying and assessing the linkages between a project's actions and various **determinants of health and well-being**. Project-induced changes to these determinants are not themselves impacts per se; rather, they signal potential pathways through which a project can impact overall health and well-being (Figure 6.2). Many health authorities, including the World Health Organization and Health Canada, have recognized the need for and benefits of addressing health and well-being as a holistic concept in EA application. Under the Mackenzie Valley Resource Management Act in Canada's Northwest Territories, for example, the consideration of the well-being of communities and individuals is an explicit requirement of EA, often focused on a combination of environmental, social, economic, and cultural factors.

Project activities or actions may trigger changes in:

Physical environments	Socio-economic & structural determinants	Community context	Individual factors
e.g., air quality; water quantity; water quality; wildlife; critical habitat	e.g., income; income equality; educational services; health and social services; food security; access to traditional lands; housing availability	e.g., community cohesion; social supports; social networks; community identity; safety; social capital	e.g., diet and nutrition; alcohol consumption; stress; attachment; self-esteem; mental health; physical health risks

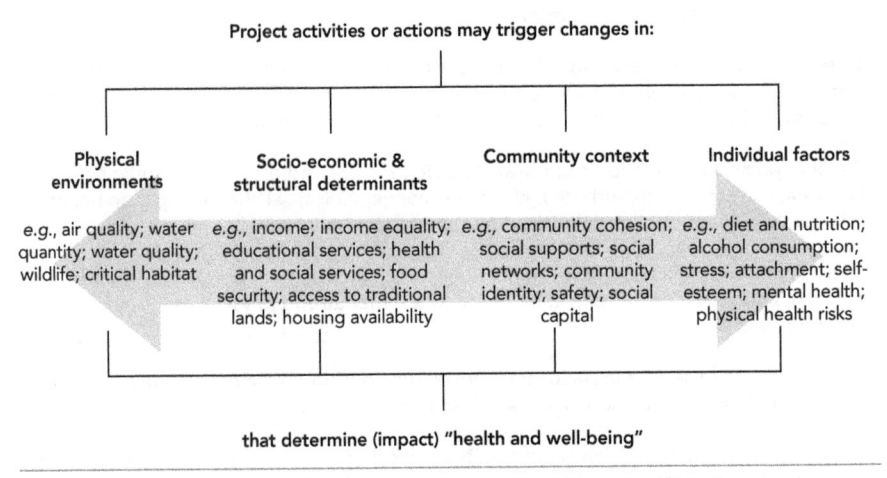

that determine (impact) "health and well-being"

Figure 6.2 Determinants of impacts on human health and well-being

Environmental Assessment in Action

Predicting the Impacts of the Lower Churchill Hydroelectric Generating Project on Community Health and Well-Being

In 2006, Nalcor Energy, a provincial Crown energy corporation of Newfoundland and Labrador, filed an application to develop the Lower Churchill Hydroelectric Generating Project on the lower Churchill River, central Labrador. The project would consist of two hydroelectric facilities totalling 3074 megawatts and associated dams and reservoirs, two interconnecting transmission lines between the two facilities and the existing grid totalling 263 kilometres, and the construction of work camps to accommodate the anticipated 2000-member workforce during peak of construction. The project's location, in the Upper Lake Melville region of central Labrador, is relatively remote. There is one larger centre in the project's primary impact area, Happy Valley–Goose Bay, with a population of approximately 7500, and three smaller centres: Mud Lake, North West River, and Sheshatshiu—the largest Innu community in Labrador, with an estimated population of about 1200 people.

The project was subject to a joint federal-provincial EA review. The EIS guidelines were prepared by the governments in 2008, and Nalcor submitted its EIS to the joint review panel in 2009. A community health and well-being baseline study and impact assessment were completed for the project. The baseline assessment indicated a history of social and community health issues in Upper Lake Melville, particularly in the Innu community of Sheshatshiu, including solvent abuse and alcoholism. Gambling activity, particularly video lottery terminal addictions, was found to have become much more prevalent in recent years, with the number of gambling addictions in the region reported as significantly higher than the provincial average. The baseline study also identified women in the Upper Lake Melville area as having children at considerably younger ages than women in the province as a whole, with births to

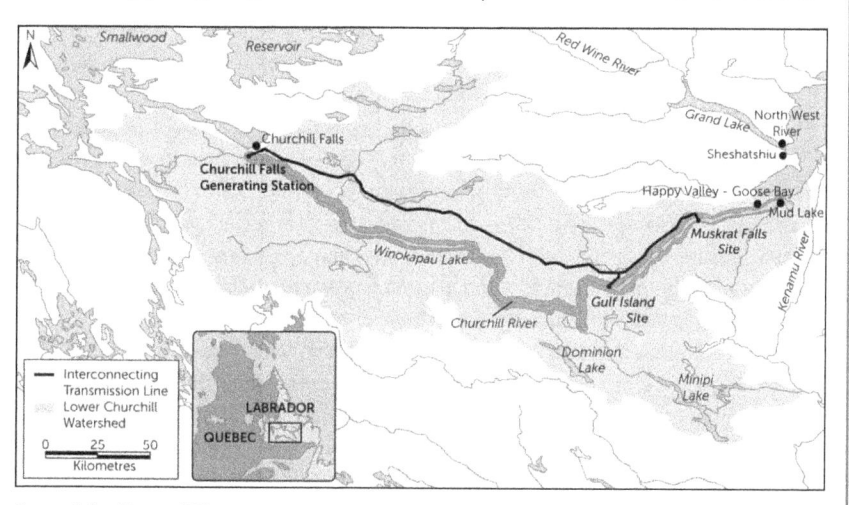

Source: Nalcor Energy, 2009

continued

teenage mothers most pronounced in Sheshatshiu. The baseline study reported an increased demand for social and mental health services and treatment facilities in the region, with current services already operating at or beyond capacity. Primary health care delivery in the Upper Lake Melville area, as measured by the family-physician-to-population ratio, was found to be substantially lower than for the province.

Some of the primary issues of concern identified in the project's community health and well-being impact assessment were issues related to the potential for increased alcohol and drug use and the implications for criminal activity, the health of the individual and the family, and the ability to provide social and health services under a project development scenario. The challenge for the proponent was that many of these were ongoing issues of concern with or without the proposed project and many of the health and social services in the region were already either at or beyond capacity. Most impacts on community health as a result of the project would be experienced indirectly through demographic change, increased disposable income, and any in-migration to and worker–community interactions within the Upper Lake Melville area.

As with most assessments, the first source of information used for predicting the impacts of the Lower Churchill project was lessons learned based on experiences elsewhere, in this case similar hydroelectric projects in Quebec, Manitoba, and British Columbia. However, results from these projects showed a complex mix of both positive and potentially adverse impacts. Further, lessons from the socio-economic environments of those projects were not easily transferable to the Innu health, cultural, and historical context of the Lower Churchill project. The proponent recognized early on that few quantitative predictions were possible, impacts on health and well-being were comprised of complex and interrelated factors, and not all impacts could be predicted. For example, most adverse effects on mental health, addictions, and counselling services would likely be associated with an increase in disposable income as a result of project employment and associated local spending patterns. An additional concern raised by the Innu community was the potential for an increased incidence of teenage pregnancies as a result of worker interaction with the community. Many such impacts depend on individual and community behaviour as well as on personal and community coping skills.

The proponent stated in its EIS that many such impacts cannot be predicted in any quantifiable way and any predictions that could be made would be subject to considerable uncertainty. The view was that it was better to plan to adapt than to make inaccurate impact predictions or ones that simply could not be verified. Given a baseline assessment that indicated that any increase in demand for community health services associated with the project was likely to be beyond the current system capacity, the proponent focused instead on what might be deemed an acceptable level of impact, resembling a threshold or maximum allowable effects levels approach (discussed later in this chapter). Emphasis was placed on coordinated monitoring programs between the proponent and the regional health authority and impact mitigation to identify and avoid potentially adverse impacts on community health and well-being in the event they should occur. The project was approved in 2012 by the governments of Canada and Newfoundland and Labrador.

Sources: Aura Environmental Research and Consulting Ltd, 2008; Nalcor Energy, 2009

Climate change
Climate change and the potential climate impacts of development projects have taken centre stage in EA discussions in recent years (Fischer & Sykes, 2009; Ohsawa & Duinker, 2014). The International Association for Impact Assessment's international best-practice principles on climate change in impact assessment (Byer et al., 2018) state:

> Impact assessment (IA) has much to contribute in assisting governments with meeting their international commitments to address human-induced climate change, and in assisting industry and the public to understand the environmental and social consequences of climate change.

In 2016, the Canadian government introduced a requirement for the assessment of upstream greenhouse gas (GHG) emissions in energy pipeline EAs carried out under federal review and approval processes (Luke & Noble, 2019). With the recent introduction of Bill C-69, an Act to enact the Impact Assessment Act and the Canadian Energy Regulator Act, to amend the Navigation Protection Act and to make consequential amendments to other Acts (Assented 21 June 2019), a key requirement of EA is to assess the extent to which a proposed project affects the Government of Canada's ability to meet its environmental obligations and commitments related to climate change—such as the Paris Agreement on climate change.

The extent to which EA is a valuable tool for assessing a project's impacts on climate change, however, is questionable. At best, project proponents can identify their likely direct emissions and develop project-specific strategies to minimize or offset those emissions (Box 6.1); they are not well positioned to assess the implications of their project on future climate change in combination with the effects of all other

---

Box 6.1 Climate Considerations in EA Practice in British Columbia's LNG sector

Some of the world's largest shale gas deposits are found in British Columbia. Since 2012, 21 liquefied natural gas (LNG) facility and pipeline projects have been proposed in the province and nine EAs completed. LNG production requires high amounts of energy for the liquefaction process. Natural gas impurities, including CO_2, are separated out, which means that LNG projects generate high levels of GHG emissions compared to typical natural gas projects (Murillo, 2012). British Columbia's Greenhouse Gas Reduction Targets Act commits to reducing emissions by 33 per cent below 2007 levels by 2020 and by 80 per cent by 2050 (Government of British Columbia, 2007). The act also requires emissions reporting every two years by the government to track and monitor progress. In January 2016, the province provided updated legislation and regulations related to the LNG sector, including a production intensity target to make British Columbia's LNG sector the "cleanest in the world." This included new emissions reporting requirements and the development of a carbon registry whereby emission offsets and credits can be issued, transferred, and tracked (Government of British Columbia 2016).

continued

> EA in British Columbia is legislated under the Environmental Assessment Act. All LNG projects, including small projects that may not trigger EA, are regulated by the province's Oil and Gas Commission. Exceptions include projects that involve the export of natural gas, which are also subject to permitting by the National Energy Board. In 2019, Luke and Noble reviewed all LNG projects (pipelines and processing facilities) that had undergone a provincial EA review over the previous two decades. Their analysis showed that all project EAs made at least some reference to climate change, with more than 2200 references to climate change across seven pipeline projects and more than 4300 references across 11 LNG facility projects. Most of these references, however, were found in the minority of projects. For example, the Eagle Mountain–Wood Fibre EA contained more than one-third of the references to climate change found among all pipeline projects, and the Fortune Creek, LNG Canada Export Terminal, Pacific Northwest, and Woodfibre projects contained 75 per cent of all references to climate change among LNG facilities reviewed. The most frequently discussed topic in relation to climate change was GHG, amounting to more than 60 per cent of references across the sample of pipeline projects and 56 per cent for LNG facility projects. Other topics such as methane, for example, typically mentioned in terms of mitigating fugitive emissions, amounted to less than 6 per cent of references to climate change.
>
> Reference to climate change does not indicate the quality of climate change consideration, though it does reveal where in the EA process climate change receives the most attention and what proponents focus on when considering climate impacts. All 10 project descriptions for LNG facilities made some reference to climate change, compared to only three of the eight project descriptions for pipelines. There was no difference between older and more recent projects. The application information requirements issued by the province, stating what the proponent must include in their assessment, focused largely on the project's GHG emissions. Proponents' impact statements thus predominately focused on GHG emissions mitigation rather than predicting total GHG emissions or the project's contributions to climate change per se. At the decision phase, climate change was addressed in all EA certificates, where issued, except the Woodfibre LNG facility. Reference to climate change throughout all stages of the EA process for most projects indicates that, at least on the surface, climate change is routinely incorporated into EA practice in the LNG sector—at least in terms of GHG emissions and mitigation strategies. The practice is consistent with the province's view that it is not possible to predict the impacts of an individual project's emissions on global climate change.
>
> Source: Luke & Noble, 2019

human actions, nor are they able to devise sector-wide strategies to mitigate climate change. Luke and Noble (2019) argue that project proponents should indeed be responsible for predicting the emissions caused by their project and for mitigating them, but modelling a project's impacts on global climate change is beyond the scope of a single project EA. According to Gray (2015), the role of proponents is to assess and mitigate their own GHG emissions; the role of governments is to determine whether the project aligns with broader climate commitments or can otherwise be justified when it does not. This is consistent with the province of British Columbia's

Environmental Assessment Office, for example, which notes that the impacts of GHG emissions must be addressed globally and that it is not possible to estimate the impacts of an individual project's emissions on global climate change. Such matters are best addressed through other policy and assessment mechanisms, such as strategic environmental assessment (Chapter 12).

Predicting Impacts of the Environment on the Project

Not only is it necessary to predict a project's impacts on the environment, but the impacts of environmental conditions on the project must also be considered. These impacts are often referred to as environmental risks, such as floods or the impacts of major storm events on project infrastructure. In 1998, for example, a major ice storm in Ontario, Quebec, and parts of Atlantic Canada caused major disruption to electricity supply to millions of people; more than 1000 steel transmission towers and thousands more wooden utility poles collapsed under the weight of the ice (Phillips, 2002). Data for predicting and mitigating such risks are often collected external to the EA process through long-term, science-based monitoring and model development.

Increasingly important is the need to consider the impacts of climate change on projects (Table 6.2), often referred to as **climate risk**, and to develop **climate-resilient projects**. Predicting and understanding climate risk is about understanding the *climate hazard, exposure,* and *vulnerability* of the project (Figure 6.3). Consider a proposed hydroelectric project, for example, and the implications of increasing temperatures

Table 6.2 Illustrative Examples of the Adverse Impacts of Climate Change on Projects in Different Sectors

	Temperature change	Sea-level change	Changing precipitation patterns	Changing storm patterns and severity
Energy projects	Impacts on peak system reliability as temperatures warm and demand increases Pipeline movement or damage as permafrost thaws	Increased risk to offshore energy production infrastructure Inundation of coastal infrastructure for energy export (e.g., LNG terminals)	Impact of low flows on hydroelectric production potential Insufficient water supplies for cooling of power stations	Damage to and reliability of electricity production (e.g., wind turbines) and distribution systems (e.g., transmission lines)
Transportation projects	Increased stress on rail and bridge joints with wide temperature fluctuations Damage to infrastructure due to thawing permafrost Loss or reduced use of ice roads in the North	Erosion of coastal transportation networks Inundation of coastal transportation infrastructure or hubs	Landslides or washouts as more winter precipitation falls as rain on saturated soils	Damage to road or rail infrastructure due to major flood events Need for increased protection of coastal infrastructure from storm surges and hurricanes

continued

Table 6.2 Continued

	Temperature change	Sea-level change	Changing precipitation patterns	Changing storm patterns and severity
Water resource projects	Increased evaporation from reservoirs (e.g., hydro reservoirs, storage ponds) Increased treatment costs Increasing and competing demands and cost	Impacts to coastal water infrastructure, including sewage treatment systems Salinization of water supplies	Increased need for storage capacity Impacts of low (or high) flow on infrastructure, extraction allowances, and treatment costs	Increased sediment loading and contamination of supplies
Forest resource projects	Increased disease outbreaks Increased wildlife risk	Loss of coastal forest habitat and resources	Increased erosion due to heavy precipitation events Shifts in species distribution and access	Damage due to hurricanes or major ice storm events

Sources: Boyle, Cunningham, & Dekens, 2013; US EPA, 2019

on flow levels and the ability to alter a project's design to keep pace with peak energy demands under a warming climate or the sustained availability of water resources to cool reactors for nuclear power production. The vulnerability of a project to climate risk is a function of *sensitivity*, or the degree to which the project and its associated infrastructure are affected by climate change, and *adaptive capacity*, or the ability to adapt project design or project characteristics to changing climatic conditions.

Projects with coastal infrastructure, such as shipping or supply terminals, for example, must be able to adapt to the impacts of accelerated sea-level rise, storm surges, and coastal erosion. For projects in the Arctic, thawing permafrost (Figure 6.4)

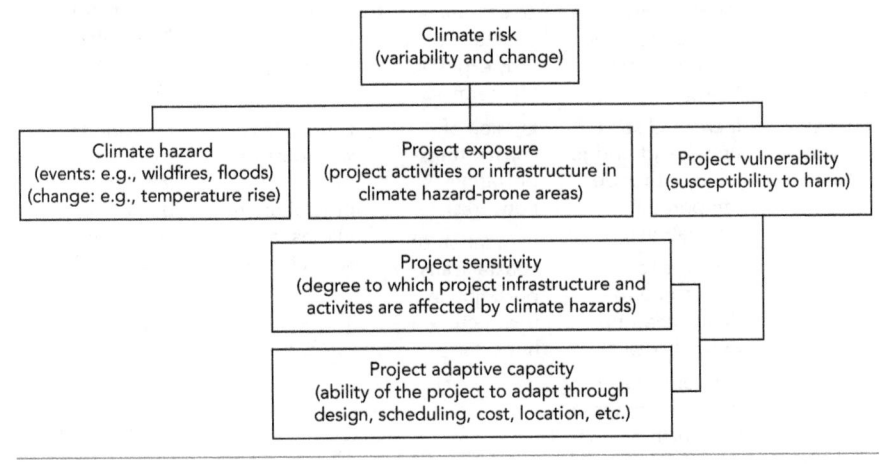

Figure 6.3 Predicting a project's climate risk

Source: Based on Boyle, Cunningham, & Dekens, 2013

Figure 6.4 **Permafrost thaw (top photo) and slumping (bottom photo) on Ellesmere Island, near the Eureka weather station.** Permafrost thawing across the Arctic is a clear indicator of changing climate conditions. Permafrost thaw can pose risk to existing infrastructure and will require innovative design in future infrastructure.

poses significant risk to infrastructure, and shortening ice road seasons means that the costs of transporting project goods, construction materials, or resupply are likely to increase over time. If the project is not adaptable, there is high risk of **stranded assets** whereby projects become no longer economically viable because of increasing climate-related costs (Boyle, Cunningham, & Dekens, 2013). In Alaska, Larsen et al. (2008) estimated that for public infrastructure, for example, climate change could add an additional 10 to 20 per cent to annualized replacement costs beyond normal "wear and tear." Such costs must be accounted for in assessing project viability.

Predicting the Cumulative Impacts of the Project

A project's impacts do not occur in isolation—they often add to, interact with, or amplify the impacts of other projects, activities, or disturbances in the project's regional environment.

Consider, for example, the total downstream effects on fish health resulting from upstream **point-source stress** and **non-point-source stress** in a watershed (Figure 6.5), such as sedimentation due to forestry and stream crossings by transmission lines, alterations in flow and increases in methyl-mercury at a hydroelectric

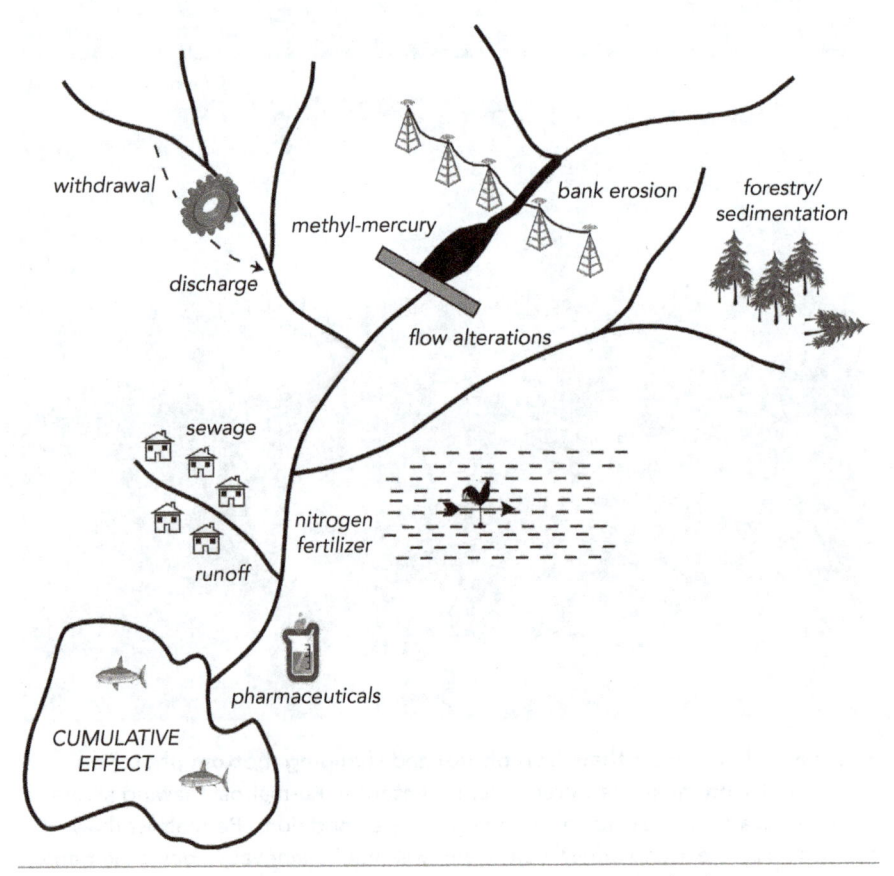

Figure 6.5 Sources of potential cumulative effects in a watershed

facility, water withdrawal and discharge from heavy industry, septic leakage and urban storm water runoff from residential areas, nutrient loadings from agricultural runoff, and pharmaceuticals and other chemicals from industry and manufacturing.

The total effect of all these activities, accumulating over time and across space, combined with larger-scale stress caused by climate change and transboundary effects acting on a single VC, is known as a cumulative effect or **cumulative impact**. The US Council on Environmental Quality (1997) characterizes cumulative effects as:

- the total effect, including direct and indirect, on a given resource, ecosystem, or human community of all actions taken;
- effects that may result from the accumulation of similar effects or the synergistic interaction of different effects;
- effects that may last for many years beyond the life of the action that caused them;
- effects that must be analyzed in terms of the specific resource, ecosystem, or human community affected and not from the perspective of the specific action that may cause them; and
- effects that must be approached from the perspective of carrying capacity, thresholds, and total sustainable effects levels.

In order to understand the *true* impact of a project, it is essential to understand cumulative impacts. In some jurisdictions, including under Canadian federal EA legislation and in some provinces and territories, project proponents are required to consider the impact of their project in combination with the impacts of other projects and activities in the project's regional environment—past, present, and future. In the absence of assessing cumulative impacts, a project's impacts may be mistaken for the cumulative impacts of *other* activities, or the significance of the project's additional impacts to a VC may be underestimated. Consider Figure 6.6, in which the condition of a VC is declining over time. Predicting a project's cumulative impact requires consideration of:

- the VC baseline condition (A) in the past (t_{-1}), present (t_0), and future (t_{+1});
- the predicted VC condition in the absence of the project but considering the impacts of other known or reasonably foreseeable future projects and disturbances; and
- the predicted VC condition considering the project's impacts in addition to the impacts of other known or reasonably foreseeable future projects and disturbances.

Although a project proponent is often responsible for mitigating only the *additional* impacts to a VC that can be attributed to their project (I), the importance or significance of those impacts must be interpreted based on the cumulative impact of *all* disturbances considering future (II), current (III), and, most important, previous (IV) VC conditions. How to determine the significance of a project's impact is discussed in Chapter 8. In some cases, should a project be predicted to impact a VC that is severely degraded, a proponent may be required to implement enhanced mitigation measures to reduce the overall cumulative impact. For example, a proponent

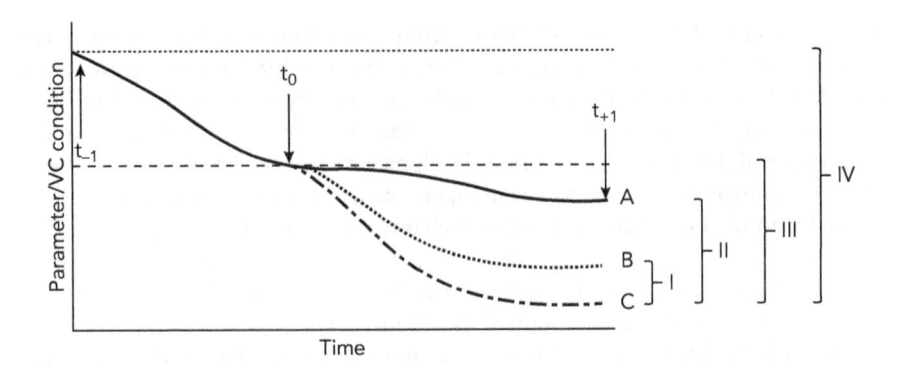

Legend

t = reference condition: past or historic condition (t_-1), present condition (t_0), predicted future condition (t_+1)

A = predicted future VC condition in absence of the proposed project and assuming no other future developments or disturbances
B = predicted future VC condition in absence of the proposed project but including other future developments or disturbances
C = predicted future VC condition including both the project and other future developments or disturbances

I = predicted project impact
II = predicted cumulative impact relative to a future VC condition
III = predicted cumulative impact relative to current VC condition
IV = predicted cumulative impact relative to past or historical VC condition

Figure 6.6 Predicted project and cumulative impacts

may be proposing to construct an oil and gas pipeline in a region that is heavily fragmented by linear features, including abandoned roads and trails. Considering the project's cumulative impact on habitat fragmentation, a condition of approval may be that the proponent contribute to the reclamation of already existing abandoned roads and trails such that the project's development will reduce the overall cumulative impact.

Although predicting the cumulative impacts of a project is an important part of EA, such predictions are complex and often require some form of regional or systems-based modelling. Many different types and sources of stress can contribute to a cumulative impact (Table 6.3). Predicting a project's cumulative impact requires baseline data that are often beyond the scope of that collected for the project under review, including data from other sectors and land uses. It also means that other project proponents must be willing to share information about their existing and anticipated project undertakings (Canter & Ross, 2010). Obtaining this information can prove to be challenging in practice (CESD, 2011). Hackett, Liu, and Noble (2018) describe the case of Cardinal River Coal's proposal for development of the Cheviot coal mine, Alberta. The proponent required information from other mining companies and land users, including the forest and tourism industries, to predict the cumulative impact of its proposed mining operation (Kennett, 2002). The proponent's argument was that it simply could not be expected to rely upon the goodwill of other mining

Table 6.3 Nature and Sources of Change That Contribute to Cumulative Impacts or Effects

Source of change	Characteristics	Example
Space crowding	High spatial density of activities or effects	Multiple mine sites in a single watershed
Time crowding	Events frequent or repetitive in time	Forest harvesting rates exceeding regeneration and reforestation
Time lags	Activities generating delayed effects	Human exposure to pesticides
Fragmentation	Changes or interruptions in patterns and cycles	Multiple forest access roads cutting across wildlife habitat
Cross-boundary movement	Effects occurring away from the initial source	Acid mine drainage moving downstream to community water supply systems
Compounding	Multiple effects from multiple sources	Heavy metals, chemical contamination, and changes in dissolved oxygen content resulting from multiple riverside industries
Indirect	Second-order effects	Decline in recreational fishery caused by decline in fish populations due to heavy-metal contamination from industry
Triggers and thresholds	Sudden changes or surprises in system behaviour or system structure	Collapse of a fish stock when persistent pressures from harvesting and environmental stress result in a sudden change in population structure

companies and resource sectors to obtain data to support their own EA application and that the data simply were not available from other projects and land users.

EA across Canada has a relatively poor track record of predicting and dealing with cumulative impacts (Duinker & Greig, 2006; Cronmiller & Noble, 2018), as evidenced by the number of legal challenges that have emerged concerning the quality of a proponent's cumulative impact assessments (Noble, Liu, & Hackett, 2017). This is not to say that a project proponent must not predict and manage the cumulative impact of their projects, but governments need to ensure that the information is available for proponents to understand their project in a larger, regional context. Chapter 11 discusses in detail the challenges to predicting cumulative impacts in project EA and the importance of regional EA initiatives.

How to Predict

Predicting the potential impacts of proposed projects is a complex and uncertain task. Many cause–effect relationships are unknown, and biophysical and human environments are moving targets. Morris and Therivel (2001) suggest that impact prediction requires five basic elements, including:

i) sound understanding of the nature of the proposed undertaking;
ii) knowledge of the outcomes of similar projects;
iii) knowledge of past, present, or approved projects whose impacts may interact with the proposed undertaking;
iv) predictions of the project's impacts on other environmental and socio-economic components that may interact with those directly affected by the project; and
v) information about environmental and socio-economic receptors and how they have in the past and might in the future respond to change.

There is no set of laws or regulations that prescribe *how* impact predictions must be made. In many cases, the quality of an impact prediction depends on the skill and experience of the practitioner, the nature and availability of baseline data, the quality of the predictive tool or **model** used, and the degree of complexity or uncertainty in the environmental or socio-economic system of concern. Greig and Duinker (2011) argue that when impacts are predicted they should, where at all possible, be based on rigorous and falsifiable null-impact hypotheses stating the relevant affected variables, impact magnitude, spatial and temporal extent, probability of occurrence, significance, and associated confidence intervals. Of course, a hypothesis-driven approach does not supplant the role of professional judgment when considering what constitutes a significant or acceptable deviation from the null and that it inevitably relies on the availability and quality of data.

Impact predictions must also consider accuracy and precision. **Accuracy** refers to the extent of system-wide bias in a prediction, or the closeness of a predicted value to its "true" value. **Precision** refers to the level of preciseness or exactness associated with an impact prediction. Predictions such as "slight reduction" or "minor effect" are of little value for understanding, monitoring, and managing the actual effects of a project if a "slight reduction" or "minor effect" is neither quantified nor qualified. It is thus impossible to generate accurate predictions when such predictions are couched in vague and imprecise language. A prediction stating "there will be a slight increase in methyl-mercury concentrations in the reservoir due to flooding" is difficult to prove—"slight increase" is a vague concept, the increase is not quantified relative to a baseline condition, and no time frame for the impact is identified. Such predictions are of limited value in terms of verifying and managing actual project effects. But as precision increases, the possibility of being wrong also increases. For example, "methyl-mercury concentrations will increase by 7.6 per cent above the current baseline level and persist for 180 days after reservoir flooding." The fact that an impact prediction is accurate does not necessarily constitute a useful prediction if the level of precision is relaxed. However, if more precise impact predictions are desired, then accuracy may be forfeited (De Jongh, 1988).

Analogue Approaches

The starting point for impact prediction is examining existing information, referred to as **analogue approaches**. The most common, and valuable, analogue approach is the examination of similar projects and learning from past project experiences. Few projects, or their impacts, are so unique that we cannot look to other projects to learn something about potential impacts and management strategies. Examination of similar projects makes impact predictions of a proposed development based on comparisons with the impacts of existing developments where conditions might be similar. Data might be collected and comparisons made based on previous impact statements, site visits, or monitoring reports. Using similar projects to understand the potential impact of a project under review is a valuable approach, particularly when multiple projects of a similar nature already exist in the development area. Difficulties do emerge, however, when transferring lessons from one context to another in that project environments must be sufficiently alike to justify a transfer of findings.

Expert Judgment

Expert judgment underlies many aspects of EA and is commonly relied upon for impact prediction. As suggested by the United Nations Environment Programme (UNEP, 2002), the successful application of many EA methods relies heavily on the nature and quality of expert judgment. There is no standard by which judgment is integrated in EA, and there are no specific criteria in EA professional practice for *who* is an expert. An expert may be a disciplinary expert with experience and credentials in a specific field or technique or a local knowledge holder with lived experience on the land. Multiple techniques are available for integrating expert judgment in EA, from unstructured and ad hoc techniques, such as roundtable discussions or storytelling, to highly structured and organized techniques, such as the **Delphi technique**. No matter the source or nature of expert judgment in EA, it must be substantiated. This means that all sources of expert judgment must be explicit about uncertainties, including confidence levels where appropriate, and assumptions. Expert judgment should also be accompanied by other sources of supporting evidence, such as demonstrated knowledge, scientific literature, or experience from similar projects or contexts.

Modelling and Extrapolation

Models are simplifications of real-world systems; they can range from simple box-and-arrow diagrams to sophisticated equations and computer simulation. A wide range of models is available, from those that deal with single issues such as air quality to complex models for evaluating ecosystems and interactions. Modelling creates simplified representations of the system under investigation, including causal mechanisms, and is based on explicit assumptions about the behaviour of the system. Modelling approaches are typically very data-demanding, requiring input from the scientific community beyond that which is normally available through EA baseline studies. There are many different types and classifications of models, but among the most common in EA are **balance models** and **statistical models**.

Balance models, or mass balance equations, identify inputs and outputs for specified environmental components, such as soil moisture volume or surface runoff, and are comprised of inputs, storage, and outputs for an environmental system. For example, inputs to a system might include water and energy, whereas outputs include wastewater and outflow. Balance models, such as surface hydrological models, are particularly useful for predicting physical changes in environmental phenomena when predicted changes equal the sum of the total inputs minus the sum of the total outputs (Glasson, Therivel, & Chadwick, 1999). An example is a site water budget, which predicts the impacts of a proposed development on the input, storage, and output of water available in a sub-basin.

Statistical models are often used for extrapolation, or trends analysis, and are based on assumptions about fundamental relationships or correlations underlying an observed phenomenon in order to project beyond the range of available data. Such models often include simple linear or multivariate regression and fitting data to mathematical functions. For example, based on hypothesized or proven statistical associations between neighbourhood housing density and quality of life, statistical models can be used to extrapolate changes in the quality of life resulting from a proposed large-scale residential housing development program. Statistical models are also used to determine whether a statistically significant difference exists between the predicted changes in impact indicators due to project influence as opposed to natural changes that would occur in the absence of project development. Some statistical models are explicitly spatial in nature. **Spatial models** often draw heavily on Geographic Information Systems and range from simple overlay approaches whereby potential land-use conflicts can be predicted to more complex spatial modelling and simulation in which the distribution of impacts or phenomena are modelled over both space and time.

Threshold-Based MAELs

Impact predictions often turn out to be inaccurate because of the mix of assumptions that must be made and the multiplicity of exogenous factors involved (Mitchell, 2002). Projects are almost always modified during the development phase, thereby making initial impact predictions less valuable. Given the nature of constantly changing and often unpredictable human environmental systems, the impacts of development can rarely be predicted with any degree of certainty. However, not all environmental effects need to be predicted per se; in many cases, especially when baseline data are inadequate or predictive models poorly developed, it may be appropriate to state that a specific outcome will not exceed a specified limit or benchmark, in which case management practices, the setting of targets, and the determination of limits become the focus of attention (Storey & Noble, 2004). These **threshold-based predictions** may rely on previous experience with similar projects, similar types of impacts in different environments, public acceptability, or regulatory standards.

When comparative examples or regulatory standards are not available, an alternative approach is to base impact predictions on a desirable level or maximum level of change. **Maximum allowable effects levels** (MAELs) reflects an approach to impact prediction in which the impact is stated as follows: "project impact i will not exceed a particular limit or desired effects level for impact indicator j." This approach is particularly useful for managing project outcomes and meeting desired sustainability objectives when baseline data and predictive models are not available. MAELs is also a useful approach for socio-economic effects, particularly for those phenomena that depend on individual and community behaviours. For example, in the case of a small community subject to temporary worker influx due to project construction, predicting additional demand on local health care services is a highly uncertain task. It depends, in part, on worker health and safety policies and practices, the health of the incoming workforce itself, and the social behaviours of the workforce in the community—such as alcohol or drug consumption. An alternative approach is to set an acceptable population-to-physician ratio based on the policy of the relevant health authority or on what the affected community deems an acceptable level of access to health care in comparison to current baseline conditions. An MAEL is then stated, indicating that the population-to-physician ratio will not be exceeded as a result of worker influx to the local community. The focus shifts from prediction per se to monitoring to ensure that adverse effects beyond the MAEL do not occur.

Scenarios

Scenarios are not predictions or forecasts per se; rather, scenarios are alternative images of the future that serve to challenge current assumptions and account for the likely possibility that the future is not a static continuation of the past. The Intergovernmental Panel on Climate Change describes a scenario as "a coherent, internally consistent and plausible description of a possible future state" whereby each scenario is an alternative vision of how the future can unfold (Figure 6.7).

Scenario development in EA is about constructing and evaluating future events or conditions through the "consideration of plausible (although not equally likely)" outcomes (Liu, Mahmoud, Hartmann, Stewart, & Wagener, 2008). This involves making assumptions about potential future conditions and drivers and often considering best- and worst-case outcomes. Scenario development is most useful when several alternative scenarios are constructed and analyzed, each providing some significant contrast from the others—but each retaining some common variables such that scenarios can be compared. Hulse, Branscomb, and Payne (2004) further add that the drivers considered in scenario development should be physically and politically plausible but that space be provided for the consideration of extreme events and circumstances. Scenarios are especially valuable when looking far into the future (e.g., 10 years or more past project development) (Duinker & Greig, 2007), for exploring the potential induced impacts of a project, for exploring a project's impacts in combination with the impacts of other reasonably foreseeable or hypothetical developments or undertakings, or when attempting to account for such wildcards as

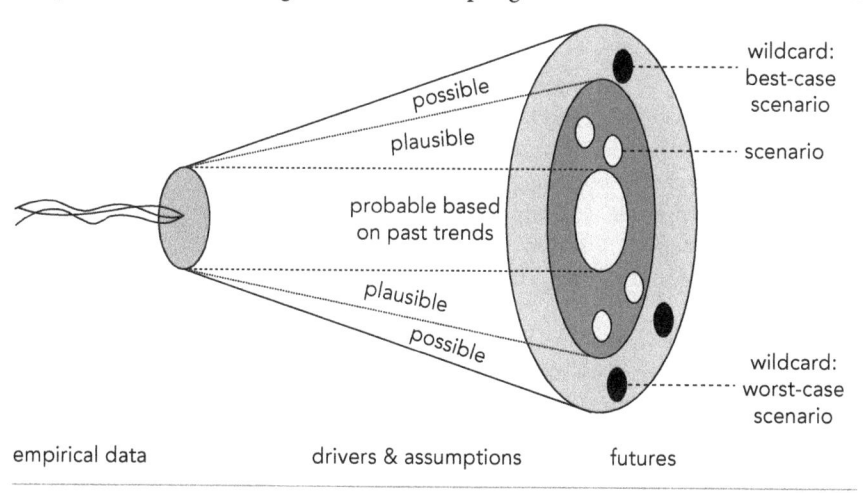

Figure 6.7 Scenarios in EA

natural disasters or climate uncertainties. Several tools are available to support scenario exploration and analysis, including spatial modelling and simulation software such as **ALCES** or **MARXAN** (Box 6.2).

Box 6.2 ALCES and MARXAN: Tools for Scenario Analysis

ALCES (A Landscape Cumulative Effects Simulator) is a system dynamics simulation tool for exploring the behaviour or response of resource systems to disturbances. ALCES was developed by Dr Brad Stelfox, founder of Forem Technologies in 1995 and an adjunct professor at the University of Calgary. ALCES is used for exploring and quantifying the relationships between different land-use sectors and natural disturbances and their potential environmental and socio-economic consequences. As a simulation model, it does not search for optimal solutions, such as an optimal land-use configuration or spatial pattern; rather, it is designed to forecast and explore alternative outcomes and conditions. ALCES is a spatially stratified model. It allows users to examine the area and length of different land uses or disturbances, such as roads or forest cut blocks, within each landscape unit. Coupled with Geographic Information Systems, ALCES can generate maps that illustrate the plausible location and extent of simulated land-use features and landscape types, including footprint types and disturbed area, and other user-defined indicators, such as water use or wildlife population parameters. ALCES has been applied in numerous resource and land-use planning and assessment exercises in western Canada, including the Upper Bow Basin Cumulative Effects Study, State of the Baptiste Lake Watershed report, and the Terasen Jasper National Park–Mt Robson Pipeline assessment. For additional information on ALCES, including the modelling tool, applications, and training opportunities, see http://www.alces.ca.

MARXAN (Marine Spatially Explicit Annealing) is a landscape and marine optimization tool. MARXAN was developed by Drs Ian Ball and Hugh Possignham at the University of Adelaide and later integrated with a Geographic Information System through a Nature Conservancy–funded project. MARXAN uses a site-selection algorithm to explore options for conservation and regional biodiversity protection. It divides a landscape into small parcels termed planning units and then attempts to minimize the "cost" of conservation while maximizing attainment of conservation goals. The objective is to select the smallest overall area needed to meet set conservation goals. It does this by using a simulated annealing with iterative improvement algorithm to select areas of high value for conservation, changing the planning units selected and re-evaluating the cost through multiple iterations. MARXAN uses data on ecosystems, habitats, and other relevant conservation features to find efficient solutions to the problem of selecting a set of conservation areas. MARXAN is a useful tool for exploring alternative land-use and conservation scenarios and management options and to facilitate discussion among stakeholders. However, MARXAN is limited in its ability to consider social and economic systems and dynamics. It has been used in several conservation planning initiatives, including rezoning of the Great Barrier Reef and the North American Wildlands Project, and by the

World Wildlife Fund to define a global network of marine protected areas. MARXAN has not been widely used for applications in impact assessment, but the Great Sand Hills Regional Environmental Study in Saskatchewan (2004–7) demonstrated the utility of MARXAN as a tool in strategic environmental assessments to support scenario-based analysis and cumulative effects assessment of human-induced disturbances due to cattle ranching, oil and gas development, and linear developments. For additional information on MARXAN, including the modelling tool, applications, and training opportunities, see http://www.uq.edu.au/marxan.

Characterizing Predicted Impacts

Impact prediction should provide insight into the characteristics of potential impacts. Impact characterization is important to prioritizing impact management actions and to determining the significance of potentially adverse impacts. Effects and impacts can be classified in several ways. There is no single best set of criteria for use in every project situation, but there are several common classifications that should be considered for any given project (Table 6.4). The following is by no means a comprehensive list; rather, it serves to illustrate classification approaches to predicted effects and impacts and to show that not all impact characteristics are independent.

Order

Effects resulting from a project can be **direct effects** (first order), **secondary effects** (resulting from change caused by direct effects), or **induced effects** (resulting from spin-off activity). For example, the flooding of land during construction of a hydroelectric project has a direct effect on the availability of terrestrial habitat. Indirectly, human health may be affected because of exposure to increased mercury levels in fish caused by mercury release and the flooding of soils and decay of organic matter. The relationship between first-order effects and second-order effects is not always straightforward. For example, runoff from an agricultural operation may cause enrichment of a local freshwater body by depositing excess nutrients. The result may be excess plant growth near the shores of the water body, which may create an aesthetically pleasing waterscape. However, the subsequent **eutrophication**—excess

Table 6.4 Classification of Environmental Effects and Impacts

Order	Nature	Magnitude	Spatial Extent	Frequency and Duration	Reversibility	Likelihood
Direct	Incremental	Size	Local	Continuous	Reversible	Probability
Indirect	Additive	Degree	Regional	Immediate	Irreversible	Risk
Induced	Synergistic	Concentration	National	Delayed		
	Antagonistic	Increasing	International	Short-term		
		Decreasing		Long-term		

> **Box 6.3 Actions, Causal Factors, Effects, and Impacts**
>
> | | |
> |---|---|
> | Action | Hydroelectric dam construction on a river system. Assume a simple development project in which a river is modified and dammed for hydroelectric power generation. |
> | Causal factor | Dredging. The project involves dredging and widening the river channel upstream of the dam site prior to its construction. |
> | Condition change | Increased sediment flow into the river downstream of the dredging site creates changes in deposition of sediments and increased turbidity. |
> | Effects and impacts | Direct: Decreased growth rates in fish |
> | | Indirect: Decline in recreational fishery |
> | | Several different impacts may occur as a result of the initial change in condition. For example, deposition of sediments and increased turbidity (project effects) may result in decreased growth rates in river fish (first-order, direct impact), leading to a decline in the region's recreational river fishery (second-order, indirect impact). |

plant decomposition absorbing dissolved oxygen at high rates—generates a negative secondary effect, making water unsuitable for aquatic life (Box 6.3).

Induced effects or impacts are those resulting from other activities or actions triggered indirectly by a project but actions that the project proponent may have little control over. For example, the development of an all-season access road to a remote northern community may create new economic opportunities for the community and attract new businesses and development opportunities that generate actional impacts. A new mining operation may hire workers from a local community, who spend a portion of their earnings on local entertainment and goods and services. This new injection into the local economy, and increased demand for goods and services, can create additional jobs and attract investment in other non-mining sectors.

Nature

Incremental effects or impacts are marginal changes in conditions that are directly attributable to the action being assessed. For example, a forestry operation may lead to an incremental loss of habitat each year should harvesting outpace regeneration. For incremental effects, it is important to consider the "rate of change" in the affected environmental parameter and to regularly assess total change against determined thresholds, standards, or specified targets.

Additive effects or impacts are the consequence of individual but separate actions that may be minor individually but together can create a significant overall impact (Figure 6.8). Additive effects can result from the progressive impacts from the same project or activity over time or from the actions of several independent projects, such as land clearing in a single region. For example, oil and gas operations may result

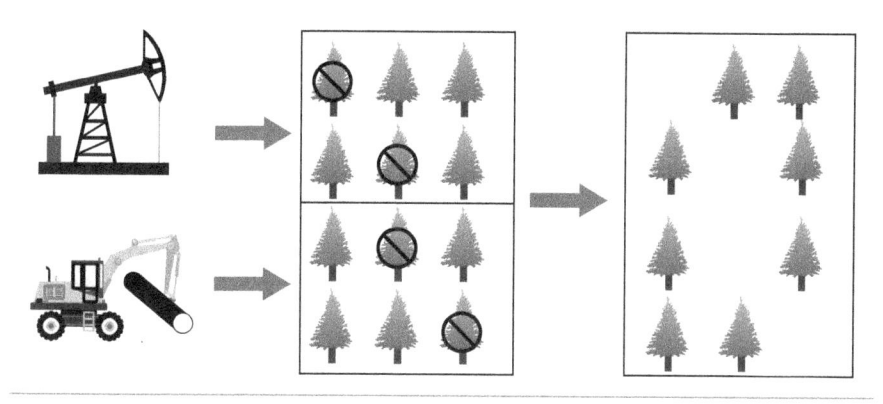

Figure 6.8 Additive effects

in a 10 per cent removal of forest cover in a region. Forest harvesting may result in the same amount of forest removal in the region. If land clearing for oil and gas operations and forestry occur during the same time period, or if clearing for oil and gas infrastructure occurs before forests can recover from forest harvesting, the additive result would be 20 per cent loss of forest cover. While perhaps individually manageable, together such operations can present significant adverse additive effects on wildlife because of habitat fragmentation. Additive effects thus result from two types of actions:

1. Where a single development action repeatedly affects the same environmental component over space or time. For example, the development of a mine site may create habitat loss, changes in surface water drainage and quality, and increased noise from vehicles and operations. In combination, they can have an additive negative effect on wildlife.
2. Where two or more of the same type of actions are affecting the same environmental component. For example, multiple industrial developments may cause additive impacts on the landscape because of vegetation clearing for project infrastructure.

Synergistic effects or impacts are the result of interactions between effects and occur when the total effect is greater than the sum of the individual effects (Figure 6.9). For example, a single industry located adjacent to a river system may alter water temperatures, change dissolved oxygen levels, and introduce heavy metals. Individually, each effect may be tolerable for fish, but the toxicity of certain heavy metals is multiplied in high water temperatures and low dissolved oxygen content. The effect on fish as a result of the interaction of these effects is thus greater than the sum of the individual project-induced changes. Alternatively, an industrial plant releasing chemical "A" into a river may result in a 10 per cent mortality in fish. A second industrial activity, releasing a different chemical "B" into the same river may also result in a 10 per cent mortality in fish over the same time period. If chemical "A" and chemical "B" are released together and interact, the resulting mortality in fish may be 30 per cent.

Antagonistic effects or impacts occur when one adverse effect or impact partially cancels out, offsets, or interrupts another adverse effect or impact. For example, when

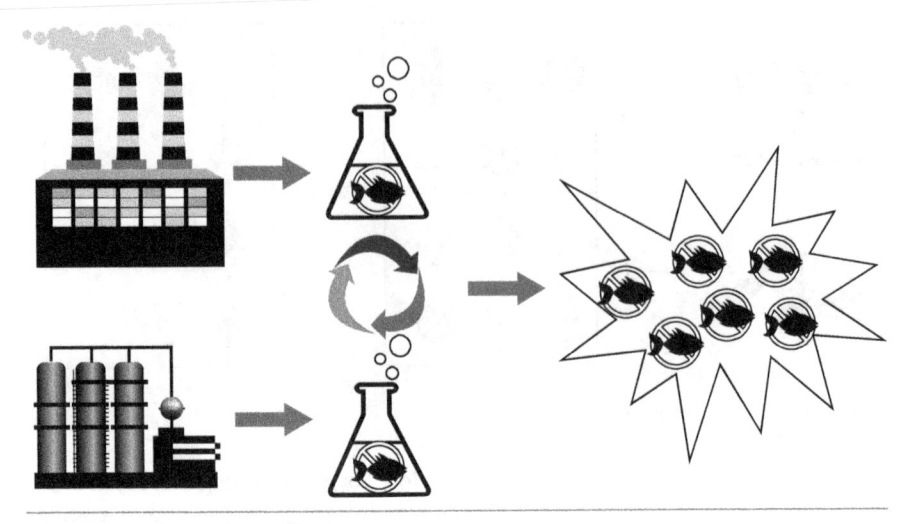

Figure 6.9 Synergistic effects

chemical "A" with a 10 per cent mortality rate in fish is combined with chemical "B," which has the same mortality rate, the antagonistic result is less than a 20 per cent mortality rate. These effects are usually less common in EA, since additional stressors to the environment more often create further disturbance and degradation. One example, however, is the reduced eutrophication of a water body receiving effluents containing both chlorine and phosphates. While each substance is individually harmful for aquatic life, together and in moderate amounts they may be beneficial for managing eutrophication.

Magnitude

The **magnitude** of an effect refers to its size or degree—for example, the specific concentration of an emitted pollutant. Many impacts can be measured in quantitative terms such as concentration or volume, but others, such as the aesthetic impact on view scape, are subjective and not easily reduced to quantitative terms. The magnitude of an effect is not necessarily related to the significance or importance of the impact. For example, removal of 5 per cent of the forest cover in an area that supports a rare or endangered species may be considered much more significant than removal of 10 per cent of the forest cover in an area that supports stable species populations.

The **direction of change** associated with an effect is closely related to its magnitude. Direction of change refers to whether the affected parameter is increasing or decreasing in magnitude relative to its current or previous state or baseline condition. For example, the construction and operation of a large-scale industry may lead to an increase in the local population if it generates employment opportunities and triggers in-migration; however, it may lead to a decrease in local population if smaller industries are displaced or if quality of life in the community decreases as a result of environmental change. In the first instance, the impact may be considered positive, whereas in the second, an adverse effect is the more likely result. The situation is not always this straightforward. In some cases, effects are multi-directional. For example, some individuals in the affected community may see an increase in local population as a positive impact and an improvement in their quality of life, whereas others may see population increase due to industrial growth as having an adverse impact on their quality of life. When characterizing environmental change, it is important to give at least some indication of direction, even if it cannot be quantified, in order to understand the magnitude of the effect and impact.

Spatial Extent

The spatial extent of an effect or impact will vary depending on the specific parameters involved and the project's environmental setting. For example, soil erosion may be highly localized and agricultural drought may be regional, while economic impacts associated with continuous drought may be national or international in extent. In the case of the Jack Pine mine project, a proposal by Shell Canada to mine bitumen deposits in the Athabasca oil sands of Alberta, the spatial extent of the assessment was defined by "local study areas" and "regional study areas." Local study areas were identified as those directly affected by project development. Regional study areas were identified as those seen from a larger geographic and ecological perspective that would experience direct and indirect impacts.

Sometimes spatial extent is classified according to **on-site impacts** and **off-site impacts**, a recognition that the impacts of a development or action may extend well beyond the specific project site. On-site actions may trigger environmental change, which in turn affects off-site environmental components. For example, chemical discharges at an industrial complex generate on-site environmental impacts because of soil contamination, which in turn may create off-site impacts through groundwater contamination. Such off-site impacts may go beyond biophysical impacts alone. **Fly-in fly-out** employment arrangements, often associated with remote mining projects, have impacts that extend well beyond the mine project site to the communities the workers come from and affect personal matters such as family life and relationships (Box 6.4). Determining the spatial extent of EA is a key element of the scoping process and is critical to ensuring that all necessary impacts are considered within the scope of the assessment.

Box 6.4 Impacts of Fly-in Fly-out Mining Operations

Fly-in fly-out is now a common practice in mining operations globally, particularly in Australia and northern Canada's mineral and energy resource sectors. Storey (2010) defines fly-in fly-out as long-distance commuting work practices whereby workers travel by air or other mode of transport to and from worksites that are typically in remote areas and often at a distance from existing communities. Storey explains that fly-in fly-out was encouraged by the expansion of mining into increasingly remote areas at a time when corporate interests were focusing on leaner and more flexible modes of production and when governments were unwilling to support the development of new single-industry communities in remote areas. There are several fly-in fly-out mining operations in northern Canada, including: the Diavik diamond mine in the Northwest Territories, located 300 km north of

continued

Yellowknife; Cameco's Key Lake Uranium Mine in Saskatchewan, located 570 km north of Saskatoon by air and 220 km by road from the nearest village; CanTung, owned by the North American Tungsten Corporation, located about 300 km northeast of Watson Lake in Yukon; and the Meadowbank gold mine in Nunavut, located 300 km west of Hudson Bay and 70 km north of Baker Lake, the nearest town.

Some of the characteristics of fly-in fly-out work practices in the mining industry include working in relatively remote locations where the mining company typically provides accommodations for the workers at or near the mine site and a work roster based on a fixed number of days at the worksite and a fixed number of days at home, such as a two-week work rotation. Although fly-in fly-out work arrangements ensure affordable accommodation and the necessary social and infrastructure services for workforces in remote areas, there are also a number of potentially adverse impacts, both on-site (in the host community or at the work camp) and off-site (in the worker's home community and household). For example, resident communities housing a fly-in fly-out workforce can experience additional demand on local social infrastructure and services. There is the potential for adverse impacts on the workers themselves, including fatigue and other health and safety risks associated with long-distance commuting to and from often remote areas. The worker's household may also experience potentially adverse impacts, including family disruption and reduced socialization with family and friends in the home community. If workers are Indigenous and still engaged in traditional hunting and fishing activities, there is the potential for cultural disruption. At the same time, fly-in fly-out operations generate new income for those finding employment, and fly-in fly-out personnel can bring external income from the mine site to the local community.

Such impacts are difficult to predict, and methods for doing so often tend to draw on experiences from other, similar projects and work arrangements. But these off-site and on-site impacts associated with fly-in fly-out work arrangements need to be identified and effectively managed as part of the impact prediction and mitigation strategies for remote mining operations.

Sources: Storey, 2010; Morris, 2012

Frequency and Duration

Frequency refers to how often an effect or impact occurs. Frequency is especially important when characterizing effects on ecological systems, since more frequent impacts or disturbances mean less time for an affected component to recover. Consider the scenario depicted in Figure 6.10, where a valued component is subjected to a series of disturbances over time. The magnitude of each consecutive disturbance decreases, but the frequency of disturbances increases, meaning less recovery time and an overall decline in VC condition. *Duration* is the length of time that the effect or impact exists—for example, short-term effects such as the noise associated with a bridge construction or the long-term effects associated with riverbank erosion and downstream loss of arable land.

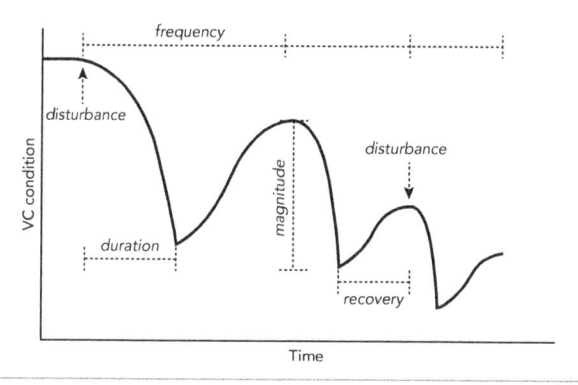

Figure 6.10 Relationship between disturbance frequency, duration, magnitude, and VC recovery time and condition

Sometimes impacts endure for long periods of time and are **continuous effects** or impacts, such as energy fields associated with transmission lines. Other effects may be discontinuous and last only for short intervals, such as the noise from blasting at a construction site. Different effects and impacts can also arise at different times during the life of a project. Immediate effects and impacts follow shortly after a change in the condition of the environment. For example, odour from the emissions of a pulp and paper mill would occur immediately after its establishment, whereas health effects due to continued exposure to poor air quality may be delayed. Consideration must also be given to the regularity of effects, since those with greater predictability can usually be dealt with more easily than those that happen irregularly or that come as a surprise. For example, the impact of low-level military flight training in Newfoundland and Labrador on caribou health and reproduction is of concern not because of the noise level itself but because of the startle or surprise effect of low-flying military jets.

Reversibility

The reversibility of an impact is important for determining the nature and effectiveness of impact management strategies. An effect or impact is considered a **reversible impact** when it is possible to approximate the pre-disturbed condition. For example, in the case of a sanitary landfill operation, the surface can be remediated to resemble its pre-project condition, or traffic congestion may return to its original level when a bridge reconstruction project ends. Reversing an environmental effect, however, does not always mean restoring a biophysical or socio-economic environment to its *exact initial* condition, since that is neither always possible nor always desirable. A mine site, for example, may be remediated and reforested based on current forest stand composition in the project area—it may be neither desirable nor feasible to return the site to its pre-project forest stand composition. Restoring disturbed environments to their pre-disturbed condition is at best an approximation of what the affected environment might look like had the development action not taken place.

Some effects and impacts are completely **irreversible impacts**. For example, the extinction of a rare plant species during site-clearing is irreversible. Other types of impacts may be reversible from a technical standpoint, but it may be impractical or economically unfeasible to attempt to do so. For example, it is common practice in large-scale mining initiatives to drain a local pond or lake for tailings disposal. While it is *possible* to restore the pond or lake following mine decommissioning, restoration is often neither practical nor economically feasible. Consequently, these effects are often considered irreversible from any practical impact management perspective. Instead, efforts are made elsewhere to compensate for the loss of aquatic habitat, either through the creation of new habitat or the restoration of damaged habitat.

Likelihood

Whether an effect or impact is considered important rests considerably on the probability of it occurring. If, for example, an impact is almost certain to occur, then it is more likely to be considered in the decision-making process than one that is determined to be highly improbable. The likelihood or probability of an effect or impact occurring is one component used to measure **risk**—an uncertain situation involving the possibility of an undesired outcome. Risk combines the probability of an adverse event occurring with an analysis of the severity of the consequences associated with that event (Figure 6.11). In other words, risk can be characterized as a function of damage potential, exposure to dangerous substances such as toxic chemicals, opportunity for exposure, and the characteristics of the population at risk. For example, the elderly and young children may be much more vulnerable to exposure than other segments of the population. **Risk assessment**, then, refers to the process of accumulating information, identifying possible risks, risk outcomes and the significance of those outcomes, and assessing the likelihood and timing of their occurrence. The acceptability of a specified level of risk depends on several factors, including the catastrophic potential associated with the risk event, scientific uncertainty, distribution of risk outcomes, and understanding of and familiarity with the

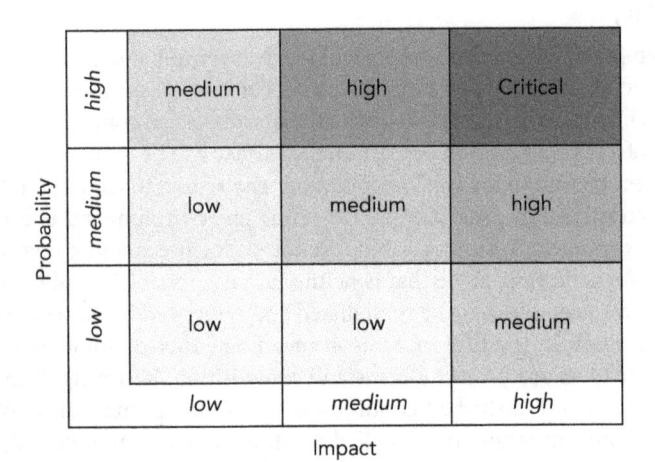

Probability		Impact		
		low	medium	high
	high	medium	high	Critical
	medium	low	medium	high
	low	low	low	medium

Figure 6.11 Simple 3 x 3 risk classification matrix

risk. In terms of the latter, for example, an individual living on a flood plain may have perceptions of flood risk significantly different from those of individuals living at a distance from the same flood plain.

Addressing Uncertainty

Practitioners of EA are expected to predict the impacts of proposed projects and communicate their findings in a way that can be used by decision-makers to make informed decisions. Yet impact predictions are often found to be wrong or inaccurate (Storey & Noble, 2004). The problem isn't that impact predictions are proved wrong per se but that a degree of confidence is presented in predictions that simply is not warranted. Tennoy, Kvaerner, and Gjerstad (2006) reviewed 22 EAs in Norway and found that uncertainty disclosure diminished over the course of the EA process—from scoping to final approval. Wood (2008) reviewed 30 EAs in the UK and found that only 23 per cent considered the uncertainty associated with prediction methods. Wood concluded that EA can impart a greater sense of certainty than is genuinely warranted, including uncertainty related to the confidence in predictions. Omitting or underestimating uncertainties can result in a systematic bias in the decisions taken and overconfidence in the assessment process, including the appropriateness of planned mitigation, almost certainly resulting in compromised protection of the environment.

There are measures that can be taken to address uncertainty in impact prediction, including **probability analysis, sensitivity analysis, confirmatory analysis**, and **uncertainty disclosure**. The first three measures assume that the level of uncertainty can be assessed and quantified. The fourth measure accounts for the reality that, often, uncertainty cannot be expressed as a probability or explained by the tools and techniques used to predict impacts. In many cases, there is inherent uncertainty in environmental and socio-economic systems.

Uncertainty disclosure requires that practitioners disclose their assumptions and uncertainties about impact predictions. However, Duncan (2008) reports that proponents, consultants, and decision-makers may have a vested interest in making EAs and their decisions appear more defensible and politically palatable, resulting in practices that systematically seek to minimize uncertainty disclosure. What has become standard practice, but arguably poor practice, is to systematically diminish uncertainty disclosure in EA, resulting in overconfidence in impact predictions. A recent cross-Canada survey of EA practice by Leung, Noble, Jaeger, and Gunn (2016), for example, reports that 80 per cent of practitioners and regulators believe that all impact assessments contain information that is uncertain but only 15 per cent believe that uncertainty is sufficiently acknowledged.

It is the responsibility of the individual providing input to the assessment process to identify uncertainties that might exist in the information they provide, including any data or knowledge gaps or assumptions on which the information is based. It is the responsibility of those receiving the information to determine how to process and use that information in their decision process. One approach to uncertainty identification and characterization is the **uncertainty matrix** (Figure 6.12). The uncertainty matrix was first developed by Walker et al. (2003) and recently

UNCERTAINTY MATRIX		Level of uncertainty			Nature of uncertainty	
Location of uncertainty		statistical uncertainty	scenario uncertainty	systemic uncertainty	knowledge-related	variability-related
Context	Assumptions about ecological, technological, economic, political, or social context					
Expert judgment	Narrative uncertainty or experience uncertainty					
Model	Model structure: relations					
	Model parameters: choice and representation					
	Model input: data, drivers					
Data	Availability, gaps, quality					

Figure 6.12 Uncertainty matrix for communicating the location, level, and nature of uncertainty

Sources: Adapted from Walker et al., 2003; Petersen et al., 2013

revised by Petersen et al. (2013); it forms the basis for uncertainty guidance developed by the Netherlands Environmental Assessment Agency. The matrix is comprised of three basic components:

- The *location of uncertainty* provides for an understanding of where the uncertainty exists and thus what type of strategies might be pursued to better understand, explore, or resolve uncertainty. For example: uncertainty related to assumptions about ecological systems or economic or political context; uncertainties arising from the predictive tools or models used to represent or conceptualize relationships; or uncertainties due to the quality and extent of available data.
- The *level of uncertainty* speaks to the degree to which one is uncertain and provides users of information with a relative understanding of decision risk. Levels of uncertainty include **statistical uncertainty** (e.g., calculated error and probabilities are known, and the decision risk can be calculated); **scenario uncertainty** (e.g., how an impacted system might change is fairly understood, but the likelihood and extent of change are not known); and

systemic uncertainty (e.g., uncertainties that cannot be estimated by any current method or technique—we simply don't know).
- The *nature of uncertainty* speaks to whether the uncertainty can be eliminated or sufficiently reduced. Knowledge-related uncertainty implies that additional research or monitoring may help to resolve uncertainty—though the uncertainty may be so significant that the knowledge cannot be obtained in a timely manner. Inherent variability, in contrast, refers to systems and values whereby "more research" (and a delayed EA or decision) will not necessarily reduce the uncertainty.

The uncertainty matrix can be used by practitioners to help communicate uncertainties and determine whether or when additional information might be required. For example, uncertainty in predictions that are largely located in data (i.e., availability or quality) may be resolved through additional data collection or supplemented by expert judgment, provided that the level of uncertainty is statistical or scenario uncertainty. In cases of systemic uncertainty or when uncertainty is variability-related, it may not be possible to resolve the uncertainty in an impact prediction, and greater emphasis must be placed on either planning for the worst-possible scenario or ensuring sufficient capacity to respond rapidly to changing circumstances or unanticipated outcomes.

Key Terms

accuracy
additive effects
ALCES
analogue approaches
antagonistic effects
balance models
climate-resilient projects
climate risk
community health and well-being
confirmatory analysis
continuous effects
cumulative impact
Delphi technique
determinants of health and well-being
direct effects
direction of change
ecosystem services
environmental change
environmental effect
environmental impact
eutrophication
fly-in fly-out

incremental effects
induced effects
irreversible impacts
magnitude
MARXAN
maximum allowable effects levels
models
non-point-source stress
off-site impacts
on-site impacts
point-source stress
precision
probability analysis
reversible impact
risk
risk assessment
scenario
scenario uncertainty
secondary effects
sensitivity analysis
spatial models
statistical models

statistical uncertainty
stranded assets
synergistic effects
systemic uncertainty

threshold-based predictions
uncertainty disclosure
uncertainty matrix

Review Questions and Exercises

1. Discuss the relationship between accuracy and precision in impact prediction.
2. What are some of the challenges to making impact predictions concerning biophysical and social change?
3. Obtain a completed EIS from the Canadian Impact Assessment Registry (https://www.canada.ca/en/impact-assessment-agency.html). Scan the document for impact predictions, and examine how these predictions are stated. Are the statements based on verifiable hypotheses? Are thresholds or maximum allowable effects levels stated? Given the nature of the predictions and the way in which they are stated, do you think that they can be followed up and verified? Compare your findings with those of others.
4. Obtain a completed EIS from the Canadian Impact Assessment Registry (https://www.canada.ca/en/impact-assessment-agency.html). Scan the document, and generate a list of the techniques used to predict, describe, or assess impacts on the biophysical and human environments. Compare your results with those of others. Is there a common set of techniques that emerge for various environmental components, such as water quality, air quality, or employment?
5. Provide an example for each of incremental, additive, and synergistic impacts.
6. Consider the following scenario: A new energy pipeline is proposed to transport oil and gas from the western Arctic to northern Alberta. What are some of the potential climate implications of such a project? Are all these implications within the scope of a project EA? What are some of the potential impacts of the environment on the project?
7. Suppose a proponent applies for the development of a large-scale open-pit gold mine operation in your region. In small groups, brainstorm the potential "on-site" and "off-site" impacts. Use the "impact characteristics" discussed in this chapter to identify the impacts you believe are most important to consider among both on-site and off-site impacts. Are there any additional criteria or characteristics that should be considered? Compare the results among groups.

References

Aura Environmental Research and Consulting Ltd. (2008). *Community health baseline study: Lower Churchill hydroelectric generating project*. Report prepared for Minaskut Limited Partnership. St John's, NL.

Boyle, J., Cunningham, M., & Dekens, J. (2013). *Climate change adaptation and Canadian infrastructure: a review of the literature*. Winnipeg, MB: International Institute for Sustainable Development.

Byer, P., Cestti, R., Croal, P., Fisher, W., Hazell, S., Kolhoffff, A., & Kørnøv, L. (2018). Climate change in impact assessment: international best practice principles. Special Publication Series no. 8. Fargo, ND: International Association for Impact Assessment.

Canter, L.W., & Ross, B. (2010). State of practice of cumulative effects assessment and management: the good, the bad and the ugly. *Impact Assessment and Project Appraisal* 28(4), 261–8.

CESD (Commissioner of the Environment and Sustainable Development). (2011). *Report of the Commissioner of the Environment and Sustainable Development*. Ottawa, ON: Office of the Auditor General of Canada.

Christensen, L., & Krogman, N. (2012). Social thresholds and their translation into socio-ecological management practices. *Ecology and Society* 17(1), 5. http://dx.doi.org/10.5751/ES-04499-170105.

Cronmiller, J., & Noble, B. (2018). Integrating environmental monitoring with cumulative effects management and decision making. *Integrated Environmental Assessment and Management* 14(3), 407–17.

De Jongh, P. (1988). Uncertainty in EIA. In P. Wathern (ed.), *Environmental impact assessment: theory and practice*. London, UK: Unwin Hyman.

Duinker, P., & Greig, L. (2006). The impotence of cumulative effects assessment in Canada: ailments and ideas for redeployment. *Environmental Management* 37(2), 153–61.

Duinker, P.N., & Greig, L. (2007). Scenario analysis in environmental impact assessment: improving explorations of the future. *Environmental Impact Assessment Review* 27, 206–19.

Duncan, R. (2008). Problematic practice in integrated impact assessment: the role of consultants and predictive computer models in burying uncertainty. *Impact Assessment and Project Appraisal* 26(1), 53–66.

Fischer, T.B., & Sykes, O. (2009). The new EU territorial agenda: indicating progress for climate change mitigation and adaptation? In S. Davoudi et al. (eds), *Planning for climate change*. London, UK: Earthscan.

Glasson, J., Therivel, R., & Chadwick, A. (1999). *Introduction to environmental impact assessment: principles and procedures, process, practice and prospects*. 2nd edn. London, UK: University College London Press.

Government of British Columbia. (2007). *Greenhouse Gas Reduction Targets Act*. Current to 5 April 2017. Victoria, BC: Queen's Printer, Government of British Columbia.

Government of British Columbia. (2016). Climate action legislation. Victoria, BC: Government of British Columbia. http://www2.gov.bc.ca/gov/content/environment/climate-change/policy-legislation-programs/legislation-regulations.

Gray, E. (2015). *Blind spot: the failure to consider climate in British Columbia's environmental assessments*. Victoria, BC: Environmental Law Centre, University of Victoria.

Greig, L., & Duinker, P. (2011). A proposal for further strengthening science in environmental impact assessment in Canada. *Impact Assessment and Project Appraisal* 29(2), 159–65.

Hackett, P., Liu, J., & Noble, B. (2018). Human health, development legacies, and cumulative effects: environmental assessments of hydroelectric projects in the Nelson River watershed, Canada. *Impact Assessment and Project Appraisal*. doi: org/10.1080/14615517.2018.1487504.

Hulse, D.W., Branscomb, A., & Payne, S.G. (2004). Envisioning alternatives: using citizen guidance to map future land and water use. *Ecological Applications* 14(2), 325–41.

Kennett, S. (2002). Lessons from Cheviot: redefining government's role in cumulative effects assessment. In *Cumulative effects management tools and approaches*. Calgary, AB: Alberta Society of Professional Biologists.

Larsen, P., Goldsmith, S., Smith O., Wilson, M., Strzepek, K., Chinowsky, P., & Saylor, B. (2008). Estimating future costs for Alaska public infrastructure at risk from climate change. *Global Environmental Change* 18(3), 442–57.

Leung, W., Noble, B., Jaeger, J., & Gunn, J. (2016). Disparate perceptions about uncertainty consideration and disclosure practices in environmental assessment and opportunities for improvement. *Environmental Impact Assessment Review* 57, 89–100.

Liu, Y., Mahmoud, M., Hartmann, H., Stewart, S., & Wagener, T. (2008). Formal scenario development for environmental impact studies. *Developments in Integrated Environmental Assessment* 3, 145–62.

Luke, L., & Noble, B. (2019). Consideration and influence of climate change in environmental assessment: an analysis of British Columbia's liquid natural gas sector. *Impact Assessment and Project Appraisal* 37(5), 371–81.

Mitchell, B. (2002). *Resource and environmental management*. 2nd edn. New York, NY: Prentice-Hall.

Morris, P., and Therivel, R. (eds). (2001). *Methods of environmental impact assessment*. 2nd edn. London, UK: Taylor and Francis Group.

Morris, R. (2012). *Scoping study: impact of fly-in fly-out/drive-in drive-out work practices on local government*. Sydney, Australia: Australian Centre of Excellence for Local Government, University of Technology.

Murillo, C. (2012). Natural gas liquids in *North America: overview and outlook to 2035*. Calgary, AB: Canadian Energy Research Institute.

Nalcor Energy. (2009). *Environmental impact statement. Lower Churchill hydroelectric generating project*. St John's, NL: Nalcor Energy.

Noble, B.F., Liu, G., & Hackett, P. (2017). The contribution of project environmental assessment to assessing and managing cumulative effects: individually and collectively insignificant? *Environmental Management* 59(4), 531–45.

Ohsawa, T., & Duinker, P. (2014). Climate-change mitigation in Canadian environmental impact assessments. *Impact Assessment and Project Appraisal* 32, 222–33.

Petersen, C., Janssen, P., van der Sluijs, J.P., Risbey, J., Ravetz, J.R., Wardekker J.A., & Hughes, H.M. (2013). *Guidance for uncertainty assessment and communication*. The Hague, Netherlands: PBL Netherlands Environmental Assessment Agency.

Phillips, D. (2002). *A closer look at a rare situation: 1998 ice storm*. Ottawa, ON: Meteorological Services of Canada.

Storey, K. (2010). Fly-in/fly-out: implications for community sustainability. *Sustainability* 2, 1161–81.

Storey, K., & Noble, B. (2004). *Toward increasing the utility of follow-up in Canadian EA: a review of concepts, requirements and experience*. Report prepared for the Canadian Environmental Assessment Agency. Gatineau, QC: CEAA.

Tennoy, A., Kvaerner, J., & Gjerstad, K. (2006). Uncertainty in environmental impact assessment predictions: the need for better communication and more transparency. *Impact Assessment and Project Appraisal* 24(1), 45–56.

UK Environmental Agency. (2002). *Environmental impact assessment: a handbook for scoping projects*. Bristol, UK: UK Environment Agency.

UNEP (United Nations Environment Programme, Economics and Trade Programme). (2002). *Environmental impact assessment training manual*. 2nd edn. New York, NY: UNEP.

US Council on Environmental Quality. (1997). Considering cumulative effects under the National Environmental Policy Act. Washington, DC: US Council on Environmental Quality, Executive Office of the President.

US EPA. (2019). Climate change impacts by sector. https://19january2017snapshot.epa.gov/climate-impacts/climate-change-impacts-sector_.html.

Walker, W.F., Harremoes, P., Rotmans, J., van der Sluijs, J., van Asselt, M., Janssen, P., & von Krauss, M. (2003). Defining uncertainty: a conceptual basis for uncertainty management in model-based decision support. *Integrated Assessment* 4(1), 5–7.

Wood, G. (2008). Thresholds and criteria for evaluating and communicating impact significance in environmental impact statements: see no evil, hear no evil, speak no evil? *Environmental Impact Assessment Review* 28, 22–38.

7 Managing Project Impacts

Impact Management

The utility of EA lies not so much in the accuracy of the predicted impacts but in the effectiveness of impact management. The EA process establishes the measures that are necessary to avoid, minimize, reduce, or offset potentially adverse impacts and, where appropriate, incorporate these into environmental management plans (de Jesus et al., 2013). Impact management is foundational to the EA process; it translates the findings from an EIS into management recommendations and practice (Tinker et al., 2005). Although procedurally impact management follows the identification of impacts and determination of their significance, identifying strategies to effectively manage impacts is not limited to any one stage of the EA process (Glasson, Therivel, & Chadwick, 1999). Impact management is inherent in all stages of EA, from pre-project planning to post-EA monitoring (Morrison-Saunders, 2018).

Mitigation Hierarchy

The adequacy of impact management strategies in EA depends on their success in reducing the significance, risk, or severity of an anticipated, adverse effect. Impact management in EA is usually referred to as **impact mitigation**. Though strictly speaking mitigation means to make "less severe," impact mitigation is often a generic term in the EA process that encompasses a hierarchy of strategies from avoiding impacts to offsetting impacts that simply cannot be avoided, minimized, or restored (Figure 7.1). For example, managing potentially adverse impacts to important species habitat should first consider options that avoid the loss of habitat, followed by options that minimize habitat loss, restoring habitat loss during project operations or at project completion, and finally, compensating for unavoidable habitat loss. Each of these strategies is discussed below.

Avoiding Impacts

Avoiding potentially adverse effects, and thus preventing them from occurring, is the most desirable approach to impact mitigation. If an impact can be avoided, then the time and financial resources required for reducing its severity or for restoration or compensation can also be avoided. It also avoids potential conflict and concerns expressed by those who may highly value the affected component (Box 7.1). Methods of avoiding potentially adverse impacts may include the consideration of alternative project locations to avoid impacts to sensitive habitat; scheduling project construction activities so that they do not conflict with the timing of wildlife migration; routing ancillary

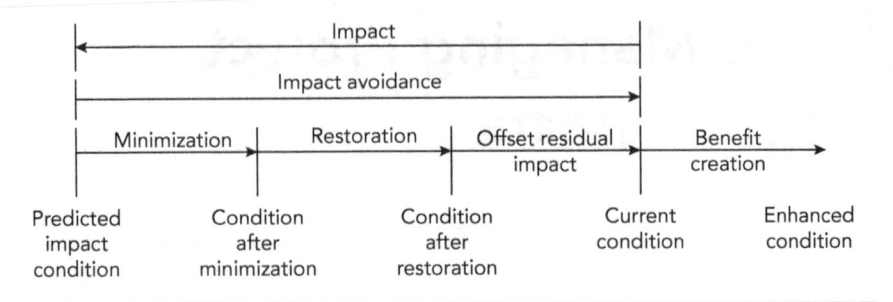

```
|<----------------------- Impact ----------------------->|
|<----------------- Impact avoidance --------------->|
| Minimization | Restoration | Offset residual | Benefit  |
|              |             |     impact      | creation |
  Predicted     Condition     Condition      Current      Enhanced
   impact         after         after       condition     condition
  condition    minimization   restoration
```

Figure 7.1 Impact mitigation hierarchy

Source: Adapted from British Columbia EAO, 2014

Box 7.1 Ecoducts for Avoiding Wildlife Collisions, Northern Sweden

The construction of ecoducts, also called wildlife bridges or wildlife overpasses, is an impact avoidance strategy in areas where wildlife collisions or disruptions to wildlife corridors or crossings are of concern. The image below shows an ecoduct near the LKAB iron ore mine in Kiruna, Norbotten County, northern Sweden. Operating for more than 100 years, the Kiruna mine is one of the world's largest underground iron ore mines. The mine's operations also have a significant surface footprint in an area that has traditionally been used by Sami Indigenous people for reindeer herding. Today, only about 10 per cent of Swedish Sami earn a living from

Ecoduct at the LKAB iron ore mine in Kiruna, Norbotten County, northern Sweden.

Photos by Bram Noble

reindeer husbandry, following the herd during its annual migration. The mine, coupled with other land uses, creates a bottleneck for herders when reindeer move between summer and winter pastures and poses high risk for collisions at road and railway crossings. This ecoduct and fence line were constructed at the mine site along the herding route to facilitate the movement of reindeer, minimize habitat fragmentation, and avoid collisions with railway traffic.

developments, such as access roads and other linear features, to avoid sensitive habitat, stream crossings, or cultural features; or the construction of self-contained work camps to avoid potentially negative socio-economic effects that might be caused by site worker–community interaction. **Impact avoidance** should enter the EA equation early, since most impact avoidance opportunities are presented early in project design processes through alternative locations of project infrastructure or project design options.

Minimizing the Severity or Extent of Impacts

Not all potentially adverse impacts are recognized in advance. Impact minimization refers to the strategies designed to reduce the severity of potential adverse impacts—often by reducing the geographic extent of the impact, the magnitude of change in baseline condition, or the duration of the impact. Minimizing impacts often, but not always, involves abatement control features (e.g., effluent treatment technologies) or physical design. For example, forest-harvesting operations can lead to soil erosion and excessive runoff, which in turn may affect the quality of aquatic environments. The use of **buffer zones** or setbacks, or areas of undisturbed riparian vegetation, is a common mitigation strategy to minimize sediment loading in streams. While buffer zones and setbacks do not fully *prevent* erosion or surface runoff from occurring, they do reduce the severity of sediment loading or contamination to aquatic environments caused by runoff.

Restoring Affected Components or Functions

Not all adverse environmental effects can be avoided or minimized. Sometimes environmental components will inevitably be temporarily damaged. **Restoration** refers to restoring environmental quality, rehabilitating certain environmental features, repairing ecological functions, or restoring environmental components to varying degrees. For example, in cases where the construction of a project requires clearing the vegetated landscape and destruction of important species habitat, impact management efforts can focus on restoring the landscape during project operations or post-operation to resemble the pre-disturbed state or function. Restoration in EA does not always mean that the resource or the land will be returned to the same pre-disturbance use or function. The objective is to return it to a more desirable condition compared to the state created by project actions. One step in restoration is **remediation**. Remediation is broadly defined as reducing contamination levels of a site to safe levels within the ecosystem to protect human health and to restore certain land uses and hydrological functions. Remediation is the end stage in the life cycle of a development project. It sometimes forms part of a project's EA application and regulatory approval; in other cases, it is treated as a separate undertaking, and the remediation activity itself is subject to EA review.

An emerging problem in Canada, and internationally, is the increasing amount of abandoned project infrastructure and contaminated project sites with no clear party responsible for remediation or reclamation (Box 7.2). In 2019, for example, the province of Nova Scotia announced that it would spend $48 million to clean up two former gold mines—Montague Gold Mines and Goldenville—as a first step in a province-wide effort to review and prioritize 69 abandoned sites on Crown land (Gorman, 2019). Perhaps the most prominent example in Canada is the Giant Mine, located in Yellowknife, Northwest Territories. The gold mine operated under various owners between 1948 and 1999, when the mine went into receivership. The Government of Canada assumed ownership of the property and sold it to Miramar Giant Mine Ltd to restart operations, create jobs in the mining sector, and ship the gold ore off-site for processing. As a condition of the sale, Canada agreed that Miramar would *not* be responsible for the existing state of the mine. The mine ceased operations in 2004, the land surface lease was returned, Giant Mine officially became an abandoned site, and responsibility for remediation now rests with the federal government. Processing at the mine site created 237,000 tonnes of arsenic trioxide dust—a by-product that is toxic to people and wildlife—that is housed in 15 underground chambers. The current remediation plan calls for the arsenic trioxide dust and the rock surrounding each chamber to be maintained completely frozen to prevent groundwater inflows and contamination (SRK & SENES, 2007). The plan also includes remediation of nearly 95 hectares of tailings and sludge, eight mine pits, surface soil remediation, and buildings and waste disposal. The total estimated cost for remediation is upwards of $1 billion.

Box 7.2 The Burgeoning Cost of Remediating Alberta's Oil and Gas Well Sites

Alberta is Canada's lead developer of conventional oil and gas. Conventional oil and gas deposits are extracted using conventional drilling methods whereby the natural pressure of the wells and pumping or compression operations force the resource to the surface. There are more than 300,000 conventional oil and gas wells in Alberta. As of mid-2019, only about 176,000 of these wells were considered active, whereas 90,000 were designated inactive wells. The number of inactive wells has been steadily increasing. An inactive well is one that has stopped producing or operating for technical or economic reasons but that may be reactivated in the future. However, some of Alberta's inactive wells date to the early 1900s, suggesting that inactive wells often remain inactive. An additional 77,000 wells in Alberta are considered abandoned, meaning that the well has been "plugged" or "capped" and is no longer operational, but the site has still not been remediated (cleaned up) or reclaimed (restored). Considering the more than 300,000 total conventional oil and gas wells in the province, the Alberta Liabilities Disclosure Project—a consortium that includes landowners and scientists—reports that cleaning up all the old and unproductive oil and gas wells will cost between $40 billion and $70 billion.

Of concern to the province is the number of orphaned wells. An orphaned well is a well that does not have a legally or financially responsible owner to deal with its closure, reclamation, and site restoration. An orphaned well is usually the result of insolvency. Given the significant and lasting drop in oil and gas commodity prices in recent years, insolvencies in the industry have been on the rise and the number of industrial operations leaving behind environmental liabilities increasing. Orphaned wells occupy land that cannot be safely used for other purposes, are human health hazards, and pose significant risks to the local environment because of leaks and soil and water contamination. A 2017 report by the C.D. Howe Institute estimated the cost to fully reclaim currently orphaned wells at between $129 million and $257 million, but given various estimates of bankruptcy rates in the industry, and considering firms that are close to being insolvent, the cost to reclaim orphaned well sites could be as high as $8.6 billion. This estimate does not include oil sands wells.

Sources: Dachis, Schaffer, & Thivierge, 2017; Riley, 2019; Orphan Well Association, 2019

Offsetting or Compensating for Unavoidable Loss

Some environmental effects cannot be avoided, mitigated, or rectified. In such cases, the typical action is **compensation** for those unavoidable, residual, or irreparable impacts that remain after other impact management options have been exhausted or for which no management alternative exists. Compensation sometimes involves monetary or other benefit payments to those affected by the damage caused by the project, while in other cases it involves measures to re-create environmental habitats at an alternative site. In Canada, for example, a federal policy on aquatic-based habitat declares that there should be "no net loss." This is not to say that projects posing a threat to aquatic habitat will not be approved; rather, any habitat that is lost must be compensated for (Box 7.3). Compensation can be a controversial form of

Box 7.3 LNG Canada Export Terminal Wetland Compensation Strategy

The LNG Canada Export Terminal, proposed by LNG Canada Development Inc., involves the construction and operation of a liquefied natural gas (LNG) processing and storage site and marine terminal for exporting LNG via shipping. The project site is near Kitimat, northern British Columbia, in the traditional territory of the Haisla Nation and within the Coastal Western Hemlock Very Wet Maritime Submontane Variant (CWHvm1) biogeoclimatic ecosystem. The project is among the largest in British Columbia's history. At its peak, the LNG project will produce 26 million tonnes per annum. The project was subject to EA review under the Canadian Environmental Assessment Act, 2012 and provincially under the British

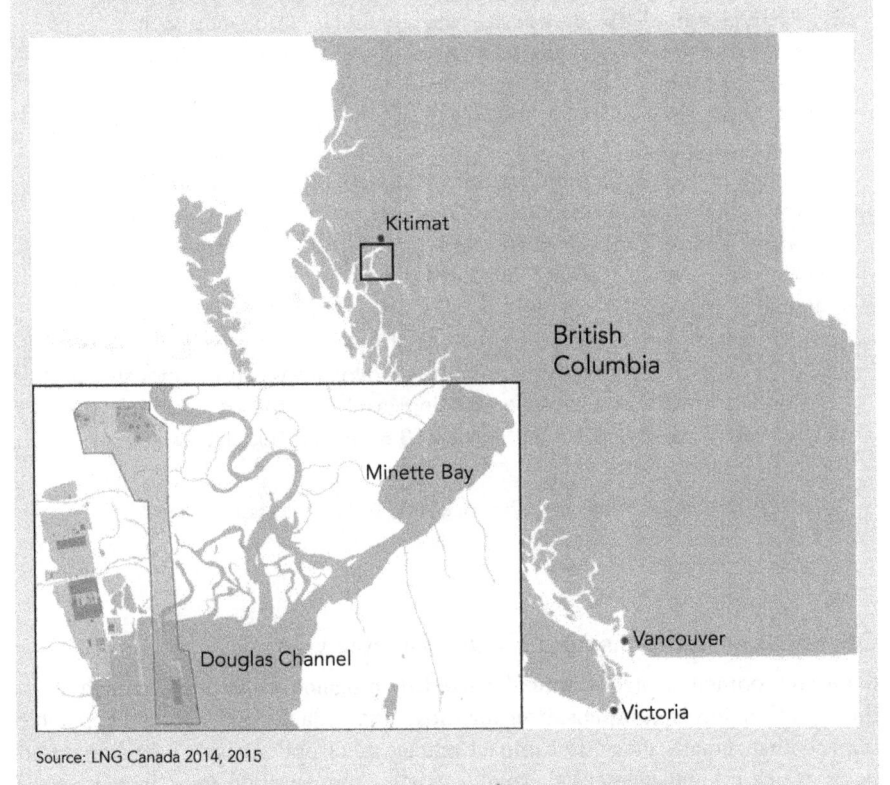

Source: LNG Canada 2014, 2015

Columbia Environmental Assessment Act. The project received federal EA approval and a provincial EA certificate of approval in June 2015.

The project footprint is 412 hectares, which will require vegetation clearing, grading, and replacement with infrastructure and the unavoidable loss of ecological functions of approximately 85 hectares of wetlands. Included among the EA conditions of approval were the development and implementation of measures to offset the residual loss of wetlands and the development of a wetland compensation strategy. The project's Wetland Compensation Plan identifies 41 hectares of provincially listed or estuarine wetlands within the project's footprint that are deemed ecologically significant and subject to a no net loss policy. Applying a 2:1 compensation ratio (wetland functions replaced: wetland functions lost), the plan identified 82 hectares of compensatory wetlands to ensure no net loss. The wetland compensation strategy includes the following components:

i. Implementation of marine fish habitat offsetting to establish 17 ha of estuarine wetlands within the Kitimat River Estuary, with similar habitat function to those lost as a result of project development.
ii. In-lieu fees to an environmental non-governmental organization to deliver land securement and the restoration, enhancement, and/or creation of 65 ha of wetlands through a legally binding agreement with LNG Canada.
iii. Development and delivery, by the environmental non-governmental organization, of a wetland monitoring program in accordance with the compensation plan and agreements with the proponent.
iv. Incorporation of traditional-use plants where appropriate and technically feasible in wetland compensation measures and providing access to those sites for the purposes of gathering traditional-use plants.

Source: LNG Canada, 2014, 2015

impact mitigation for many reasons: the mitigation action is often delayed into the future; the affected component itself is not being replaced *in situ*; compensating for a physical impact does not necessarily compensate for function; and there can be skepticism that the proponent will actually follow through on compensation measures or payments.

Checklist for Management Prescriptions

Environmental management plans (EMPs), also referred to as **environmental protection plans** or environmental mitigation plans, are often a mandatory requirement in EA. EMPs are usually prepared by the proponent and detail the specific impact mitigation strategies and the ways in which they are to be implemented (Box 7.4). EMPs vary in their nature and requirements from one jurisdiction to the next and by industrial sector. In Saskatchewan, for example, the Ministry of Environment has implemented guidelines under the Environmental Assessment Act for the preparation

> **Box 7.4 Typical Components of an Environmental Management Plan**
>
> 1. Introduction
>   a. Project's regulatory context
>   b. Project description
>   c. Assessment area and affected components
>   d. Description of potential impacts
> 2. Impact Management Strategies
>   e. Affected component "X"
>   i. Avoidance
>   ii. Minimization
>   iii. Restoration
>   iv. Offset or compensation
>   v. Residual impacts
> 3. Monitoring and Reporting
>   f. Targets, indicators, or benchmarks
>   g. Roles and responsibilities
> 4. Conclusion

of EMPs specific to oil and gas projects. Oil and gas projects that have a potential to trigger a full EA review require a detailed EMP. The EMP is prepared by the proponent and describes how the project will be undertaken, including such aspects as a description of the project environment, potential conflicts, and environmental protection measures that the proponent will take to avoid or minimize those conflicts.

The primary focus of impact management should be on potentially significant, adverse impacts. Management options should reflect a hierarchy from avoidance to compensation. Ensuring high-quality impact management design in EA also requires a degree of precision in the way that strategies or recommendations are formulated. In other words, decision-makers must understand what proponents are proposing as management strategies; proponents must clearly understand what they are being asked to implement; and regulators must be able to follow up and verify what was implemented. Vague generalizations such as *"the proponent will exercise supervision and control during construction to prevent bank erosion," "the proponent should give special consideration to use of machinery in sensitive riparian zones," "construction noise will be minimized,"* or *"the project will be carried out in such a way as to ensure as minimal disturbance as possible to sensitive habitat"* are of little value to sound impact mitigation design, implementation, and verification. Though such statements often have good intention, they are ambiguous. They cannot be verified in terms of their implementation or their effectiveness for reducing potentially significant adverse effects. There are no specified targets, management objectives, desired conditions, or expected outcomes.

Good practice requires that impact management prescriptions are verifiable such that subsequent inspections or follow-up studies can verify that the measure was implemented and effective and thus a valuable option for subsequent projects with similar impacts (Tinker et al., 2005). Of course, the level of detail and precision in management prescriptions should be commensurate with the risk associated with the potential effect being managed (i.e., the likelihood of significant adverse effect) and the degree to which the proposed management action has been proven effective—based either on scientific or technical literature, expert-based knowledge, or the same or similar applications elsewhere (Marshall, 2001; British Columbia EAO, 2013). Some management prescriptions are routine or well-accepted industry standards, whereas others may be more innovative or less certain in terms of their efficacy. Scientific and technical feasibility must also be given some consideration—i.e., is it possible for the proponent to implement the action?—as well as assurance that the prescription does not contradict existing legislation or regulations.

Based on guidance developed by Aura Environmental (2018), the following minimum standards must apply for ensuring good-practice impact management prescriptions in the EA process:

- ☑ Management actions clearly demonstrate the *nexus* between the proposed mitigation and the adverse effect of concern.
- ☑ Management actions focus on the potentially significant adverse effects.
- ☑ Management prescriptions reflect a hierarchical consideration of viable options, including:
  - Options to avoid
  - Options to minimize or reduce
  - Options to restore
  - Options to compensate
- ☑ Consideration is given to the known or anticipated efficacy of the prescriptions, including uncertainties, and any potentially adverse side effects:
  - Based on previous, similar projects or assessments
  - Based on scientific or technical literature
  - Based on expert judgment, with appropriate substantiating evidence
- ☑ Management prescriptions set out targets, benchmarks, desired conditions, or objectives against which the efficacy of a prescribed mitigation action can be evaluated.
- ☑ Consideration is given to the scientific and technical (design) feasibility.
  - Based on the significance of the adverse effect
  - Based on the size or scope of the project under consideration
  - Based on the available science and technology
- ☑ Prescribed management actions do not contradict existing regulations, land-use plans, or established management objectives.
- ☑ Where there are uncertainties *and* the potential for significant adverse effects, project management actions are part of a larger process of adaptive management.

Adaptive Management

Fundamental to impact management, especially where uncertainty exists about the nature of potential impacts or how the affected environment might respond to mitigation options, is an **adaptive management (AM)** strategy. At a 2002 US Department of Defense news briefing, then–Secretary of Defense Donald Rumsfeld made the following statement:

> There are known knowns. These are the things we know that we know. There are known unknowns. That is to say, these are the things that we know we don't know. But there are also unknown unknowns. These are the things we don't know we don't know.

Identifying programs and actions to mitigate the predicted impacts of a project yet to be developed is full of "known unknowns" and "unknown unknowns." The EA process is full of uncertainties about such factors as the project's ultimate design and timetable, inherent uncertainties due to the complexity of environmental and socio-economic systems, and uncertainties caused by exogenous factors such as the emergence of new technology or changing political or market conditions. Thus, a management system to deal with potentially significant and adverse effects should be designed to expect the unexpected, adapt to changing or surprise conditions, and learn from less than effective mitigation designs.

The success of a prescribed impact management measure is never truly known until it is implemented and tested in practice. As such, it should be expected that any management measure prescribed in an EA could be wrong or less than effective. Although often described as "learning by doing," AM is not a haphazard approach—i.e., "if X doesn't work, we'll try Y." AM is a structured, well-planned approach to environmental management that treats management prescriptions or mitigation as experiments to test hypotheses, monitor the outcomes, and subsequently adapt actions as new knowledge and understanding are gained (Taylor, Kremsater, & Ellis, 1997) (Figure 7.2). Adaptive management is a deliberative process that explores alternative management actions and makes explicit forecasts about their outcomes, carefully designs monitoring programs to provide feedback and understanding of the reasons underlying outcomes, and then adjusts objectives or management actions based on this new understanding (Canter, 2008). Diduck, Fitzpatrick, and Robson (2012) explain that AM treats environmental management actions as hypotheses, or questions rather than answers; management actions thus become treatments, in an experimental sense, with AM structured to make learning both deliberate and more efficient.

Within the context of EA, AM is a structured approach for proceeding with management actions despite uncertainties about the best course of action (Schreiber et al., 2004). Even the most tested and thought-out management prescriptions can fail; AM encourages project proponents, EA practitioners, and regulators to approach management with an expectation that a mitigation prescription may be wrong but that new knowledge can be gained, and management actions improved, through careful, planned monitoring and evaluation of management programs. AM should not be used as an excuse for not committing to specific impact management measures or

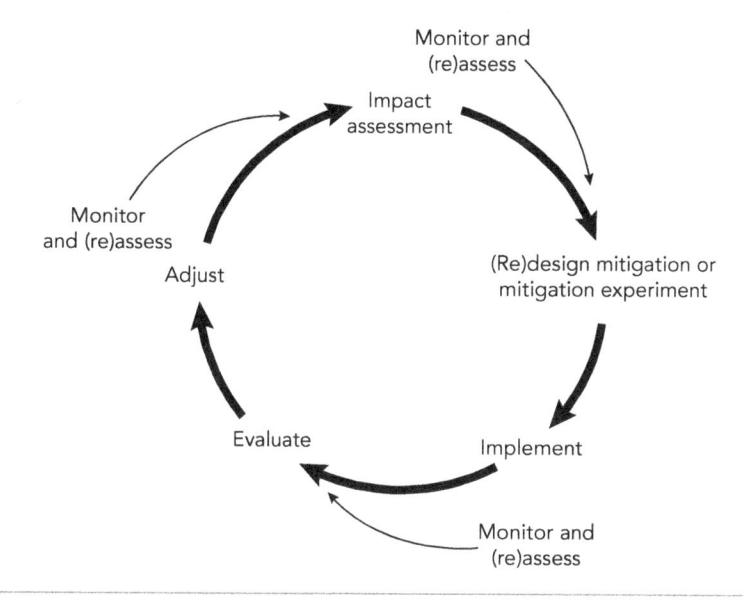

Figure 7.2 Adaptive management cycle

as an attempt to cover a situation in which a proponent is not sure how to mitigate a negative environmental impact but is committed to finding the technology or science in the future if a problem arises (Box 7.5). As Olszynski (2010) argues:

> Simply put, without clear and measurable objectives, indicators, hypotheses, thresholds, and commitments with respect to monitoring, follow up and adjustment, which is to say rigorous AM, there would appear to be little to no basis for concluding that the uncertainty associated with proposed mitigation measures will actually be reduced, let alone that these measures will prove effective and that significant adverse environmental effects will be mitigated.

Box 7.5 Tales of Adaptive Management from Two Mining Projects

Although most commonly applied in the context of resource management, AM has become commonplace in impact mitigation and management plans for major resource development projects in Canada—including the Bipole III Transmission Line, the Energy East Pipeline, and the Pacific Northwest LNG project, to name a few. But not all AM initiatives proposed by project proponents, or required by review panels and regulatory authorities, live up to the nature and intended objectives of AM. Indeed, what is often labelled adaptive management in EA is too often

continued

haphazard management, conventional management inappropriately labelled, or a means to defer responsibility for solving a problem.

The Diavik diamond mine is located on a 20 km^2 island in Lac de Gras, Northwest Territories, approximately 300 km from Yellowknife. The Diavik project received approval for permitting and licensing in 1999, and mine production commenced in 2003. Water licences issued for mine operation required that Diavik include an AM strategy as part of its aquatic effects mitigation and monitoring program. The *Diavik diamond mine adaptive management plan for aquatic effects* (Diavik Diamond Mines Inc., 2007) provides a framework for how the mine's aquatic effects monitoring program will be used to identify additional mitigation strategies to minimize the project's impact on the aquatic environment. The plan defines AM as "a systematic process for continually improving mine operation practices by learning from the outcomes of performance monitoring and review programs . . . a cyclical process of plan, monitor, review, revise plan, monitor etc." A 2008 review of Diavik's AM plan, however, concluded that AM "is being viewed with much less rigor than required to be done properly, or is being misunderstood as managing adaptively" (Murray & Nelitz, 2008, p. 5). Diavik's approach does not conform to the experimental design of AM, and the aquatic effects monitoring program was never intended to be undertaken to improve management goals or objectives. Rather, the program was established to manage impacts—both known and unexpected—through monitoring for management. Neither the mine itself nor the impacts caused by its operations was designed as an experiment but rather as development activities with environmental effects to be monitored and managed. This approach does not mean that Diavik's aquatic effects monitoring program is not effective or that monitoring for management is not a worthwhile activity, but it should not be confused with AM (Murray & Nelitz, 2008). Diavik's approach is typical of how AM is often used in EA, but AM is more than monitoring and responding.

In the case of Imperial Oil's Kearl oil sands mine project, however, the use of AM is much more controversial. The Kearl oil sands mine is located in the Athabasca oil sands region, Alberta, approximately 70 km north of Fort McMurray. Kearl is an open-pit oil sands mine and tailings management facility. The joint federal–provincial panel appointed to review the project issued its recommendation in 2007 and found no adverse environmental effects from the project. The panel's recommendation, and the project authorization, were legally challenged by environmental organizations, based in part on the uncertainty about the effectiveness and technical and economic feasibility of "end-of-pit-lake" technology (Kwasniak, 2010). An "end-of-pit-lake" is a mined pit that will receive mine tailings near the end of the mining operation; in Kearl's case, this is about 60 years after project start-up. The proposed concept is that, after covering the last of the tailings and filling the pit with fresh water, it will be possible to create a lake that will again support fish populations.

Uncertainty relating to the effectiveness and technical and economic feasibility of an end-of-pit-lake mitigation program was a key AM issue. Imperial Oil proposed an adaptive management approach to its end-of-pit-lake plan. The Federal Court reviewing the legal challenge indicated that some uncertainty existed with respect to end-of-pit-lake technology but that the level of uncertainty was not enough to deny project approval. The uncertainty was not the complexity of the environmental system per se but rather whether the proposed mitigation would work given the state of knowledge about the technology. Kwasniak (2010, p. 427) argues that AM was being used as a substitute for committing to specific impact management measures and that AM "cannot be used to attempt to cover a situation where a proponent is not sure how to mitigate a negative environmental impact but commits to finding the technology or science in the future, if a problem arises." What Imperial Oil proposed is not AM but managing adaptively—i.e., if the proposed technique does not work, we will try something else.

Source: The above cases are summarized from Noble, 2015

Creating and Enhancing Positive Impacts

Making impacts less severe is not good enough. As Gibson (2011) argues, "ultimately, the enhancement we need to deliver through environmental assessment is confidence that every approved undertaking will move us positively towards a desirable and durable future." Development projects can often create as many positive impacts, particularly economic ones, as they can negative impacts. Thus, an important management strategy is to create new benefits, enhance existing benefits, and maximize the duration of those impacts—especially for those communities most adversely affected by the project. One of the primary instruments for ensuring that communities affected by development also receive substantial benefits from development is negotiated agreements.

Negotiated agreements, often referred to as **impact benefit agreements (IBAs)**, are external to the regulatory EA process but often occur in conjunction in EA. IBAs are private, bilateral arrangements between an industry proponent and (usually) an affected community or Indigenous group. With few exceptions, governments are not directly involved in the development and negotiation of these agreements. IBAs typically include negotiated measures to mitigate adverse project impacts beyond those included in the EA and to ensure that affected communities will benefit from project contracting and employment opportunities (O'Faircellaigh, 2017). For Indigenous communities, IBAs can provide, among other things, assurance of benefits from development they may not otherwise receive. For proponents, IBAs provide some level of guaranteed support for a project, especially during the licensing process. With more than 300 agreements negotiated across Canada since the 1970s for mining projects alone (Mining Association of Canada, 2018), IBAs have become an accepted add-on to the regulatory EA process.

The growth in IBAs in recent years may be attributed, in part, to the deficiencies of EA in negotiating community socio-economic issues, impacts, and benefits at the time of the development proposal and impact assessment. In a review of IBAs in the Mackenzie Valley of the Northwest Territories, for example, Galbraith, Bradshaw, and Rutherford (2007) found that IBAs addressed many issues of concern to the local community, including community–industry relationships and benefits-sharing, that project EAs did not. Of course, the rise in IBAs may also be attributed to the frustration of project proponents when community grievances are aired about a project as EA approvals and certificates are being sought. The focus of IBAs is most often on cultural, social, or economic aspects, such as guaranteeing employment, local contracting, infrastructure investment, or revenue-sharing opportunities, but increasingly IBAs are also addressing biophysical impact mitigation, such as wildlife or habitat protection (e.g., Mary River Project Inuit Impact and Benefit Agreement; Meadowbank Mine Inuit Impact and Benefit Agreement), as well as longer-term environmental monitoring provisions (e.g., Athabasca Working Group, 2004) (see Environmental Assessment in Action: Negotiated Agreements and EA Governance in the British Columbia and Saskatchewan Mining Sectors).

The timing of negotiated agreements relative to the EA process varies, but timing can have significant implications for EA and its role as a deliberative process for impact identification and management for informed, public decision-making. There are three basic windows when agreements are negotiated: prior to EA commencement during the early pre-project planning stages; during the EA process, in parallel with impact assessment, mitigation planning, and decision-making; and post-EA, focused primarily on impact management and monitoring (Figure 7.3). Each window of negotiation has its advantages and limitations:

Pre-EA Negotiation: Agreements are often negotiated before EA commencement or very early in the project design and application process. The most significant benefit to pre-EA negotiation is *certainty* provisions. For project proponents, IBAs provide certainty (often legally binding) that the affected group will consent to the project or at least not publicly challenge it or air grievances during the EA process. For communities, early negotiation provides an opportunity to exert influence over project design and mitigation strategies and offers greater leverage to secure more favourable benefits from a project. However, negotiations in advance of EA occur under considerable uncertainty—the community is less informed about the project's final design, its potential impacts, and the nature and efficacy of the proponent's impact mitigation strategies. Negotiations unfold without the benefit of the EA review process, including knowledge about baseline trends, thresholds, and predicted or modelled

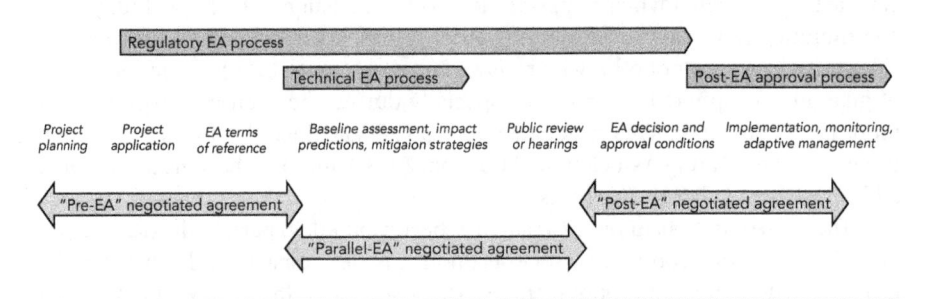

Figure 7.3 Timing of IBA negotiation in the EA process

future conditions, and in the absence of knowledge about regulatory standards or requirements for impact management. Of course, certain terms of an agreement may be open to renegotiation during the EA process, or there may be additional compensation to cover any unforeseen impacts, but this is not always the case (Gogal, Riegert, & Jamieson, 2005).

Parallel-EA Negotiation: Simultaneous EA and IBA negotiation provides an opportunity for both processes to leverage information and input. In principle, the knowledge and concerns raised during the EA process can inform and influence agreement-making and vice versa. In practice, however, the two processes may be in conflict given that they serve two different functions. Parallel EA and IBA may allow a proponent to make an undisclosed commitment to address concerns—commitments that have yet to be bounded by an agreement—and avoid putting those commitments to the public acceptability test during the EA process. In principle, the processes should be mutually supportive. In practice, the confidential nature of agreements may prevent the open flow of information such that each process equally benefits from the other.

Post-EA Negotiation: After a proponent has filed its EIS but before impact management plans are implemented, negotiations can proceed based on much greater certainty about anticipated impacts and mitigation. Communities may be more informed about a project and its potential impacts and benefits, including knowledge about baseline conditions and limits, regulatory approval conditions, and proponent commitments. Post-EA agreements can also strengthen the requirements for impact management through development of participatory monitoring or impact management programs. However, negotiated post-EA, the IBA is late in the project planning cycle—there may be less opportunity for communities to leverage benefits. The EA process itself also cannot benefit from any issues that may have emerged when negotiating the agreement, assuming such information is shared openly and there is limited ability through the IBA to influence the types of mitigation or impact compensation measures that may become part of EA regulatory approval conditions.

IBAs are not meant to replace project EA but to complement the existing regulatory process and with a focus on ensuring that economic benefits from project development accrue to local communities. These agreements are negotiated for different reasons depending on the community's interests or Indigenous land and resource rights, the regulatory framework in place, and the relationship that exists between the community and the company (Sosa & Keenan, 2001). A major challenge to the impact assessment community in understanding the value added by these agreements to the EA process is that IBAs and similar negotiated agreements occur outside the public realm and little is known about their content, benefits-sharing and impact management details, and overall efficacy.

The Canadian experience is in sharp contrast to the Greenlandic context, where impact assessment leads to an IBA with communities before a project can gain approval. Greenlandic IBAs are trilateral agreements between a company, community, and government and are required for projects subject to social impact assessment under Greenland's Mineral Resources Act. Negotiations do not commence until after a draft impact assessment report has been produced by the proponent, along with a white paper documenting public comments and concerns and the proponent's responses. The proponent's draft EIS contains a benefit and impact plan, detailing how mitigation and strategies to enhance local opportunities from project development will be implemented. This information informs the IBA negotiation process. The IBA must then be signed prior to project approval, but the agreement may require modifications to the proponent's proposed benefit and impact plan identified in its impact statement.

Environmental Assessment in Action

Galore Creek Project and Athabasca Uranium Mining Agreements

Galore Creek Project, British Columbia

The Galore Creek Project is a proposed open-pit porphyry-copper-gold-silver mine in northwest British Columbia. The Galore Creek property was acquired by NovaGold Resources Inc. in 2007 and is now a joint venture between Newmont Mining Corporation and Teck Resources Ltd. The project is situated within the traditional territory of the Tahltan First Nations, a group with a long history of involvement with the mining sector. Before commencement of an EA for the project, a participation agreement was signed between the Tahltan and NovaGold, defining how the Tahltan and NovaGold would collaborate to achieve EA approval—specifically, how the Tahltan and the proponent would work together in identifying the potential environmental effects of the project on the Tahltan Nation and determining how best to avoid or minimize such impacts.

The Galore Creek Project Agreement, although voluntarily negotiated, was a legal framework and enforceable contract between the mining company and the community to which the government was not privy. Its purpose was to create certainty, for both parties, for investment, access, extraction, and ownership of mineral rights. The agreement was pivotal in defining how the Tahltan and NovaGold would collaborate during the mine's life cycle, specifically through the EA approval phase. The agreement sought to create a framework for communication and partnership to ensure benefits to the Tahltan from the project and to guarantee the Tahltan's support of the project. The agreement captured many of the same issues protected under environmental legislation but went further to provide additional investment security for the signatories and benefits to the Tahltan, including environmental monitoring, heritage resources, an ongoing review of the closure plan, scholarships, employment training, and contracting opportunities. The EA for the project was approved, but project construction has been suspended since 2008 because of unfavourable economic conditions. The proponents are currently reviewing optimal project designs and feasibility assessments but are still obligated to meet commitments to the Tahltan based on the participation agreement.

Athabasca Basin Uranium Mining Operations, Saskatchewan

The Athabasca Basin of northern Saskatchewan is one of the world's most productive uranium mining regions, contributing approximately 20 per cent of global uranium supply. The region is also home to seven Indigenous communities. Uranium mining operations are "fly-in fly-out," and local economies are based primarily on traditional hunting, trapping, fishing, and guiding activities. Cameco Corporation and AREVA Resources Canada are the two uranium producers in Saskatchewan. In 1991, the Canadian federal and Saskatchewan provincial governments established a Joint Federal–Provincial Panel on Uranium Mining Developments in Northern Saskatchewan to review industry proposals for uranium mine development and existing mine expansion in the Athabasca Basin. The joint panel concluded that, among other things: (i) many of the communities in the Athabasca Basin wanted to participate in, and receive benefits from, uranium mining activities; (ii) although monitoring systems were in place that met regulatory requirements, there were enduring concerns about their effectiveness in determining the impacts of mining activities; and (iii) because of the proximity of mining operations to communities, community involvement should extend beyond consultation.

In response to the joint panel, the uranium industry established the Athabasca Working Group (AWG) in 1993—a private partnership between the two uranium mining companies and the seven communities of the Athabasca Basin. The rationale for forming the AWG was largely the recognized need to build relationships with the local communities and to build an environment of corporate policy and responsibility to ensure that stakeholders are involved in and familiar with mining operations. The AWG itself has no legal authority. It was created by the industry to address the concerns of the Athabasca communities about the impacts of uranium mining; to ensure the engagement of communities in mining-related activities, including impact management; and to facilitate discussions about the sharing of benefits of mining activity. An Impact Management Agreement (IMA) was later signed between Cameco, AREVA, and the local communities. The IMA is focused on local employment, training and business development, benefit-sharing, and environmental protection. As part of the agreement's commitment to environmental protection, and in part because of the joint panel's recommendation for more direct community involvement, a community-based environmental monitoring program was established to monitor the "off-site" impacts of uranium mining operations. Community members, appointed by the community, are responsible for identifying monitoring locations and for monitoring data collection.

Lessons Learned

In the Galore Creek case, the agreement was established to formalize a process of working together through the EA process. In the Saskatchewan case, the agreement was established post-EA to engage communities in monitoring and impact management programs. Both cases illustrate going beyond the prescribed

continued

requirements of EA and, in doing so, reveal several lessons for improving the governance of EA as a tool for environmental management:

- Successful agreements hinge on community trust, the result of which is enhanced corporate image and a social licence to operate. Enhanced community trust can lead to increased community confidence in the mining company and in the EA process.
- Building this level of trust requires genuine community participation pre- and post-EA in the design, implementation, and evaluation of impact management strategies. There is an opportunity through negotiated agreements to enhance the capacity of Indigenous communities to become engaged in resource development by, for example, providing funding for participation in the EA process, ensuring employment and ongoing skills development and training opportunities, and involving communities in post-EA monitoring and management programs.
- There is a danger with negotiated agreements of blurring the lines of responsibility and accountability between the mining company and government with respect to Indigenous communities and commitments to impact management. Agreements negotiated prior to or parallel with the EA process provide an opportunity to chart out how regulatory approval can be achieved but can be criticized for undermining the public EA process. When agreements are negotiated post-EA, communities may have much less control over impact management and benefit enhancement strategies as conditions for granting their approval of the project.

Sources: Based on Noble & Fidler, 2011; Noble & Birk, 2011

Key Terms

adaptive management (AM)
buffer zone
compensation
environmental management plan (EMP)
environmental protection plan

impact avoidance
impact benefit agreement (IBA)
impact mitigation
remediation
restoration

Review Questions and Exercises

1. Obtain a completed project EIS from your local library or access one online through a government EA registry. Select a sample of valued components from the table of contents, and then identify and classify the impact management measures proposed based on the mitigation hierarchy. Are most impact management measures based on avoidance, mitigation, restoration, or compensation? Are there any impact management measures that emphasize creating or

enhancing positive project impacts? Generate a list for your project, and compare your results with those of others.
2. From orphaned wells to mine sites, the amount of resource infrastructure that is abandoned in Canada and requiring reclamation and restoration is increasing. Although for current projects there are mechanisms to ensure that project proponents are financially liable for post-project clean-up (e.g., requirements for insurance policies or bonds), this does not resolve the challenges associated with project sites that are already abandoned. Who pays? Through discussion groups, identify a range of policy options or solutions for addressing the financial cost associated with site clean-up.
3. Suppose you are responsible for negotiating an IBA for a small, remote community (population less than 5000) that is about to be the recipient of a large mining operation. What items might you want to negotiate with the proponent for inclusion in the IBA? What obstacles might you face in negotiating with the proponent? How might the timing of the negotiation relative to the EA process influence your ability to negotiate for maximum benefits?

References

Athabasca Working Group. (2004). *Athabasca Working Group annual report*. Wollaston Lake, SK: Athabasca Working Group.
Aura Environmental. (2018). *High quality assessment input: current challenges, key principles, and recommendations*. Whitehorse, YT: Major Projects Management Office.
British Columbia EAO (Environmental Assessment Office). (2013). *Guidance for best practices for the selection of valued components and the assessment of potential effects*. Victoria, BC: BC EAO.
British Columbia EAO (Environmental Assessment Office). (2014). *Mitigation policy*. Victoria, BC: BC EAO.
Canter, L. (2008). Adaptive management for integrated decision-making: an emerging tool for cumulative effects management. Paper presented at Assessing and Managing Cumulative Environmental Effects, Special Topic Meeting of the IAIA, 6–9 November, Calgary.
Dachis, B., Schaffer, B., & Thivierge, V. (2017). *All's well that ends well: addressing end-of-life liabilities for oil and gas wells*. Commentary no. 492. Calgary, AB: C.D. Howe Institute.
de Jesus, J., Bingham, C., Canter, L., Partidario, M., Cashmore, M., Croal, P., . . . & Keshkamat, S. (2013). Mitigation in impact assessment. *Fastips* 6. Fargo, ND: International Association for Impact Assessment.
Diavik Diamond Mines Inc. (2007). *Diavik diamond mine adaptive management plan for aquatic effects*. Prepared for Wek'èezhii Land and Water Board, NT.
Diduck, A., Fitzpatrick, P., & Robson, J. (2012). *Guidance from adaptive environmental management, monitoring, and independent oversight for Manitoba Hydro's upcoming development proposals*. Winnipeg, MB: Public Interest Law Centre of Legal Aid Manitoba.
Galbraith, L., Bradshaw, B., & Rutherford, M. (2007). Towards a supraregulatory approach to environmental assessment in northern Canada. *Impact Assessment and Project Appraisal* 25(1), 27–41.
Gibson, R.B. (2011). Application of a contribution to sustainability test by the Joint Panel for the Canadian Mackenzie Gas Project. *Impact Assessment and Project Appraisal* 29(3), 231–44.
Glasson, J., Therivel, R., & Chadwick, A. (1999). *Introduction to environmental impact assessment: principles and procedures, process, practice and prospects*. 2nd edn. London, UK: University College London Press.

Gogal, S., Riegert, R., & Jamieson, J. (2005). Aboriginal impact and benefit agreements: practical considerations. *Alberta Law Review* 43, 129.

Gorman, M. (2019). Nova Scotia to spend $48 million to clean up two former gold mines. *CBC News online*. Posted 25 July. https://www.cbc.ca/news/canada/nova-scotia/gold-mining-remediation-environment-government-funding-1.5224766.

Kwasniak, A.J. (2010). Use and abuse of adaptive management in environmental assessment law and practice: a Canadian example and general lessons. *Journal of Environmental Assessment Policy and Management* 12(4), 425–68.

LNG Canada. (2014). *Environmental assessment certificate application*. LNG Canada Export Terminal. Section 2, project overview. Vancouver, BC: LNG Canada.

LNG Canada. (2015). *Wetland compensation plan*. LNG Canada Export Terminal. Vancouver, BC: LNG Canada.

Marshall, R. (2001). Application of mitigation and its resolution within environmental impact assessment: an industrial perspective. *Impact Assessment and Project Appraisal* 19(3), 195–204.

Mining Association of Canada. (2018). Aboriginal affairs. http://mining.ca/our-focus/aboriginal-affairs. Accessed 21 March 2019.

Morrison-Saunders, A. (2018). *Advanced introduction to environmental impact assessment*. Cheltenham, UK: Edward Elgar.

Murray, C., & Nelitz, M. (2008). *Review of Diavik and EKATI adaptive management plans*. Prepared by ESSA Technologies Ltd, Vancouver, for Fisheries and Oceans Canada, Western Arctic Area, Central and Arctic Region, Yellowknife, NT.

Noble, B.F. (2015). Adaptive environmental management. In B. Mitchell (ed.), *Resource and environmental management in Canada*, 5th edn. (pp. 87–111). Toronto, ON: Oxford University Press.

Noble, B.F., & Birk, J. (2011). Comfort monitoring? Environmental assessment follow-up under community–industry negotiated environmental agreements. *Environmental Impact Assessment Review* 31, 17–24.

Noble, B.F., & Fidler, C. (2011). Advancing Indigenous community–corporate agreements: lessons from practice in the Canadian mining sector. *Oil, Gas and Energy Law Intelligence* 9(4), 1–30.

O'Faircellaigh, C. (2017). Shaping projects, shaping impacts: community-controlled impact assessments and negotiated agreements. *Third World Quarterly* 38, 1181.

Olszynski, M. (2010). Adaptive management in Canadian environmental assessment law: exploring uses and limitations. *Journal of Environmental Law and Practice* 21(1).

Orphan Well Association. (2019). OWA orphan lists. https://www.alberta.ca/upstream-oil-and-gas-liability-and-orphan-well-inventory.aspx.

Riley, S.J. (2019). Regulator projects Alberta's inactive well problem will double in size by 2030, documents reveal. *The Narwhal*, 8 April. https://thenarwhal.ca/regulator-projects-albertas-inactive-well-problem-will-double-in-size-by-2030-documents-reveal.

Schreiber, E.S.G., et al. (2004). Adaptive management: a synthesis of current understanding and effective application. *Ecological Management and Restoration* 5(3), 177–82.

Sosa, I., & Keenan, K. (2001). Impact benefit agreements between Aboriginal communities and mining companies: their use in Canada. http://www.cela.ca/publications/impact-benefit-agreements-between-aboriginal-communities-and-mining-companies-their-use.

SRK Consulting Inc. & SENES Consulting. (2007). *Giant Mine remediation plan*. Yellowknife, NT: Department of Indian Affairs and Northern Development.

Taylor, B., Kremsater, L., & Ellis, R. (1997). *Adaptive management of forests in British Columbia*. Victoria, BC: British Columbia Ministry of Forests.

Tinker, L., et al. (2005). Impact mitigation in environmental assessment: paper promises or the basis of consent conditions? *Impact Assessment and Project Appraisal* 23(4), 265–80.

8 Significance Determination

The fact of an environmental impact is the change itself, its magnitude, direction, units, and the estimated probability that it will occur. The meaning of an environmental impact is the value placed on the change by different affected interests. It is the answer to the question: If this impact occurs, so what? The "so what?" determines how important or "significant" an environmental issue is, and to whom. (Haug et al., 1984, p. 18)

Impact Significance

Are the project's impacts "significant," and if so, are they justified? This is perhaps the most challenging and controversial aspect of the EA process. What is considered a significant impact is dynamic, contextual, political, and uncertain (Lyhne & Kørnøv, 2013). There is no universal standard that defines a "significant" effect; there will always be regulatory, social, political, and site-specific issues to take into consideration, and what is considered significant in one context may not be so in the next. Consider a mining project that will involve the discharge of effluent to a lake system, leading to a decline in fish population. Is the effect of the mining operation on the lake system a significant adverse effect? It might depend on the magnitude of the effect (i.e., how much will the population decline?) or the duration and reversibility of the effect (i.e., how long will it last, and can it be corrected?), but it also depends on context (i.e., whether the lake is a highly valued source of traditional foods or whether the fish is a rare, threatened, or protected species).

Determinations of **impact significance** begin at the outset of the EA process when a decision is made as to whether the proposal requires a formal assessment and extends throughout the scoping, prediction, mitigation, and follow-up stages (Table 8.1). As discussed in Chapter 4, screening decisions are often based on specified thresholds (e.g., project size, disturbed area) or contexts (e.g., rare species) that are deemed in some way to be "significant." The dominant focus of significance determination in EA, however, is on the significance of potential **residual effects**—the effects that remain after proposed mitigation measures are taken into consideration.

There is no single definition of a "significant" impact in EA. The Canadian federal Impact Assessment Act, for example, makes 20 references to "significant" in the context of project impacts and the assessment of those impacts and includes various factors to be considered, such as Indigenous and local knowledge, public input, and the feasibility of mitigation measures, but it provides no definition of a "significant"

Table 8.1 Interpretations of Significance in the EA Process

EA Activity	Significance Interpretation
Screening	If (and what) EA requirements are to be applied Criteria and procedures for making screening decisions
Scoping	Alternatives that are reasonable and criteria for comparing them Analysis of boundaries and components to focus on Public and agency issues Proposal characteristics most likely to induce significant effects Proposal characteristics that warrant mitigation and monitoring
Baseline analysis	Criteria for determining environmental significance and sensitivity Choice of valued components
Impact analysis	Potential impacts to analyze and at what level of detail and interpretation Impact magnitude and impact significance criteria and thresholds Impact acceptability and significance interpretations
Cumulative effects analysis	Cumulative effects criteria and thresholds effects analysis Acceptability of cumulative effects Cumulative effects significance interpretations
Decision-making	Proposal acceptability and compliance with standards, policies, etc. Basis for decision-making
Mitigation	Impacts that warrant mitigation, compensation, or benefits Choice of measures and significance of residual impacts
Monitoring and auditing	Impacts that should be monitored and choice of monitoring methods Effectiveness of mitigation and monitoring

Source: Based on Lawrence, 2004

impact. There are, however, several basic principles and concepts that characterize significance and are more or less accepted among the community of EA scholars and practitioners:

- Significance determination is not solely a scientific exercise.
- What is significant is subjective and varies based on the values and perceptions of different stakeholders.
- What is significant in one context or at one place and time may not be so at another.
- Significance determinations are made based on incomplete information and under uncertain conditions.
- There is no standard method for significance determination that will work for all projects or for all impacts.
- A determination of a significant adverse impact does not mean that a project should be rejected, but if the project is approved, then the impacts must be justified.
- Significance determinations and the justification of projects with significant adverse impacts must be transparent.

Measurement and Meaning: Components of Significance

Significance determination is one the most critical elements of EA (Duinker & Beanlands, 1986), yet it remains one of the most contested components of EA practice (Wood, 2008). Reviewing significance determinations under the US NEPA, for example, Haug et al. (1984) observed that regulations "provide no clear definition of significance that can be applied objectively and uniformly to environmental issues and the consequences of man's activities." Little has changed. Lawrence (2005) suggests that clear and unequivocal good-practice significance determination standards are unlikely to emerge in the foreseeable future and argues that a necessary first step is greater clarity regarding the basic characteristics of significance determination activities.

There are two main components to significance and thus significance determination in EA: **impact measurement** and **impact meaning** (Table 8.2). Measurement refers the characteristics of the impact (e.g., magnitude, spatial extent, duration). Meaning refers to the context within which those characteristics are viewed and interpreted (e.g., regulatory, social, ecological, sustainability). Both measurement and value judgment play an important role in significance determinations in EA. Impacts are ultimately measured on the yardstick of human values, and any comprehensive definition of a significant impact in EA must reflect this value judgment (Beanlands & Duinker, 1983). As Ehrlich and Ross (2015) explain, subjective judgment, informed by a body of evidence and reflective of societal values, is not only credible but a mainstay of some of the most important decisions made in society; the same principles lie at the heart of significance determinations.

Measurement: Effect Characteristics

For each prediction in EA, several characteristics of the anticipated environmental effect or impact are typically examined (Table 8.2). These characteristics were introduced in Chapter 6. Although the characteristics vary from one jurisdiction to the next and from one assessment to another, they have found their way into international and Canadian EA systems as fundamentals in significance determination. For example, the US Army Corps of Engineers (see Canter & Canty, 1993) have adopted these characteristics into a suite of standardized review questions for characterizing impacts and determining significance under the US National Environmental Policy Act.

Table 8.2 Measurement and Meaning of Impact Significance

Impact measurement characteristics	Impact meaning characteristics
Order of impact	Regulatory designations and standards
Nature of impact	Benchmarks or limits
Impact magnitude	Vulnerability
Spatial extent	Culture and recognized and expressed rights
Frequency and duration	Contributions to sustainability
Reversibility	Social values and acceptability
Likelihood	Political context and public interest

Under Nova Scotia's Environmental Assessment Regulations, "significant" is defined as an adverse effect that could occur as a result of its magnitude, geographic extent, duration, frequency, reversibility, or probability of occurrence (sec 2(1)(l)(i–vi)).

There are many other examples of EA guidance that identify impact characteristics for significance determination. Impact characterization has become the boilerplate for significance in EA, yet the approach has brought little resolution to the challenges and complexities associated with what, exactly, constitutes significance. Antunes, Santos, and Jordão (2001) report that when an impact scale is used for significance determinations, the impacts are often classified on that scale by experts, considering simultaneously factors such as magnitude, spatial extent, duration, and so on, but without making explicit the rules used for that classification. The result is that significance determinations can be difficult, if not impossible, to validate or replicate.

Wood (2008), for example, reports on how significance is typically addressed and communicated in EAs. Often, practitioners characterize impacts based on some combination of impact measurement and then roll those measurements up into descriptors such as "moderate" or "major" but with insufficient information as to what is meant by, for instance, a "moderate" impact or "moderate" value and to whom. Wood suggests that when relying upon a single language or term in the final assessment of significance, variance in meaning and interpretation remain deeply entrenched. This results in an impact significance determination that is not only open to multiple interpretations but one that is inherently simplistic and with no benchmarking to the project's context or environmental setting or the expert assessor's professional frame of reference. Wood concludes that the sole use of such criteria for characterizing impacts has the merit of simplicity and benefit of scale of measurement, but this "can come at a considerable cost in terms of the degree of transparency achieved."

The challenges described by Wood (2008) are evident in Canadian EA practice. In 2013, for example, the Joint Review Panel for Ontario Power Generation's deep geologic repository project, a proposal for the long-term management of low- and intermediate-level nuclear waste, was highly critical of the lack of transparency in significance determinations and required that Ontario Power Generation revisit its significance determination process. The panel specifically directed that Ontario Power Generation avoid the use of arbitrary categorization of impacts (e.g., low/medium/high) based on assigned numerical scores in favour of narrative reasoning and context.

This is not to say that impact and interaction matrices are not useful for identifying and communicating information about impacts, but it is important not to rely solely on impact characterizations. Briggs and Hudson (2013) report that the determination of significance based solely on impact characteristics depicted in simple matrices is in danger of becoming too prescriptive and simply becoming a "handle-cranking" exercise for consultants.

Meaning: Effect Context

The complexity of significance is exacerbated by context, comprising issues of social and cultural values, ecological sensitivity, economic goals, and institutional and political interests. Baker and Rapaport (2005) suggest that the evaluation of significance based strictly on scientific data (e.g., species populations, habitat metrics, emissions levels) is inadequate in many cases because technical and quantitative approaches often do not capture issues of social or cultural significance. Rowan (2009) agrees, suggesting that characterizing impacts to attribute major, moderate, or minor levels of significance all make sense and is useful but it is important that there be a human element. As Ehrlich and Ross (2015, p. 88), explain:

> In our experience, we have observed that technical experts are usually engaged in analyzing impact characteristics such as impact geographic extent, magnitude, etc. (typically described as the technical bases for significance determinations). For example, a biologist may predict that a valued component may be affected to a certain degree, over a certain area, over a certain time, with a certain probability. We suspect, however, that if you were to ask that biologist the crucial question of whether or not the predicted change is acceptable, the biologist should respond that the answer is not a strictly scientific judgement.

This human element, or subjectivity, cannot be avoided in significance determinations. Subjectivity arises from the value placed on a receptor (species or habitat) of an impact; it is dependent on the value society places on it (Briggs & Hudson, 2013). In other words, knowing whether indicator "X" is likely to change by "Y" units from baseline conditions, with a "Z" per cent probability, is useful information, but it stops short of interpreting the meaning of that change. Does it matter? Why? To whom? Effects characterization is essential to significance determination, but it does not indicate whether the characterized effect is acceptable. The significance of an effect is dependent on the specific context in which the effect is occurring (Briggs & Hudson, 2013) (Figure 8.1).

There are several common context-based considerations that inform significance determinations (see Table 8.2). These may not be the *only* considerations, and in any given case the relative importance of these themes may vary. Haug et al. (1984), however, do suggest a priority in the types of criteria used to provide context for significance determination—namely, legal context (e.g., laws, regulations), functional context (e.g., science, ecological limits), and normative context (e.g., social values, acceptable levels).

Legal or regulatory designations or standards
Predicted effects or impacts following mitigation are often compared against environmental standards or regulations—in essence, specified thresholds. The use of standards and regulations is the most common and arguably robust, context-based criterion in significance determination. Impacts within specified standards or that do not exceed certain regulatory limits are deemed to be insignificant in comparison to impacts that do exceed standards or limits. Such standards or limits may include, for example, Canadian Council of Ministers of the Environment (CCME) water quality guidelines for the protection of aquatic life (CCME, 2002) or critical habitat thresholds for caribou populations (Environment Canada, 2012). Lawrence

Figure 8.1 Placer gold mine in the Indian River valley, Yukon, south of Dawson. The mine is on a site where permafrost and peatlands once were. Understanding the significance of the project's impacts to peatlands requires understanding the regional context of peatland disturbance in the watershed, the ecological services they provide, how conditions have changed over time, regulatory standards and policy objectives, and the importance of peatlands to Indigenous use and culture.

(2004) suggests three types of standards or regulatory limits typically used for significance determinations:

- *Exclusionary* – leads to automatic rejection of a proposal
- *Mandatory* – leads to a mandatory finding of significance
- *Probable* – normally significant but subject to confirmation

An impact should always be identified as significant if it exceeds a government-determined limit or does not meet a specified regulatory standard, but the corollary is not necessarily true—an impact may well be within a determined objective, regulation, or standard yet still be significant for other reasons (Ehrlich & Ross, 2005). This was the case argued by Noble and Gunn (2013) in their analysis of Manitoba Hydro's Keeyask project, a proposed 695 MW hydroelectric generation project on the lower Nelson River, northern Manitoba. The project's contribution to critical habitat loss for caribou in the project's study area was considered to be within the critical habitat limits as identified by Environment Canada, but any such loss of habitat caused by the project was considered by local Indigenous populations a significant impact because of potential changes in the distribution of caribou and access to local caribou populations.

Benchmarks or limits

There is a tendency for some proponents and EA practitioners to compare the *relative* impact of a project to the impacts of other activities as a basis for significance determination. If a project's impacts are small compared to the impacts of other activities, a case is then presented that the project's impacts are not significant. This is poor practice (Box 8.1). Rather than assess the significance of a project's impact relative to the impacts of others, impacts should be assessed relative to defined benchmarks or limits. These benchmarks or limits must be linked explicitly to a decision-making process so that if a limit is approached or exceeded, then proponents, managers, or regulators know the action they will take.

Limits also need to be consistent with those monitored by the scientific community (Dubé et al., 2013). One approach is to benchmark project-induced change against the range of natural variability (RNV) for the affected environmental component (Figure 8.2). The RNV refers to the spectrum of natural conditions possible in ecosystem structure, composition, and function when considering both temporal and spatial scales (Swanson, Jones, Wailin, & Cissel, 1994). In other words, it is the spectrum of states and processes encountered in an ecological system (e.g., habitat

---

Box 8.1 Debunking the "Compared-to" Approach for Significance Determination

The EA of the Ajax mine project, a proposed open-pit copper and gold mine in British Columbia, estimated that the project's annual GHG emissions would constitute only 0.048 per cent of the province's annual emissions and 0.016 per cent of Canada's annual emissions. The EA concluded that the project's effects on GHG emissions were therefore negligible and overall not significant. Although a common practice used in EA to justify a project's impacts, this "compared-to" argument is a misrepresentation of the significance of a project's impacts and holds little merit. Joseph (2019), in a letter to Impact Assessment and Project Appraisal journal, turns the tables on the "compared-to" argument. Using climate change and the Ajax mine as the example, Joseph explains that for adverse impacts, proponents will often compare the magnitude of their project's GHG emissions to the emissions of other projects or to the emissions of an entire sector or state and then present the case that the project's impacts are thus insignificant. Joseph presents a counter-scenario, arguing that a project proponent would never use the same flawed logic to assess and communicate a project's positive impacts. Employment opportunities created by the project, for example, or the project's contribution to gross domestic product, are often presented as "significant" positive impacts based on project-specific values alone—without comparison to the resource sector or state at large. However, if one adopted the same reasoning used for characterizing adverse impacts, then the contribution of a single project to employment or gross domestic product would likely be highly "insignificant" compared to the entire resource sector. Joseph describes this approach as "faulty logic."

Source: Based on Joseph, 2019

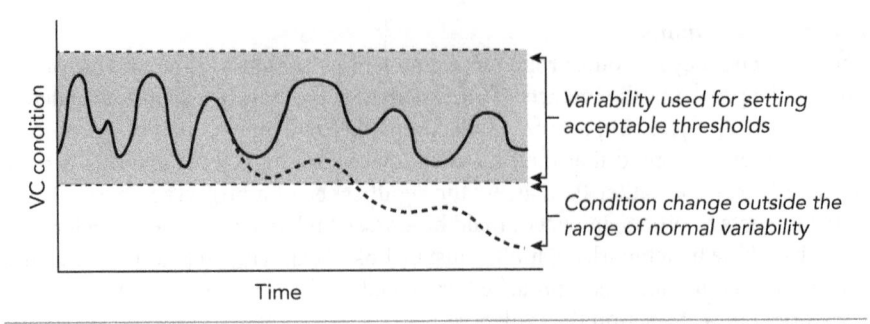

Figure 8.2 Using range of normal or natural variability for setting acceptable thresholds

condition, species population) over a long period of time (Gayton, 2001). Deviations close to, or outside of, the RNV caused by development actions may be considered significant adverse impacts.

Defining ecological limits is not an easy task. It usually implies that enough scientific knowledge exists to understand when the assimilative capacity of an ecological receptor has been exceeded and that there is agreement on what is "natural." This means that even if defining RNV is possible for the environmental component of concern, most definitions of ecological limits still require the use of a non-ecological standard against which to interpret the severity of the impact (Duinker & Beanlands, 1986). For example, an impacted wildlife population may still be functioning within its RNV. To assert that the impact, which may indeed be ecologically insignificant, is therefore not significant may not be an acceptable conclusion because it excludes the societal values that a local human population may place on the species. According to Ehrlich and Ross (2015), while ecological significance must play an important role in determining the significance of an impact, it must not be the only determinant, since societal values play an important role in determining what is significant in the overall assessment of a project. Traditionally, limits have not been based on ecological limits per se but more commonly on public perceptions of risk (Piper, 2002).

Vulnerability
Impacts to irreplaceable or vulnerable systems or functions are often deemed to be significant impacts. Toro, Requena, Duarte, and Zamorano (2012) explain that vulnerability can be measured by a series of parameters related to the stress to which a valued component or human or ecological system is exposed because of pre-existing impacts or sensitivities or another type of stressor that reduces the resilience of the system. In the context of human systems, this may include vulnerable populations or segments of the population, or aspects of cultural traditions and practices that have been declining or under existing pressure, or other factors that affect a society's capacity to cope with change. The basic notion of vulnerability in social systems is that some groups can adapt more quickly and make use of opportunities arising from infrastructure projects and other groups are less able to adapt and will bear more of the negative consequences of change (Slootweg, Vanclay, & van Schooten, 2003). Rowan (2009) thus recommends the need to consider whether impacts will

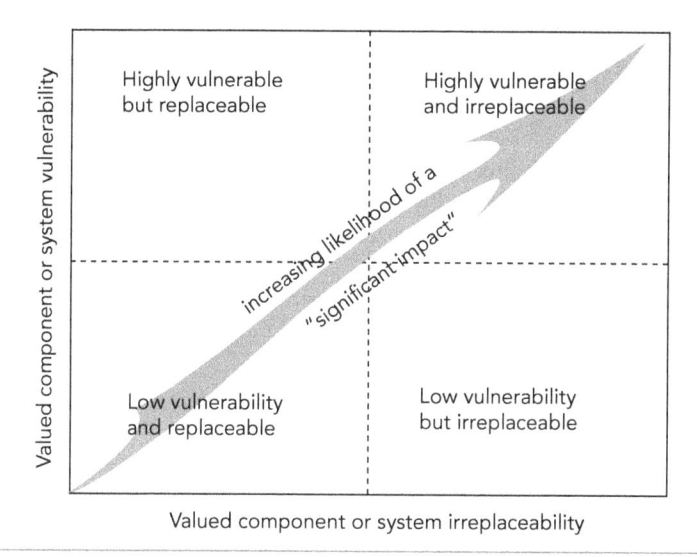

Figure 8.3 Vulnerability and irreplaceability as context for significance determination

increase or reduce vulnerability by the creation or deletion of landlessness, joblessness, homelessness, marginalization, food insecurity, and so on.

Scholarship focused on ecological vulnerability captures similar concepts. Wood (2008), for example, suggests that the ecological context plays an important role in that a relatively small impact in an ecologically sensitive environment may be considered as having a more significant impact than a far larger impact located in a more robust setting. This relationship is often characterized as one of vulnerability versus irreplaceability (Figure 8.3), suggesting that effects on rare or threatened plants, animals, and their habitat, or on particularly vulnerable species or components of the natural environment that are also irreplaceable in terms of ecological functioning (or in terms of their use value to local communities or cultures), are more likely to be deemed significant impacts than impacts to those components that are not vulnerable or whose functions or roles can be replaced or substituted.

Cultural context and recognized and expressed rights
Impacts to Indigenous culture and to formally recognized or asserted rights can constitute significant adverse impacts. Section 35 of the Constitution Act, 1982 entrenches Indigenous and treaty rights, imposing an obligation on governments to justify any interference with those rights that poses hardships on the rights-holders or impacts their means of exercising their rights. Justifying such impacts often means sufficiently demonstrating that Indigenous rights have been prioritized over those of other resource users, that the infringement on rights is minimal, and that sufficient compensation is provided. Even in cases where rights are asserted but remain unproven, there is a legal obligation to accommodate for those rights and impacts where appropriate.

Several policy and guidance documents, and volumes of scholarly research, address Indigenous rights in the context of EA. In some cases, such as under land claims agreements, more explicit (yet broad) guidance is provided for interpreting significance in the context of Indigenous culture and rights. Aboriginal Affairs and Northern Development ministerial guidance for EA under the Nunavut Land Claims Agreement, for example, specifically includes under guidance for significance determinations whether a project or its impacts will:

- pre-empt the use of a natural resource in such a way that will adversely affect the well-being and self-reliance of future generations of the Inuit and other residents of the designated area; or
- affect ecosystem functions in such a way that will adversely affect the well-being and self-reliance of future generations of the Inuit and other residents of the designated area.

However, aside from legal scholarship and select case law, there is limited practical guidance that *specifically* focuses on significance determination in EA in relation to cultural context and recognized and expressed rights.

There are instances in which projects have been rejected in part because of concerns about the incompatibility of a project with Indigenous values of the cultural landscape—such as the Ur-Energy Inc. Screech Lake uranium exploration project, Northwest Territories (Ehrlich, 2010) (see Chapter 10). However, these examples are more the exception than the norm. Booth and Skelton (2011) report on the challenges faced by Indigenous peoples in the EA process. Reflecting on the West Moberly First Nation's experience with First Coal Corporation in northwest British Columbia, Booth and Skelton argue that not only are the project's potential impacts on caribou of significant biodiversity concern but impacts to caribou and caribou habitat are significant to Moberly identity as a Mountain Dunne-za people. No specific reference is made to cultural context as a significance determination factor, but there is a connection between impacts on resources, the cultural value placed on those resources, and interpretations of significance. Similarly, in an analysis of Spectra Energy's (now Enbridge Inc.) proposed Westcoast Connector Gas Transmission project, an approximately 850-kilometre natural gas pipeline corridor from northeast British Columbia, traversing the traditional territory of the Blueberry River First Nations, terminating at Ridley Island near Prince Rupert, Noble (2014) concluded that impacts (e.g., disruption to riparian habitat, terrestrial habitat, traditional use and access constraints) that potentially affect the First Nation's ability to use their traditional lands and derive benefit (e.g., sustenance, cultural, spiritual, economic) constitute significant adverse effects on the First Nation.

Contributions to, or detractions from, sustainability

Some scholars have suggested that significance determination must consider broader, longer-term sustainability-based criteria. Gibson, Hassan, Holtz, Tansey, and Whitelaw (2005), for example, provide a package of sustainability-based criteria and trade-off rules for evaluating significance. These criteria draw on Gibson's earlier work, which is perhaps the most frequently cited guidance for sustainability-based significance determinations. Gibson et al. (2005) suggest that significance determination is about asking whether a project's impacts make a positive contribution to or detract from sustainability. The 12 generic sustainability-based questions proposed by Gibson et al. for evaluating the significance of environmental impacts are as follows:

- Could the effect add to stress that might undermine ecological integrity that could damage life-support functions?
- Could the effect contribute to ecological rehabilitation and/or otherwise reduce stress that might otherwise undermine ecological integrity?
- Could the effect provide more economic opportunities for human well-being while reducing material and energy demands and other stresses on socio-ecological systems?
- Could the effect reduce economic opportunities for human well-being and/or increase material and energy demands and other stresses on socio-ecological systems?
- Could the effects increase equity in the provision of material security, including future as well as present generations?
- Could the effect reduce equity in the provision of material security, including future as well as present generations?
- Could the effect build government, corporate, and public incentives and capacities to apply sustainability principles?
- Could the effect undermine government, corporate, or public incentives and capacities to apply sustainability principles?
- Could the effect contribute to irreversible damage?
- Are the relevant aspects of the undertaking designed for adaptation if unanticipated effects emerge?
- Could the effect contribute positively to several or all aspects of sustainability in a mutually supportive way?
- Could the effect in any aspect of sustainability have consequences that might undermine prospects for improvement in another?

A sustainability-based context was used by a federal-provincial joint review panel in its evaluation of the significance of the Kemess North copper-gold mine project, British Columbia. This was the first mining project in Canada for which a joint federal-provincial review panel recommended outright rejection based on the potential for significant adverse impacts (see Environmental Assessment in Action: Sustainability-Based Review of the Kemess North Copper-Gold Mine Project). In some respects, the sustainability-based context simply captures the considerations identified above under the previous context-based considerations—but placing it under a more attractive label. However, Barnes, Marquis, and Yamazaki (2012) express serious reservations about the use of sustainability-based approaches for significance determination. They argue that it is not reasonable for a project proponent to be subjected to policy debates around the broader sustainability of a project. They use the illustration of an application for a natural gas development project, arguing that it is unfair to evaluate or prohibit a proponent from proceeding with a project on the basis of some discussion or evaluation related to the sustainability of hydrocarbon as an energy source or to debate energy policy in general. Barnes et al. argue that project EA is not the forum for such policy-level debate.

Environmental Assessment in Action

Sustainability-Based Review of the Kemess North Copper-Gold Mine Project

In 2004, Northgate Minerals Corporation proposed to develop the Kemess North copper-gold mine in the Peace River Regional District, 250 kilometres northeast of Smithers. The project is an expansion of the existing Kemess South mine, located just six kilometres south. The proposed new mine site would make use of existing infrastructure at the south mine site, including the work camp, airstrip, access roads, and power lines. Current ore milling capacity would be increased from 55,000 tonnes per day to up to 120,000 tonnes per day, with an estimated 397 million tonnes of tailings and 325 million tonnes of waste rock produced—most of which would be disposed of at Amazay Lake (Kemess North Mine Joint Review Panel, 2007). The project was subject to EA under the British Columbia Environmental Assessment Act and the Canadian Environmental Assessment Act.

The proposed Kemess North mine was located in the traditional-use territory of Kwadacha First Nation, Takla Lake First Nation, and Tsay Keh Dene (collectively referred to as the Tsay Keh Nay), who were opposed to the mine site. Given the project's potential to result in significant adverse environmental effects, the use of a lake for mined waste disposal, and the potential effects to Indigenous people, both the provincial and federal government determined that an independent

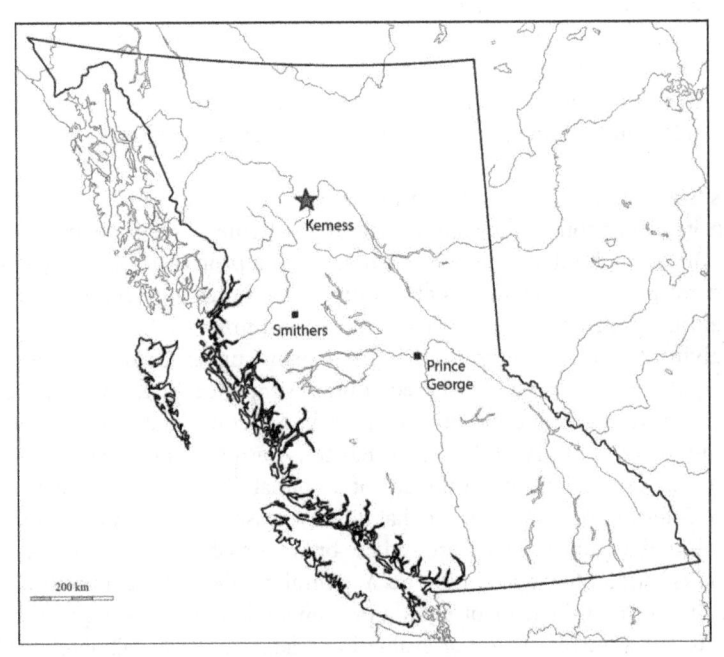

Source: Kemess North Mine Joint Review Panel, 2007

review panel be established to carry out a public review of the project in accordance with the Canada–British Columbia Agreement on Environmental Assessment Cooperation.

In conducting its assessment, the panel consulted mining sector sustainability initiatives, as well as the Government of British Columbia's 2005 Mining Plan. The panel examined the significance of the project's effects from five key sustainability perspectives: environmental stewardship, economic benefits and costs, social and cultural benefits and costs, fairness in the distribution of benefits and costs, and present versus future generations. In its report, the panel noted that the economic and social benefits provided by the project, on balance, were outweighed by the risks of significant adverse environmental, social, and cultural effects, some of which might not emerge until many years after mining operations cease.

The panel identified several key concerns in relation to the significance of the project's potential environmental effects:

- Environmental stewardship: The creation of a long-term site management legacy was a significant environmental concern, and there were doubts about how much assurance could be provided that Northgate Mineral Corporation's site management regime would remain effective over the long term.
- Economic benefits and costs: Although significant benefits would accrue to mine workers and suppliers, government, and shareholders, there were significant concerns with respect to the short duration of the incremental economic benefits—two years of construction and up to only 11 years of mining production.
- Social and cultural benefits and costs: The project would likely make a significant contribution to social well-being and community stability in communities where workers live and service suppliers operate, but the socio-cultural implications of the project for Indigenous people, and obstacles to their participation in project benefits, was a significant concern.
- Fair distribution of benefits and costs: There was concern about the likelihood for inequities in the distribution of benefits and costs between those interests that receive most of the benefits (workers, suppliers, governments, company shareholders) and those people who incur most of the costs (locally based, primarily Indigenous people).
- Present versus future generations: The creation of a long-term legacy of substantial mine site management and maintenance obligations, lasting for thousands of years, was found to be a major imposition on future generations.

The review panel noted that Indigenous communities were unlikely to accept the project or the compensation offered by the project proponent and that "to do so would entail accepting the loss of the spiritual values of Duncan (Amazay) Lake,

continued

and Aboriginal groups have said that these values are beyond price" (Kemess North Mine Joint Review Panel, 2007, p. 16). The panel recommended to the federal and provincial governments that the Kemess North mine not be approved. The recommendation was accepted and the project rejected.

In 2011, AuRico Gold purchased Northgate and acquired the Kemess property. AuRico revived the Kemess project but proposed an underground mine versus the previous open-pit design. The Kemess project was once again subject to EA review and this time received approval from both the British Columbia Environmental Assessment Office and the Canadian Environmental Assessment Agency in 2017.

Sources: Noble, 2016; Kemess North Mine Joint Review Panel, 2007

Social (public) values and acceptability

Significance determination in EA uses judgments and values equal to, or to a greater extent than, science-based criteria and standards, and the adaptability of the concept of significance to socio-political contexts (including values) is an important part of the success of EA (Sippe, 1999). Social (public) values is somewhat of a *catch-all* consideration—recognizing that all context-based considerations inherently involve some consideration of values. Much of the literature addressing the importance of values in significance determination (e.g., Kotchen & Reiling, 2000; Rowan, 2009; Weston, 2000) does so with reference to the notion of *acceptability*—what is considered by an affected group, community, or segment of society to be an acceptable impact, level of change, or risk.

Noble and Storey (2005), in their review of follow-up and thresholds for the Hibernia offshore oil project, report that *maximum allowable effects levels* were determined as part of the social impact assessment. Communities were involved in setting such levels, and for local crime rates, for example, the community established a 0 per cent increase above the current baseline as the maximum allowable increase in crime due to increased temporary worker–community interactions. Any predicted increase above this level was thus deemed a significant impact and required appropriate mitigation. For larger, more catastrophic events, such as dam failure, the issue is not necessarily whether the event is significant—this is of little value to a decision-maker because the extent of harm is almost always considered high; rather, the issue is whether the probability of event occurrence is within a socially acceptable level.

The consideration of values or *what is acceptable* is consistent with the International Association for Impact Assessment's Principles of Environmental Impact Assessment Best Practice, which states that the evaluation of significance involves determining the importance and acceptability of impacts (Senecal, Goldsmith, Conover, Sadler, & Brown, 1999). Wood (2008) explains that when determining significance, it is thus important to consider the diversity of values that exist—not relying simply on the values of the regulator, the practitioner, or the decision-maker. Social values are characterized by plurality and not simply in terms of the different perspectives of individuals and agencies regarding the desirability of change but also with respect to values that surround different ethical positions. Perceptions about what is significant can differ greatly depending on the values and attitudes of the interested parties (Toro, Requena, Duarte, & Zamorano, 2012), but the consideration of this diversity assists the interpretation and understanding of significance and lends credibility to significance determinations.

Political context and public interest
The application of political context or importance in significance determination is tricky. Wood (2008), for example, suggests that for political and economic reasons, a community dominated by high unemployment may be more supportive of controversial development proposals with significant adverse impacts than comparable areas with full employment. Although political context is a *known* factor influencing significance determination, there is limited guidance on how it should be properly applied in reaching significance determinations for potentially adverse impacts. There are some exceptions found in government decision-support guidance. Ministerial guidance prepared for Aboriginal Affairs and Northern Development Canada for decisions under the Nunavut Land Claims Agreement (Aura, 2004), for example, identifies several factors that comprise significance determinations that may be considered *political* determinations, including whether the project or its impacts:

- adversely affects matters of sovereignty;
- poses a risk to national security, including national energy security;
- contravenes commitments to international treaties and conventions on such matters as biodiversity or climate change;
- causes transboundary pollution;
- violates or interferes with Indigenous treaty rights or land title agreements;
- is inconsistent or incompatible with state environmental policies, strategies, or priorities; or
- affects or is likely to affect a structure, feature, place, or area of provincial, territorial, or national significance.

Most often, however, political context or importance is applied to *justify* a project's adverse impacts. Toro et al. (2013) refer to the "importance of the project" as one such political factor in justifying the approval of a project even though it may be found to generate significant adverse environmental effects. In some instances, such projects are referred to as "in the national interest" for economic reasons, notwithstanding significant adverse effects locally or regionally. Often, such justification is used in the context of large national infrastructure projects or programs, such as major energy pipelines or transportation corridors. In other instances, however, the test is whether the project is "in the public interest." This is the case for projects reviewed under the Canadian Impact Assessment Act (Box 8.2), where the "public interest" must be justified based on a series of related factors or considerations. One of the factors to be considered in determining the public interest concerns the significance of project impacts.

> **Box 8.2 Factors to Be Considered in Determining Public Interest under the Impact Assessment Act**
>
> Under section 63 of the Impact Assessment Act, ministerial determinations about the acceptability of a project and the determination of the Governor in Council must be based on the impact assessment report, which must specify the extent to which a project's impacts are significant (e.g., sec. 28, 33, 51), and a consideration of the following factors:
>
> (a) the extent to which the project contributes to sustainability;
> (b) the extent to which the adverse effects within federal jurisdiction and the adverse direct or incidental effects that are indicated in the impact assessment report are significant;
> (c) the implementation of the mitigation measures that the minister or the Governor in Council considers appropriate;
> (d) the impact that the project may have on any Indigenous group and any adverse impact that the project may have on the rights of the Indigenous peoples of Canada recognized and affirmed by section 35 of the Constitution Act, 1982; and
> (e) the extent to which the effects of the project hinder or contribute to the Government of Canada's ability to meet its environmental obligations and its commitments in respect of climate change.

Approaches to Significance Determination

Determining significance is a highly subjective process, yet significance can and should be determined in a consistent and systematic fashion. Several approaches can be used for significance determination, including technical, collaborative, reasoned argumentation and composite approaches, and various methods and techniques are available to support such approaches, including Geographic Information Systems, simulation modelling, and **statistical significance** tests; data scaling and screening procedures, such as threshold analysis and constraint mapping; qualitative and quantitative aggregation and evaluation procedures, including concordance analysis, multi-criteria analysis, ranking and weighting, and risk assessment; and formal and informal public interaction procedures, such as open houses, workshops, and advisory committees (Lawrence, 2004). As with many other aspects of EA practice, there is no specific set of methods and techniques for determining significance.

Technical Approach

The technical approach to significance determination typically involves adopting one or more standardized scaled or quantified methods, each of which is based on some characterization of the various dimensions of the anticipated impacts of development, such as impact scale, severity, reversibility, probability, and duration. The technical model captures the most widely used set of standardized methods for significance determination and adopts such commonly used methods and tools as weighting and scoring, **cost-benefit analysis**, **impact magnitude matrices**, multi-criteria evaluation, and **fuzzy sets**, to name a few. Hajkowicz, McDonald, and Smith (2000) provide a more detailed discussion of weighting methods and techniques.

The United Kingdom Institute of Ecology and Environmental Management's Guidelines for Ecological Impact Assessment provides a framework within which to assess significance and factors that should be considered (Briggs & Hudson, 2013). The guidelines propose placing a value on the ecological receptor at a geographic frame of reference, such as a regional or local value. This value is determined using several factors, including designations, biodiversity value, habitat value, species value, potential value, secondary or supporting value and social value, and economic value. The impact on the receptor is then predicted considering magnitude, extent, duration, reversibility, integrity, timing, and frequency of the impact. The impact and value are then combined to establish significance at a geographic level alongside the probability of the predicted impact.

However, no matter how sophisticated the methodology, at the heart of many of these approaches is some assignment or distribution of weights or priorities to score the impact or indicate the importance of the affected environmental component and thus the relative significance of the impact. Thompson (1990) concludes that approaches to significance determination that proceed from assigning scores or values through to full aggregation of impacts to derive a final significance score should be avoided. Such approaches may be useful for a practitioner's "in-house" evaluation, but they are *not* recommended for communicating impact significance. Thompson goes on to note that arbitrary weighting schemes that are the product of assessment team preferences should be avoided, unless they are complemented by a similar weighting scheme that is the product of affected public interests. Barnes et al. (2012), writing from the perspective of consultants and practitioners of EA in Canada, are quite clear on the limitations of "scoring" approaches for significance determinations:

> There are some practitioners who have and continue to base significance determination on a numeric system or ordinal score. . . . There is great risk in such efforts as the scores are often qualitative and poorly supported with rationale. Further, that an environmental effect is of high magnitude (e.g., a 3 on a scale of 1 to 3) or frequent (e.g., continuous and therefore a 5 on a scale of 1 to 5) may not be an indication that it is in any way unacceptable. The summing of such rankings can result in a meaningless high score having little bearing on the significance of the environmental effect. Efforts to codify or template environmental assessment and significance determination in this way are distressing. In the hands of novices these practices can lead to incorrect EA conclusions and consequent regulatory or project risk.

Collaborative Approach

The collaborative approach, explains Lawrence (2005, p. 16), "starts from the premise that subjective, value-based judgments about what is important should result from interactions among interested and affected parties." Under this approach, there are no predefined thresholds or criteria for judgment—rather, judgments of what is important, what is acceptable, and what are the limits of allowable change emerge only through consultation with the publics. Issues pertaining to regional and community context are integrated directly into significance determination, and the compatibility of the proposal and its impacts regarding goals and objectives are central to the decision-making process. Because the approach is collaborative, the key role of regulators is to ensure the involvement of the most directly affected communities or stakeholders and that all stakeholder concerns are assessed in significance determination.

Numerous methods and approaches are available to facilitate collaborative approaches to significance determination, including open houses and community forums, interactive web-based forums, key informant interviews, community or regional profiling, rapid rural appraisal, nominal group processes, and intervener funding. The overall objectives of using such methods are to fully integrate community, technical, and traditional knowledge in significance determination and to forge a stronger relationship with the public throughout the EA process. It is important to note, however, that public concerns about environmental effects are not always the same as actual environmental effects resulting from project actions.

Reasoned Argumentation

The **reasoned argumentation** approach views significance determination as based on reasoned judgments supported by evidence. Lawrence (2005, p. 19) explains that reasoned argumentation starts from the premise that technical and collaborative models are "too narrow to provide an adequate foundation for value-based significance judgments about what is important and what is not important." Usually expressed qualitatively, the reasoned argumentation model is evident at the regulatory level in priorities and objectives of EA legislation or regulation, often defining "matters of significance," which are used as triggers during the screening process and further expressed in project-specific guidelines and requirements (Box 8.3). At the practical level, reasoned argumentation, similar to a court decision process, involves sifting through information, data, perspectives, and expressed values using structured methods (e.g., decision support aids, matrices, network diagrams) to focus on matters of most importance to decision-making and to build "reasoned" arguments that support significance determination—hence the importance of complete and clear documentation of the evaluation process and reasons for the decision (Kontic, 2000). The reasoned argumentation model is flexible and responsive to context; however, a well-reasoned argument for significance does not ensure that full consideration has been given to scientific data, public values, or existing technical information.

Composite Approach

The choice of approach to significance determination will vary with context, taking into consideration the nature of the project, the receiving environment, interests, and the political setting. In a study of methods used to determine significance in EIA practices in England, for example, Wood, Glasson, and Becker (2006) found that professional judgment and experience, consultation, and simple checklists of impacts were the methods most commonly used by local planning authorities and

> **Box 8.3 Issues of National Significance**
>
> New Zealand's Resource Management Act 1991 considers issues of national significance in EA. The act does not define what "national significance" means, but it does provide indicative guidance as to the factors that may be considered in an EA decision, including whether the proposal:
>
> - has aroused widespread public concern or interest regarding its actual or likely effect on the environment (including the global environment);
> - involves or is likely to involve significant use of natural and physical resources;
> - affects or is likely to affect a structure, feature, place, or area of national significance;
> - affects or is likely to affect or is relevant to New Zealand's international obligations to the global environment;
> - results or is likely to result in or contribute to significant or irreversible changes to the environment (including the global environment);
> - involves or is likely to involve technology, processes, or methods that are new to New Zealand and that may affect its environment;
> - is or is likely to be significant in terms of section 8 (referring to certain existing uses of land that is protected);
> - will assist the Crown in fulfilling its public health, welfare, security, or safety obligations or functions;
> - affects or is likely to affect more than one region or district; or
> - relates to a network utility operation that extends or is proposed to extend to more than one district or region.

consultants. In other words, a composite model consisting of various combinations of technical, collaborative, and argumentative approaches is desirable. Such a model may consist of technical analysis using conventional significance determination methods, supported by public consultation or traditional knowledge systems, which together, based on existing EA regulation or land-use plans, comprise a reasoned argument for significance. Lawrence (2005) suggests that at minimum, effective impact significance determination relies upon:

- a variety of technical methods and analytical techniques;
- a range of public consultation methods, including methods that facilitate stakeholder interaction, to support collaborative significance determination;
- existing regional, community, or land-use plans or local social surveys to identify values and aspirations against which to compare the proposed development;
- literature analysis and case study reviews of previous, similar proposals and outcomes in comparable environments and situations; and
- exploring the uncertainties and acceptable risks associated with the significance determination.

This suggests that significance determination requires clear *operational frameworks* (Beanlands & Duinker, 1983)—as opposed to standard, one-size-fits-all criteria or methods. Several composite approaches or supporting frameworks have been proposed for making significance determinations, including the following.

Range of acceptability
Ehrlich and Ross (2015) suggest that the significance determiner's judgment should be informed by a reasonable weighing of the evidence and by the values of society and, for social and cultural impacts, should consider the rights of, and impacts on, the affected public. The authors suggest four steps to significance determination using this model (Figure 8.4):

- decide where on the spectrum of potential impacts to place the threshold of significance for that particular valued component;
- weigh the evidence (impact predictions);
- decide which side of the threshold the predicted adverse impact falls on; and
- if the impact falls on the unacceptable side, decide if additional mitigation measures will shift the predicted impact to the acceptable side.

Decision tree for determining acceptability of impacts
Sippe (1999) proposes a conceptual decision tree for determining acceptability when significance decisions must be made (Figure 8.5). When a potential impact is identified, the framework includes two dimensions that must be considered—the impact (the predicted, measurable change and its various characteristics, such as magnitude, duration, extent, reversibility) and some consideration of how important or significant it is (based on, for example, the character of the receiving environment, resilience of the biophysical and social environment to cope with change, and public interest). Then, a third dimension must be considered—that these factors need to be interpreted and given further context in terms of what is acceptable, whereby a decision-maker applies threshold tests for acceptability—whether qualitative or quantitative. Sippe also suggests the use of establishing consistent "levels of acceptability" to help make such determinations. For example, an *unacceptable* impact may be one that exceeds a legal threshold or quality standard or that results in loss of critical habitat for a species. An impact may be *normally unacceptable* when it conflicts with existing environmental policies or land-use plans, results in loss of species of social importance, or results in large-scale loss of the productive capacity of renewable resources. Other impacts may be deemed *acceptable only with minimization, mitigation, or management*.

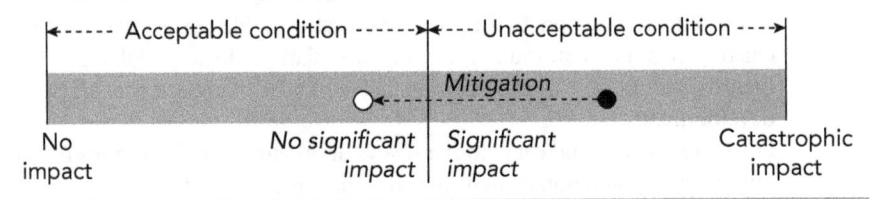

Figure 8.4 Range of acceptability for significance determination
Source: Redrawn, based on Ehrlich & Ross, 2015

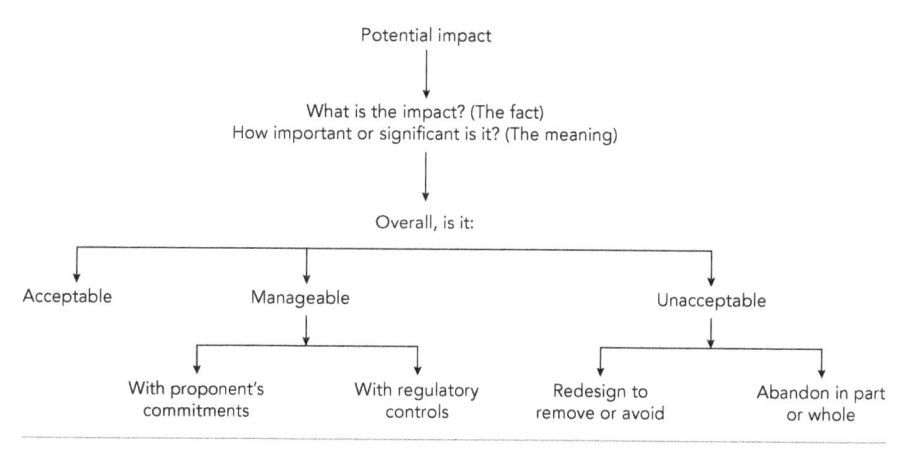

```
                        Potential impact
                              │
                              ▼
              What is the impact? (The fact)
           How important or significant is it? (The meaning)
                              │
                              ▼
                        Overall, is it:
      ┌───────────────────────┼───────────────────────┐
      ▼                       ▼                       ▼
  Acceptable             Manageable              Unacceptable
                    ┌────────┴────────┐       ┌────────┴────────┐
                    ▼                 ▼       ▼                 ▼
             With proponent's   With regulatory  Redesign to    Abandon in part
              commitments          controls      remove or avoid   or whole
```

Figure 8.5 Decision tree for determining acceptability of impacts
Source: Redrawn, based on Sippe, 1999

Key Principles for Determining Significance

When determining whether the effects of a proposed development are significant, consideration should be given not only to the characteristics of the effects but also to the contextual factors of public concern, conditions and sensitivity of the receiving environment, specified thresholds and objectives, significance scale, cumulative change and induced effects, contributions to sustainability, and the nature of proposed mitigation measures. Significance is a highly subjective concept, and differences of opinion are likely to continue as to what constitutes a significant environmental effect, how contextual factors are to be considered, what represents an appropriate level of detail for regulatory requirements and standards, and how to make significance determinations that are transparent and robust. The challenges for significance determination do not lie simply in the realms of improved science and the pursuit of objective expert evaluations but in the clarity of communication of the assessment to decision-makers and the stakeholder community as well (Wood 2008). Significance is not so much the search for objectivity as it is the adequate substantiation of any subjectivity (Lawrence, 1993). Good-practice significance determination is thus underpinned by three foundational principles:

- Transparency in *how* significance determinations are made. This is as important as (if not more important than) the methods or approach used to make that determination.
- Inclusiveness and diversity in how significance determinations are made. Significance determination processes should seek to devise a significance appraisal framework and associated criteria in conjunction with a range of stakeholders and through dialogue conducted early in the scoping phase of the EA—prior to impact predictions and mitigation proposals. Wood

(2008) suggests that such a process can serve to enhance the credibility and legitimacy of significance determinations across the diversity of stakeholders and interests involved in EA.
- Consideration of uncertainty in the factors that inform significance determinations, especially the effectiveness of proposed mitigation measures and the probability of catastrophic or irreversible loss or damage. Not all uncertainties can be reduced through more and better knowledge (Walker et al., 2003). Uncertainty rooted in the inherent variability of social and ecological systems often cannot be reduced by more research, additional baseline data, or additional information requests about the nature of a project's design or its potential impacts. Significance determinations must err on the side of caution.

Key Terms

cost-benefit analysis
fuzzy sets
impact magnitude matrices
impact meaning
impact measurement

impact significance
reasoned argumentation
residual effects
statistical significance

Review Questions and Exercises

1. Determining the significance of environmental effects is often a subjective process, particularly when issues of social concern are involved. Suppose, for example, that a proposed development is likely to result in the closure of a local outdoor public recreational area. How might you assess the significance of such an impact? Identify the types of criteria that you would consider.
2. Obtain a completed project EIS from your local library or government registry, or access one from the Canadian federal Impact Assessment Registry (https://www.ceaa.gc.ca/050/evaluations).
    a. Determine whether the EIS describes impacts as "major," "moderate," or "minor" and so on.
    b. Determine whether, if any, definitions are provided to explain the meaning of those concepts.
    c. Are quantifiable measures, thresholds, or benchmarks used to describe impacts?
    d. Are any impacts identified as "significant"?
    e. Is there evidence of "sustainability criteria" used or discussed in the EIS?
    f. Compare your findings to those of others. Are there noticeable similarities among the significance criteria across impact statements?
3. Who should be responsible for determining impact significance? Are there certain criteria that would apply to practically all proposed developments in your area? What is the role of the public in determining impact significance?

References

Antunes, P., Santos, R., & Jordão, L. (2001). The application of Geographical Information Systems to determine environmental impact significance. *Environmental Impact Assessment Review* 21(6), 511–35.

Aura Environmental. (2004). *Criteria for ministerial decisions under sections 12.4.7 and 12.4.9 of the Nunavut Land Claims Agreement*. Report prepared for the Department of Indian and Northern Development. Ottawa, ON.

Baker, D., & Rapaport, E. (2005). The science of assessment: identifying and predicting environmental impacts. In K. Hanna (ed.), *Environmental impact assessment practice and participation*. Toronto, ON: Oxford University Press.

Barnes, J., Marquis, D., & Yamazaki, G. (2012). Significance determination in energy project EA in Canada. Moncton, NB: Stantec Consulting.

Beanlands, G.E., & Duinker, P.N. (1983). Lessons from a decade of offshore environmental impact assessment. *Ocean Management* 21(3/4), 157–75.

Booth, A., & Skelton, N. (2011). Industry and government perspectives on First Nations' participation in the British Columbia environmental assessment process. *Environmental Impact Assessment Review* 31, 216–25.

Briggs, S., & Hudson, M.D. (2013). Determination of significance in ecological impact assessment: past change, current practice and future improvements. *Environmental Impact Assessment Review* 38, 16–25.

Canter, L.W., & Canty, G.A. (1993). Impact significance determination—basic considerations and a sequenced approach. *Environmental Impact Assessment Review* 13, 275–97.

CCME (Canadian Council of Ministers of the Environment). (2002). Canadian water quality guidelines for the protection of aquatic life: total particulate matter. Winnipeg, MB: Canadian Council of Ministers of the Environment.

Dubé, M., Duinker, P., Greig, L., Carver, M., Servos, M., McMaster, M., . . . & Munkittrick, K. (2013). A framework for assessing cumulative effects in watersheds: an introduction to Canadian case studies. *Integrated Environmental Assessment and Management* 9(3), 363–9.

Duinker, P.N., & Beanlands, G.E. (1986). The significance of environmental impacts: an exploration of the concept. *Environmental Management* 10(1), 1–10.

Ehrlich, A. (2010). Cumulative cultural effects and reasonably foreseeable future developments in the Upper Thelon Basin, Canada. *Impact Assessment and Project Appraisal* 28(4), 279–86.

Ehrlich, A., & Ross, W. (2015). The significance spectrum and EIA significance determinations. *Impact Assessment and Project Appraisal* 33(2), 87–97.

Environment Canada. (2012). *Fifth national assessment of environmental effects monitoring data from pulp and paper mills subject to the Pulp and Paper Effluent Regulations*. Public Works and Government Services of Canada. ISBN 978-1-100-21715-4.

Gayton, D.V. (2001). *Ground work: basic concepts of ecological restoration in British Columbia*. Southern Interior Forest Extension and Research Partnership, Kamloops, BC. SIFERP Series 3.

Gibson, R., Hassan, S., Holtz, S., Tansey, J., & Whitelaw, G. (2005). *Sustainability assessment criteria, processes and applications*. London, UK: Earthscan.

Hajkowicz, S.A., McDonald, G.T., & Smith, P.N. (2000). An evaluation of multiple objective decision support weighting techniques in natural resource management. *Journal of Environmental Planning and Management* 43(4), 505–18.

Haug, P.T., et al. (1984). Determining the significance of environmental issues under the National Environmental Policy Act. *Journal of Environmental Management* 18, 15–24.

Joseph, C. (2019). Problems and resolutions in GHG impact assessment. *Impact Assessment and Project Appraisal*. doi: 10.1080/14615517.2019.1625253.

Kemess North Mine Joint Review Panel. (2007). *Panel report: Kemess North copper-gold mine project*. CEAA Registry Doc 04-07-3394. Ottawa, ON: Kemess North Mine Joint Review Panel.

Kontic, B. (2000). Why are some experts more credible than others? *Environmental Impact Assessment Review* 20, 427–34.

Kotchen, M., & Reiling, S. (2000). Environmental attitudes, motivations, and contingent valuation of non-use values: a case study involving endangered species. *Ecological Economics* 32, 93–107.

Lawrence, D.P. (1993). Quantitative versus qualitative evaluation: a false dichotomy? *Environmental Impact Assessment Review* 13(1), 3–12.

Lawrence, D.P. (2004). *Significance in environmental assessment*. Research supported by the Canadian Environmental Assessment Agency's Research and Development Program for the Research and Development Monograph Series, 2000. Gatineau, QC: CEAA.

Lawrence, D.P. (2005). *Significance criteria and determination in sustainability-based environmental impact assessment*. Report to the Mackenzie Gas Project Joint Review Panel. Langley, BC: Lawrence Environmental.

Lyhne, I., & Kørnøv, L. (2013). How do we make sense of significance? Indications and reflections on an experiment. *Impact Assessment and Project Appraisal*. doi: 10.1080/14615517.2013.795694.

Noble, B.F. (2014). *Review of the approach to cumulative effects assessment in Spectra Energy's environmental assessment certificate application for the Westcoast Connector Gas Transmission Project*. Commissioned report by Ratcliff and Company, on behalf of the Blueberry River First Nation, BC.

Noble, B.F. (2016). *Learning to listen: snapshots of Aboriginal participation in environmental assessment*. Aboriginal People and Environmental Stewardship Series. Ottawa, ON: Macdonald-Laurier Institute.

Noble, B.F., & Storey, K. (2005). Toward increasing the utility of follow-up in Canadian EIA. *Environmental Impact Assessment Review* 25(2), 163–80.

Noble, B.F., & Gunn, J. (2013). *Review of KHLP's approach to the Keeyask Generation Project Cumulative Effects Assessment*. Prepared for the Public Interest Law Centre of Manitoba, Winnipeg, MB.

Piper, J.M. (2002). CEA and sustainable development: evidence from UK case studies. *Environmental Impact Assessment Review* 22, 17–36.

Rowan, M. (2009). Refining the attribution of significance in social impact assessment. *Impact Assessment and Project Appraisal* 27(3), 185–91.

Senecal, P., Goldsmith, B., Conover, S., Sadler, B., & Brown, K. (1999). Principles of environmental impact assessment best practice. Fargo, ND: International Association for Impact Assessment.

Sippe, R. (1999). Criteria and standards for assessing significant impact. In J. Petts (ed.), *Handbook of environmental impact assessment*, vol. 1, *Environmental impact assessment: process, methods and potential*. London, UK: Blackwell Science.

Slootweg, R., Vanclay, F., & van Schooten, M. (2003). Integrating environmental and social impact assessment. In H. Becker & F. Vanclay (eds), *The international handbook of social impact assessment* (pp. 56–73). Cheltenham, UK: Edward Elgar.

Swanson, F.J., Jones, J., Wailin, D., & Cissel, J.H. (1994). Natural variability—implications for ecosystem management. In USDA Forest Service General Technical Report PNW-GTR-318 (pp. 85–99).

Thompson, M.A. (1990). Determining impact significance in EIA: a review of 24 methodologies. *Journal of Environmental Management* 30(3), 235–50.

Toro, J., Requena, I., Duarte, O., & Zamorano, M. (2012). A qualitative method proposal to improve environmental impact assessment. *Environmental Impact Assessment Review* 43, 9–20.

Walker, W.F., Harremoes, P., Rotmans, J., van der Sluijs, P., van Asselt, M.B., Janssen, P., & von Krauss, M.P. (2003). Defining uncertainty: a conceptual basis for uncertainty management in model-based decision support. *Integrated Assessment* 4(1), 5–7.

Weston, J. (2000). EIA, decision-making theory and screening and scoping in UK practice. *Journal of Environmental Assessment Policy and Management* 43(2), 185–203.

Wood, G. (2008). Thresholds and criteria for evaluating and communicating impact significance in environmental statements: "See no evil, hear no evil, speak no evil"? *Environmental Impact Assessment Review* 28(1), 22–38.

Wood, G., Glasson, J., & Becker, J. (2006). EIA scoping in England and Wales: practitioner approaches, perspectives and constraints. *Environmental Impact Assessment Review* 26(3), 221–41.

9 Follow-Up and Monitoring

Follow-Up

Much of what has been covered in this book thus far has focused on what happens before a decision is made to approve or not to approve a project. In this context, EA is a predictive process with no mechanism to verify predictions or evaluate the effectiveness of measures proposed to manage actual project impacts. **Follow-up** occurs after a project's approval and involves determining whether a project has had or is continuing to have environmental effects (Kilgour et al., 2007), but follow-up plans are developed as part of the EA process. Follow-up can transform EA from a static impact prediction tool to a dynamic environmental management process. It has been described by Arts, Caldwell, and Morrison-Saunders (2001) as a link between EA and the **life-cycle assessment** of a project (Figure 9.1). The underlying reason for follow-up in EA is to address uncertainties—whether they are associated with predicted impacts, mitigation measures, the commitments of proponents, or project outcomes. For some EA practitioners, follow-up is about ensuring that management measures committed to in EA have been implemented and are effective. Other practitioners approach follow-up as an umbrella concept for the variety of post-project approval activities that take place, including monitoring or auditing and ongoing learning and project improvement.

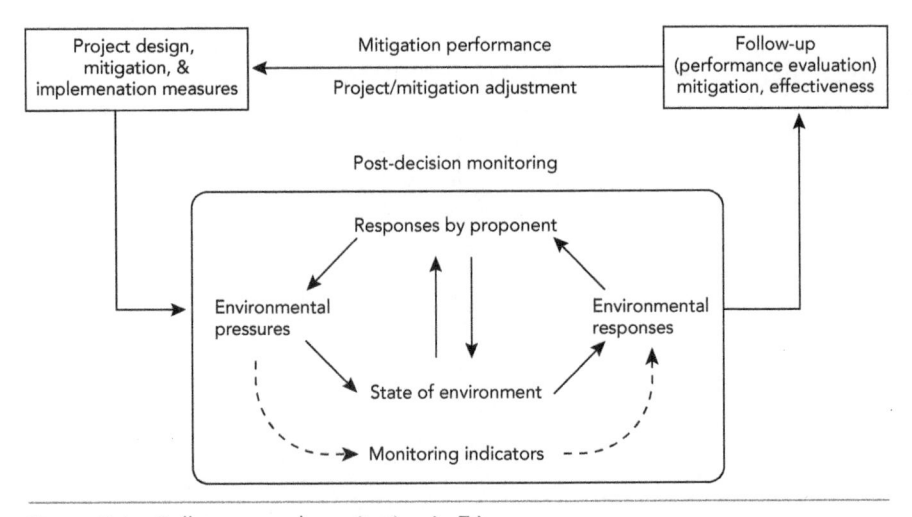

Figure 9.1 Follow-up and monitoring in EA
Source: Based on Ramos, Caeiro, & de Melo, 2004

Follow-Up Requirements

Follow-up is foundational to managing the *actual* impacts of development projects. Mitigation in EA is sometimes a series of non-binding proposals (Morrison-Saunders et al., 2001), and as such, conditions and recommendations need to be monitored and enforced to ensure their implementation and effectiveness (Tinker et al., 2005). It would seem intuitive that follow-up is a mandatory and routine part of any EA system. However, the nature of follow-up requirements and its implementation vary considerably by EA jurisdiction. At the federal level, for example, the Impact Assessment Act defines follow-up as a program for verifying the accuracy of an impact assessment and determining the effectiveness of mitigation. A stated purpose of the act itself is to encourage improvements to impact assessment through follow-up programs, and information on follow-up results is to be made publicly available.

At provincial and territorial levels, follow-up is required to various degrees under the respective laws and regulations of the provinces, with each jurisdiction placing more or less emphasis on monitoring and post-decision impact management activities. In Yukon, for example, the Yukon Environmental and Socioeconomic Assessment Act allows for a recommendation that project effects monitoring be conducted, but reviews of EA practice in Yukon have indicated that follow-up programs are rarely required as a condition of project approval. There has been little change since the act's five-year review (SENES Consultants Ltd, 2009), which showed that of a sample of 18 EA decision documents, only once was there a recommendation that an effects monitoring program be included among the follow-up measures for a project.

Follow-Up Components

Follow-up consists of three interrelated components: monitoring, auditing, and ex-post evaluation. **Monitoring** is about identifying the nature and cause of change and tracking change in valued components or indicators over time. Monitoring is often undertaken to generate knowledge about the specific condition or functioning of valued components or indicators (Bisset & Tomlinson, 1988). Such indicators can be indicators of human- or project-induced stress (e.g., changes in road densities, measures of habitat fragmentation) or indicators of environmental effects (e.g., changes in water sediment/chemistry). Monitoring programs normally consist of repetitive observation, measurement, and recording over a defined period of time to answer specific questions (Arts & Nooteboom, 1999), the objective of which is to determine whether change in an indicator has occurred and to understand the nature, cause, and magnitude of that change.

Auditing is the assessment of observations against predetermined objectives or criteria. There are a variety of types of audits (Box 9.1), but auditing generally is a periodic exercise focused on comparing the results generated from monitoring activity against specified standards or expectations and public reporting of results. Whereas monitoring is a frequent or ongoing process, auditing is a periodic or single event. Monitoring results can be of limited value if they are not subjected to comparative analysis or audit against standards or expectations (Arts & Nooteboom, 1999).

A third and related concept is **ex-post evaluation**, or the collection and appraisal of information about a project's impacts and making decisions on remedial actions and communicating the results (Arts, 1998, p. 75). The differences and relationships among these three components are illustrated in Figure 9.2.

> **Box 9.1 Types of EA Audits**
>
> **Draft EIS audit:** Review of the project EIS according to its terms of reference
>
> **Project impact audit:** Determination of whether the actual environmental impacts of the project were those that were predicted
>
> **Decision-point audit:** Examination of the role and effectiveness of the EIS based on whether the project is allowed to proceed and under what conditions
>
> **Performance audit:** Examination of the proponent's environmental management performance and ability to respond to environmental incidents
>
> **Predictive technique audit:** Comparison between actual and predicted effects of the project
>
> Source: Tomlinson & Atkinson, 1987

Rationale for Post-decision Monitoring

There are several, often overlapping, reasons for undertaking monitoring activities. However, three broad reasons for decision monitoring can be identified: compliance, management, and understanding.

Compliance Monitoring

The primary purpose of **compliance monitoring** is to determine a project's compliance with regulations, mitigation commitments, agreements, or legislation. Compliance monitoring serves to ensure that a project is operating within the specified guidelines and that mitigations have been implemented as committed to by the proponent or as required under the conditions of project approval. Compliance monitoring alone does not fulfill the requirements of an EA follow-up program. It is simply a means of ensuring that what a proponent said would be done in the EIS has actually been done once the project has been implemented. There are several types of compliance monitoring, including implementation, regulatory permit monitoring, and agreement monitoring.

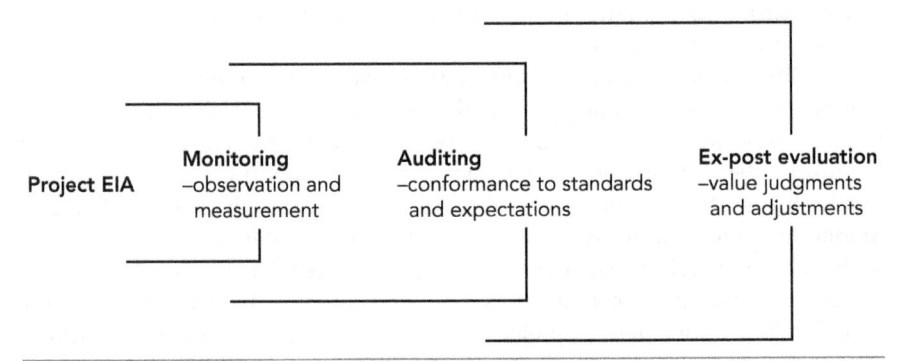

| Project EIA | Monitoring –observation and measurement | Auditing –conformance to standards and expectations | Ex-post evaluation –value judgments and adjustments |

Figure 9.2 Follow-up components
Source: Based on Arts, 1998

Implementation monitoring, the simplest form of monitoring, involves checking to ensure that operating procedures are being followed and commitments or conditions are being met. Implementation monitoring may involve facility visits and reporting of project activities. Implementation monitoring is about checking for compliance with agreed-upon procedures and ensuring that project operations are meeting expected standards. **Regulatory permit monitoring** involves the tracking of conditions that may be required for a proponent's maintenance or renewal of a project permit, such as permits for the operation of water treatment facilities at a project site or waste disposal systems. The **monitoring of agreements** between project proponents and affected groups, such as impact benefit agreements or socio-economic agreements, is becoming commonplace, particularly in northern Canada where such monitoring may be required under Indigenous land claims and self-government agreements. Such monitoring is often focused on tracking socio-economic commitments regarding such matters as housing, employment, and infrastructure demands (Box 9.2).

Monitoring for Management

The purpose of monitoring for management is to confirm anticipated outcomes, verify the effectiveness of mitigation actions, and alert managers to unanticipated outcomes (Box 9.3). In this sense, follow-up allows managers to measure project

Box 9.2   Monitoring of Socio-economic Agreements: Diavik Diamond Mine

The Diavik diamond mine is located approximately 300 kilometres northeast of Yellowknife, Northwest Territories, on a 20-square-kilometre island in Lac de Gras. The mine started operations in 2003 following the discovery of four kimberlite-bearing deposits in the mid-1990s. Since operations started in 2003, the mine has produced more than 110 million carats. As part of the mine's EA approval process, and to ensure that project commitments were implemented and outcomes meeting stakeholder expectations, Diavik entered into a Socio-Economic Monitoring Agreement with the Government of the Northwest Territories, later ratified by the Dogrib Treaty 11 Council (now the Tłı̨chǫ Government), the Yellowknives Dene First Nation, the Lutsel K'e Dene First Nation, the Kitikmeot Inuit Association, and the North Slave Metis Alliance. The agreement specifies that Diavik and the Government of the Northwest Territories report twice a year on issues relating to employment and spending in the North during the construction and operation of the Diavik mine. Included among the socio-economic agreement monitoring requirements are northern employment, employment by the community, employment by job category, employment by contractor group, northern outsourcing, northern purchasing, northern business participation initiatives, site-based training opportunities, Indigenous leadership development, summer student placements, and community investment. The Diavik Socio-Economic Monitoring Agreement and annual reports are available on the project's website at www.diavik.ca.

Box 9.3 Following up for Impact Management: Bison Reintroduction and Experimental Grazing in Grasslands National Park, Saskatchewan

Grasslands National Park is located in southwest Saskatchewan, near the Saskatchewan–Montana border in the Great Plains grassland biome. It covers approximately 906 square kilometres of mixed-grass prairie and is divided into two separate blocks. The park is a semi-arid ecosystem that has evolved over time with migratory and sedentary bison grazing, a disturbance considered to have contributed to the heterogeneity and ecological integrity of the landscape. However, bison were extirpated from the grasslands area by the turn of the nineteenth century, and most of the region outside the park has been cultivated for crop production. The remaining native vegetation has been subjected to grazing by domestic animals.

In 2002, the Grassland National Park Management Plan identified grazing as an important ecosystem function for maintaining mixed-grass prairie and necessary for restoring ecological integrity. The management plan proposed implementing a grazing prescription to represent regimes more consistent with historical patterns of migratory and sedentary bison herds. In 2005, the park prepared environmental impact assessment screening reports for two large-scale experiments: a grasslands grazing experiment and the reintroduction of plains bison after 120 years of absence. The grasslands grazing experiment was to involve the introduction of

Bison in Grasslands National Park, Saskatchewan.

livestock to the park as part of a broader adaptive management process for restoring ecological integrity. The primary objective of the grazing experiment was to determine how grazing intensity alters spatial and temporal heterogeneity of the mixed-grass prairie community. Over a 10-year period, approximately 2664 hectares of native prairie was to be set aside for the experiment, which was based on a "before-after-control-impact" design process. The bison reintroduction project would involve an initial introduction of 50 to 75 animals to the park, transported from Elk Island National Park in Alberta. The long-term stocking-rate objective was to develop a bison herd size of up to 358 animals over approximately 185 square kilometres.

Both EAs were prepared by way of the Canadian Environmental Assessment Act. As the proponent of both projects, Grasslands National Park was also responsible for the design and implementation of a follow-up program. The two projects were different from most development initiatives in that their overall purpose was to introduce disturbance to the environmental system in order to learn how it responds to management actions. At the same time, however, several unintended, potentially adverse environmental effects were also recognized.

A total of 45 potentially adverse impacts were identified in the impact statements, as well as 64 prescribed mitigation measures. Impacts and mitigation concerned a range of biophysical and socio-economic issues, including visitor experience, vegetation, wildlife, aquatics, and cultural resources. An integrated follow-up program, encompassing both projects, was designed and focused on following up for mitigation effectiveness. The objectives of the follow-up program were to: (i) ensure that the mitigation commitments identified in the impact statements were implemented; (ii) establish systems and procedures for this purpose; (iii) monitor the effectiveness of mitigation; and (iv) take any necessary action when unforeseen impacts occurred or when mitigation measures were not performing as expected.

The follow-up framework consisted of four core components: (i) an implementation audit or one-time check or verification that proposed mitigation measures had been implemented; (ii) compliance monitoring and regular inspection to ensure that agreed-upon impact management procedures and best-management practices identified in the impact statements were being adhered to; (iii) effects monitoring and evaluation to track and evaluate changes in a range of biophysical, economic, and social variables so as to compare data collected during project implementation and operation against baseline conditions; and (iv) the development of effectiveness indicators, targets, and thresholds so as to determine whether the mitigation measure was working to avoid or minimize potentially adverse effects.

Bison were successfully reintroduced to Grasslands National Park. Currently, there are approximately 310 adult bison in the park.

Source: Updated from Noble, 2006

management and environmental progress and to respond to adverse environmental change in a timely fashion when necessary. The basic components of monitoring for management include ambient environmental quality monitoring, effectiveness monitoring, and cumulative effects monitoring.

Ambient environmental quality monitoring focuses on the effects of a project on the surrounding environment. This type of monitoring is based on data collected about environmental conditions pre- and post–project implementation or at the project site and at control sites. In most cases, ambient environmental quality monitoring focuses on the biophysical environment, such as air and water quality. However, socio-economic environments, such as changes in quality of life pre- and post-project and overall community health and well-being, can also be the focus.

Effectiveness monitoring is concerned with the mitigation actions implemented to manage anticipated impacts and whether those actions are working to hold impacts to acceptable levels. Effectiveness monitoring is important for those impacts associated with new or innovative projects or technologies, when impacts are occurring in highly sensitive environmental and socio-economic contexts, and when there is uncertainty about the effectiveness of a proposed mitigation measure and the risk to environmental or socio-economic values is considered moderate to high (Table 9.1). In those cases where a mitigation is considered standard practice, or is implemented for a routine impact, monitoring is more likely to focus on implementation than on effectiveness.

Cumulative effects monitoring focuses on the accumulated state or indicators of stress associated with developments in a region, including those of the project. Monitoring for cumulative effects has been an ongoing challenge in project EA, given that the interest of project proponents is foremost on tracking and effectively managing the impact of their own projects. The focus on project-specific monitoring is also entrenched in the conditions of approval normally attached to EA decisions. In a review of seven EA decision reports for mining, roads, and hydroelectric projects in the James Bay territory, northern Quebec, issued between 2006 and 2015 under the James Bay and Northern Quebec Agreement, Noble, Martin, and Olagunju (2016) report that all EA authorizations contained requirements for monitoring and follow-up of effects, but only one authorization contained a requirement for cumulative

Table 9.1 Effectiveness and Implementation Monitoring Based on Risk and Uncertainty about Mitigation

Uncertainty about the effectiveness of proposed mitigation	Risk to environmental values after avoidance measures have been considered or implemented		
	high	moderate	low
high (no to limited supporting evidence or experience)	effectiveness monitoring	effectiveness monitoring	implementation monitoring
moderate (some supporting evidence or experience)	effectiveness monitoring	implementation monitoring	implementation monitoring
low (well supported by evidence or experience)	implementation monitoring	implementation monitoring	implementation monitoring

Source: British Columbia Environmental Assessment Office, 2014

effects monitoring—the Whabouchi spodumene mine. Given the diversity of interests involved in monitoring for cumulative effects, such monitoring programs are most appropriate when carried out by an organization mandated with such monitoring responsibilities. The nature of cumulative environmental effects is discussed in greater detail in Chapter 11.

Monitoring for Understanding

A third reason for monitoring is to understand the relationships between human actions and environmental and social systems. Monitoring for understanding serves a *learning* function whereby new knowledge and understanding that can be applied to the methods and approaches are used to assess and manage future projects. Monitoring for understanding includes experimental monitoring and monitoring for knowledge. **Experimental monitoring** generates information and knowledge about environmental systems and their impacts through research methodologies guided by questions to test specific hypotheses. Experimental monitoring is science-driven rather than motivated by impact management per se. **Monitoring for knowledge** is a type of data collection and reporting that often takes place well after impacts occur. Rather than being used for impact management purposes, the data are used to provide insights for the management of future projects.

Effective Follow-Up and Monitoring

In Canada and elsewhere, post-decision follow-up has been described as less than satisfactory. While weak or non-existent legal requirements and institutional support mechanisms may in part be contributing to the current state of practice, a number of substantive and procedural elements are also required to facilitate effective post-decision EA follow-up programs.

Clearly Defined Follow-Up Program and Objectives

Perhaps the most important requirement for effective follow-up is the clear articulation and identification of follow-up and monitoring program objectives and priorities at the outset of program design. What are the key components of the follow-up program? What are the questions to be answered, or objectives satisfied, through follow-up and monitoring activities? Monitoring programs must provide the information needed to assess, evaluate, and make decisions regarding the impacts of projects, mitigation effectiveness, and regulatory compliance (Box 9.4). If a clear set of questions or objectives for follow-up and monitoring does not exist, whether for monitoring for compliance, management, or knowledge, then one cannot adequately determine whether the follow-up or monitoring program itself is effective. The specific objectives and questions asked of follow-up and monitoring programs vary from one project to another and may include such objectives as verifying impact predictions or ensuring regulatory compliance for project licensing. Not stating program objectives and priorities can lead to confusion about what questions the monitoring program is intended to answer and what information needs it is supposed to serve. Data are collected, but they may be of little use for informing

Box 9.4 Follow-Up Program Objectives for the Hardrock Project, Ontario

Greenstone Gold Mines GP Inc.'s Hardrock project is a proposed open-pit gold mine and processing plant located approximately 275 kilometres northeast of Thunder Bay, Ontario. At full capacity, the project would produce up to 30,000 tonnes of ore per day. The project was subject to EA under both the Ontario Environmental Assessment Act and the Canadian Environmental Assessment Act, 2012. Under the federal act, the EA was required to consider the need for, and the requirements of, follow-up and monitoring activities. The project's EIS, submitted in 2017, identifies the following objectives for its follow-up and monitoring program:

1. to verify the accuracy of the EA;
2. to determine the effectiveness of mitigation measures;
3. to evaluate compliance with environmental approvals, permits, and authorizations;
4. to provide for adaptive management measures should impacts differ from what was predicted or should new information become available;
5. to support project environmental management plans.

The first two objectives reflect the definition and scope of follow-up as required under the federal act. The remaining objectives were identified by the proponent and expand on the basic requirements as set out under the legislation. The project's EIS identifies 14 components that are the focus of the follow-up and monitoring program.

For each component, the objectives of the follow-up program are described. The objectives of the water quality monitoring program, for example, are to confirm

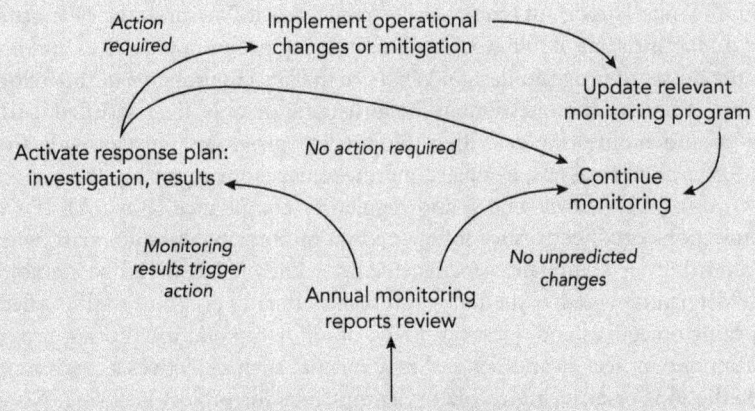

[Flow diagram showing: Action required → Implement operational changes or mitigation → Update relevant monitoring program → Continue monitoring; No action required; Activate response plan: investigation, results; Monitoring results trigger action; Annual monitoring reports review; No unpredicted changes]

Atmospheric Acoustic Groundwater Surface Water Fish and Fish Habitat
Geochemistry Upland Vegetation and Wetlands Wildlife and Wildlife Habitat
Heritage Resources Land and Resource Use Traditional Land Use
Human and Ecological Health Labour and Economy
Community Services and Infrastructure

predicted effects of effluent discharges and runoff on water quality, confirm changes in drainage patterns and surface water flow, determine if additional monitoring or mitigation measures are required, and meet regulatory requirements related to operating permits or conditions of approval. Specific monitoring variables, locations, and sampling designs are proposed, including a monitoring program to satisfy specific requirements under the Fisheries Act and regulatory standards set out under environmental effects monitoring by the Ministry of Mines, Energy and Resources.

Source: Greenstone Gold Mines GP Inc., 2017

impact management or regulatory decisions. Consequently, consideration should be given to follow-up and monitoring objectives and priorities from the outset of EA—during the scoping process, when important issues are first identified, since that will allow pre-development baseline monitoring and formulation of the questions most important to managing project impacts and informing regulatory processes.

Targeted Approach to Data Collection

Not everything of interest or importance in an EA can be monitored and effectively followed up. Further, because of the complexity of most stressor–response relationships, it is impossible to characterize all variables. Monitoring programs must be targeted and focused on the indicators that are most useful for understanding, or at least correlating, stressor–response relationships and informing impact management and meeting the information needs of regulatory decision-makers. Such indicators must be comparable over space and time, and in order to differentiate project-induced change from natural change, these variables and indicators and the data derived from them must be comparable to previous, current, and forecasted baseline conditions.

One approach to monitoring is to focus on **early warning indicators**. Measurable parameters of either a non-biological or a biological nature, early warning indicators serve to indicate stress on particular valued components before these valued components themselves are adversely affected. Early warning indicators might include, for example, changes in worker productivity and number of sick days taken as warnings of worker stress, changes in nitrogen and phosphorous concentrations as early warning of adverse effects to water quality, or changes in benthic invertebrates as early warning of adverse effects to fish communities (Table 9.2). Benthic community surveys are a core part of Environment Canada's Environmental Effects Monitoring programs for pulp and paper and mining operations and are required as part of their licences to operate if they are releasing effluent to surface waters (Kilgour et al., 2007). Early warning indicators must be:

- directly or indirectly related to the valued component;
- physically (and economically) possible to monitor;
- amenable to quantitative analysis; and
- indicative of change in valued component condition before the onset of change.

Table 9.2 Selected Valued Components and Monitoring Indicators for the Cold Lake Oil Sands Project, Alberta

Environmental component	Issues of concern	Valued components	Warning indicators
Air systems	Acidic deposition, odour, GHG emissions	Air quality	Emitted NO_x, SO_2 transported over long distances
Surface water	Lowering of lake levels, water contamination	Water quality and quantity	Combined water volume withdrawals, quality based on drinking-water standards
Groundwater	Depletion of aquifers	Potable well water	Combined water volume withdrawals
Aquatic resources	Fish contamination, harvest pressures	Sport fish species	Northern pike populations and health
Vegetation	Loss of vegetation by land clearing, airborne deposition	Vegetation ecosites	Low bush cranberry, aspen, white spruce
Wildlife	Loss, sensory alienation, habitat fragmentation, mortality due to increased traffic and hunting	Hunted and trapped species	Moose, black bear, lynx, fisher populations/health

Source: Based on Hegmann et al., 1999

Hypothesis-Based or Threshold-Based Approaches

For each indicator, significance levels and probability levels should be specified. Monitoring to inform management action requires that monitoring programs be formulated as testable hypotheses whereby analyses of significance are made against an a priori null hypothesis (Bisset & Tomlinson, 1988) or formulated with specific thresholds or benchmarks in mind. Squires and Dubé (2012) suggest that "benchmark" is the more acceptable terminology, particularly in the regulatory environment, because "threshold" implies that sufficient knowledge exists to understand when the assimilative capacity of an environmental system has been exceeded. Refer to Chapter 5 for discussion on benchmarks and cautionary, target, and critical thresholds set during baseline studies that serve as input to ongoing monitoring programs. Thresholds or benchmarks need to be linked to specific management actions or decision-making processes. Monitoring of these indicators then helps to verify that the management measures implemented by a proponent are effective, that adverse effects are not occurring, and that the project is in compliance with regulatory or industry standards.

Effects-Based Monitoring

Stress-based monitoring, or monitoring project actions or stress, is common practice in EA follow-up. For example, a project proponent for a mining operation needs to know the concentration and volume of water discharged from their operation to any lakes or containment systems to remain within the regulatory or permitted limits. Similarly, to comply with Environment and Climate Change Canada's (2018) voluntary code of practice for the management of atmospheric emissions, pulp and paper mills must monitor sulphur dioxide (SO_2) and total particulate matter (TPM) emissions and remain within the stated limits of 4.0 kilograms and 2.0 kilograms per tonne of production, respectively, for chemical facilities. Under the stress-based approach, the nature and magnitude of changes in, or the risk to, a particular valued component is assessed on the basis of the conditions imposed on the local environment by project development (Kilgour et al., 2007). Although important for ensuring that project emissions or discharge remain within the permitted limit, the assumption is often that if a project remains within its permitted limit, then no adverse effects are occurring because of the project. This assumption may be correct for simple projects or in certain environments and contexts, but it often fails to account for the complexity of environmental systems and multiple sources of stress. Stress-based monitoring should be accompanied by effects-based monitoring.

Effects-based monitoring focuses on the condition or performance of the receiving environment. Effects-based monitoring is based on the premise that measuring change in environmental indicators, or early warning indicators of potentially affected valued components, is the most direct and relevant means of assessing change (Munkittrick et al., 2000). Exceedances of stress-based limits or benchmarks, such as habitat loss or emissions levels, may be insignificant if there are no effects in the receiving environment. In contrast, habitat loss or emissions may be within stated limits or benchmarks, but the environmental conditions, or the conditions of a valued component, may be at an unacceptable level, meaning that a project's emissions, even though within the regulatory limit or standard, may be having an adverse impact. In Canada, effects-based monitoring programs are required under the federal Fisheries Act (1985) by way of the Metal and Diamond Mining Effluent Regulations and the federal Pulp and Paper Effluent Regulations. Effects-based monitoring is designed to determine whether changes in environmental quality have produced important effects. This requires that: the indicators selected for monitoring are responsive to changes attributed to the project and can provide some early warning of a potentially adverse effect; and suitable reference sites can be established, or historical data are available, to understand temporal trends.

Control Sites

Where possible, especially for effects-based monitoring, **control sites** should be established as reference locations to compare with project-affected locations. This will help to differentiate between project impacts and natural or other sources of change. The most basic monitoring design for comparing project-exposed sites to reference or control sites is the **control-impact design** (Figure 9.3). Kilgour et al. (2007) explain that differences in environmental indicators between any two sites are naturally possible, so differences in conditions between control sites and sites exposed to a project's actions may not reflect a project effect. Differences in conditions before versus after a project may indicate an effect, but natural differences from before to after are also possible. Kilgour et al. suggest that a project can be implicated for effects best when changes from before to after differ in reference and exposure sites.

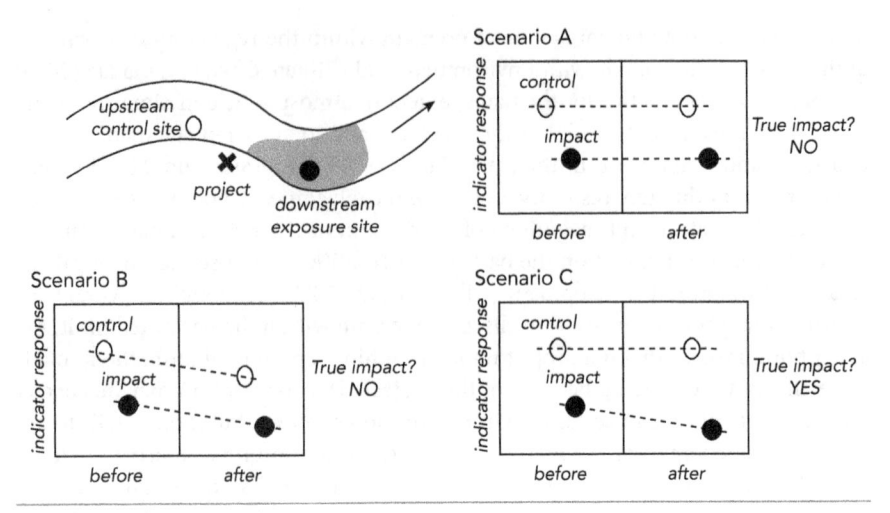

Figure 9.3 Control-impact monitoring
Source: Based on Kilgour et al., 2007

In cases where a well-established control site is not possible, it may be possible to create an artificial one using a **gradient-to-background monitoring** approach. The gradient-to-background approach assumes a clearly defined, localized impact source, such as soil pollution, and that effects can be detected at increasing distances from the source. As distance from the point source increases, effects should decrease and eventually reach ambient or background conditions (Figure 9.4).

Continuity in Data Collection

There should be continuity in data collection and handling procedures to ensure that monitoring data are transferable and comparable. Failure to achieve continuity may lead to problems in comparing monitoring results and the significance of those results from one monitoring period to the next. Such quality control problems in the continuity of data collection were evident in the Rabbit Lake uranium mine project in Saskatchewan, the province's oldest operating uranium mining facility. In 1993, following a joint federal–provincial panel review of the environmental, health, and socio-economic effects of uranium mining in northern Saskatchewan, the panel noted that monitoring met regulatory requirements but there was concern about data

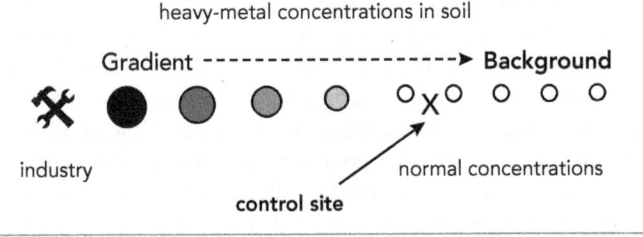

Figure 9.4 Gradient-to-background monitoring

quality and changes in the methods used for testing for radionuclides and trace elements in fish. Throughout the 1980s, testing procedures had changed several times, and some data were discarded because of quality problems, raising uncertainty about the effectiveness of impact management measures. After decades of biophysical monitoring and data collection, the proponent was unable to make a direct connection between project actions and the impacts of project-induced environmental change.

Open Data and Data Sharing

Availability of and accessibility to monitoring data are essential to supporting good-practice EA (Morrison-Saunders, Baker, & Arts, 2003) and to meeting the needs of project managers, other proponents operating in the same region as the project being assessed, and regulatory decision-makers. In principle, comprehensive monitoring data from project EAs should be available in public registries—but this is rarely the case. In the North Sea, for example, Murray et al. (2018) report that many EAs and monitoring programs have been completed or are ongoing but the data cannot be accessed, are difficult to collate, or are available in formats that are unusable. Similarly, in a recent assessment of project EAs and water licences in the Northwest Territories, Wong, Noble, and Hanna (2019) report that proponents rarely make their project monitoring data publicly available; when they do, it is often provided in protected document format, making the data available but not easily usable. Understandably, project proponents are concerned that their monitoring data could be misused or misinterpreted by those opposed to the project (Murray et al., 2018), and that monitoring data may contain information that is proprietary to project operations or sensitive to certain social conditions. To address these concerns, open and shared data may be limited to processes versus raw data, include only non–commercially sensitive data, and even be limited to a suite of agreed-upon indicators established in the project EA terms of reference that are standard for all projects operating in certain regions (Wong, Noble, & Hanna, 2019).

Community Engagement

Active engagement of the local community in follow-up and monitoring programs can lead to increased capacity at the local level to deal with environmental change and the impacts of project development (Austin, 2000) as well as improved industry–community relations. Noble and Birk (2011) report that the benefits of involving communities in the EA process have long been recognized and range from ensuring that the project itself carries more legitimacy to enhancing the effectiveness of monitoring programs and impact management measures. In practice, however, meaningful community involvement in monitoring has not frequently occurred. Follow-up programs are often criticized for not effectively engaging communities in the monitoring and management of project impacts or building trust and capacity among stakeholders. According to Hunsberger, Gibson, and Wismer (2005), it is important for communities to play a role in determining the purpose, scope, and priorities of follow-up and monitoring activities, but practice has traditionally been weak in the areas of community engagement in monitoring programs.

Although local communities are becoming more involved in resource management processes, Lawe, Wells, and Mikisew Cree First Nations Industry Relations

Corporation (2005) report that "they often have limited influence over EA follow-up including long-term monitoring programs that determine effectiveness of mitigation." There are exceptions and some promising practices, including the long-standing Wood Buffalo Environmental Association, formed to monitor air quality impacts of oil sands operations in the Regional Municipality of Wood Buffalo, Alberta (see Environmental Assessment in Action: Wood Buffalo Environmental Association), and more recent initiatives, such as the Trans Mountain Indigenous Advisory and Monitoring Committee (IMAC) (https://iamc-tmx.com). The Trans Mountain Expansion (TMX) project was approved in 2019, subject to 156 conditions set out by the National Energy Board. The project involves the twinning of an existing 1150-kilometre-long pipeline between Strathcona County (near Edmonton, Alberta) and Burnaby, British Columbia, to create a pipeline system with the capacity of moving approximately 890,000 barrels of oil per day. In 2016, a collective of Indigenous leaders along the pipeline route approached the federal government requesting Indigenous oversight of the TMX project. The IMAC was developed by a working group comprised of representatives of the federal government, including the National Energy Board and Indigenous communities. The IMAC is intended to provide for meaningful engagement of Indigenous communities in the review and monitoring of the environmental, safety, and socio-economic issues related to the project's activities over its life cycle. This includes, but is not limited to, providing advice to regulators on the project's environmental and safety performance, the adequacy of reports and filings made by Kinder Morgan, and identifying monitors to accompany and advise National Energy Board inspection officers on project compliance verification visits.

Environmental Assessment in Action

Wood Buffalo Environmental Association

The Wood Buffalo Environmental Association (WBEA) monitors the environment of the Regional Municipality of Wood Buffalo in northeastern Alberta. The Regional Municipality of Wood Buffalo extends from north-central Alberta to the Saskatchewan and Northwest Territories borders, approximately 68,500 square kilometres. The regional municipality includes the communities of Fort McMurray, Fort Chipewyan, Anzac, Janvier, Fort McKay, and Conklin and is home to the Athabasca oil sands. The WBEA originated in the mid-1980s from concerns raised by First Nations about air quality in the region associated with a growing oil sands industry. The government of Alberta and the oil sands industry subsequently formed a joint task force to explore air quality issues and establish monitoring priorities for the Wood Buffalo region. Included among the recommendations of the task force was the need for a consensus-based approach to addressing air quality, which led to the establishment of the Regional Air Quality Coordinating Committee (RAQCC).

In 1997, the RAQCC was transformed into the WBEA, an independent monitoring collaboration of industry, community groups, Indigenous interests, government, and environmental organizations. The WBEA was responsible for the operation of an air quality monitoring network in the regional municipality. Initial costs for the association were covered by the oil sands industry, with Alberta Environment providing long-term equipment and capacity-building support. Monitoring activities are focused on compliance monitoring, air quality baseline monitoring, community health monitoring, and a terrestrial ecological monitoring program to determine if emissions from oil sands operations are having long-term effects. According to the WBEA, it operates the most extensive ambient air network in Alberta, with 17 air monitoring stations and 23 passive monitoring stations, and reports ambient air quality data, in real time, on its website (https://wbea.org).

Using environmental tracers, such as stable isotopes and chemical signatures, WBEA's terrestrial effects monitoring program is designed to identify, characterize, and report impacts that atmospheric emissions have had, or may have, on terrestrial ecosystems and traditional land resources in the Wood Buffalo region. For example, epiphytic lichens have been used to map the extent of sulphur and nitrogen deposition. Lichens were collected at more than 350 sites and analyzed for trace elements, sulphur and nitrogen, stable isotopes of mercury and lead, and polycyclic aromatic hydrocarbons. Most of the contributions to environmental concentrations in lichen were found to be attributed to combustion (e.g., upgrading—removal of contaminants such as a heavy metals, nitrogen, and sulphur to turn bitumen into synthetic crude oil), fugitive tailing sand (i.e., wind-blown particles after oil sands are processed), and wind-blown mining and limestone road-building dust.

In 2010, WBEA was approached by the Fort McKay First Nation with concerns about observed changes in the quantity and quality of berries (blueberries and cranberries) on their traditional lands. Some berry patches close to the community of Fort McKay were less preferred as a food source because of the proximity of oil sands operations, while those farther from oil sands were considered a clean and acceptable food source. The Fort McKay Berry Focus Group was formed to design a community-based project to share knowledge about berry locations and quality and to couple traditional knowledge with Western science–based monitoring of berry patches. The community group harvested berries for testing of trace elements, which would be compared to store-bought berries, and worked with WBEA to install air-monitoring equipment. Results were reassuring to community members in that sampled berries in areas of concern were found to have equal or higher levels of antioxidants and other health-promoting attributes and low concentrations of trace elements.

WBEA has been subject to several scientific reviews and has maintained an effective relationship with stakeholders and communities (Lott & Jones, 2010). WBEA monitoring activities currently inform Alberta Environment's regulatory compliance initiatives and the air quality management framework under the province's Lower Athabasca Regional Plan.

Sources: Cronmiller & Noble, 2018; Wood Buffalo Environmental Association, 2016; 2017

Monitoring Methods and Techniques

As with impact prediction, there is no single set of monitoring techniques for all projects and components. The type of monitoring technique selected to provide data and to evaluate environmental change depends on the nature of the environmental components, the purpose of the data, and particular monitoring program objectives. Thus, for both biophysical and socio-economic monitoring, a variety of quantitative and qualitative techniques, or combinations thereof, are available. For example, socio-economic impacts can be monitored using annual surveys of residents' perceptions of quality of life and local police and hospital records. Selected examples of techniques for monitoring biophysical components are listed in Table 9.3. When selecting a technique for monitoring, as for any other EA procedure, it is important to keep in mind the nature and resolution of the data required.

Table 9.3 Selected Biophysical Monitoring Components, Parameters, and Techniques

Components	Parameters	Techniques
wildlife habitat	fragmentation	remote sensing
rangeland health	vegetation cover vegetation composition soil structure	remote sensing field transects trend transects
lake biology	lake benthos	Ekman grab stratified depth sampling and species counts
stream biology	stream benthos fish	Hester-Dendy sampler and frequency counts tagging electro fishing age-length keys
water quality	metals nutrients physical parameters	stream gauges and go-flow samplers chemical analysis
hydrology	water level	staff gauges pressure transducers

Key Terms

ambient environmental quality monitoring
auditing
compliance monitoring
control-impact design
control sites
cumulative effects monitoring
draft EIS audit
early warning indicators
effectiveness monitoring
effects-based monitoring
experimental monitoring
ex-post evaluation
follow-up
gradient-to-background monitoring
implementation monitoring

life-cycle assessment
monitoring
monitoring for knowledge
monitoring of agreements
performance audit

predictive technique audit
project impact audit
regulatory permit monitoring
stress-based monitoring

Review Questions and Exercises

1. What provisions exist for post-decision monitoring and auditing under your national, provincial, or territorial EA system? Are these provisions mandatory or voluntary?
2. What is the value added to EA from follow-up and monitoring activities?
3. Who should be responsible for monitoring the environment after project approval?
4. Why are socio-economic effects difficult to monitor post–project implementation? How might we address these difficulties?
5. What is the role of the public in environmental monitoring?
6. Obtain a completed project EIS from your local library or government registry, or access one online from the Canadian federal Impact Assessment Registry (https://www.ceaa.gc.ca/050/evaluations). Is there a monitoring component to the impact statement? What environmental aspects are included in the monitoring program? Are methods and techniques for monitoring identified? Compare your findings with those of others.

References

Arts, J. (1998). *EIA follow-up: on the role of ex-post evaluation in environmental impact assessment*. Groningen, Netherlands: Geo Press.

Arts, J., Caldwell, P., & Morrison-Saunders, A. (2001). Environmental impact assessment follow-up: good practice and future directions. *Impact Assessment and Project Appraisal* 19(3), 175–85.

Arts, J., & Nooteboom, S. (1999). Environmental impact assessment monitoring and auditing. In J. Petts (ed.), *Handbook of environmental impact assessment*. London, UK: Blackwell Science.

Austin, E. (2000). Community participation in EIA follow-up. Paper presented at the 2000 International Association for Impact Assessment annual meeting. Hong Kong: IAIA.

Bisset, R., & Tomlinson, P. (1988). Monitoring and auditing of impacts. In P. Wathern (ed.), *Environmental impact assessment: theory and practice*. London, UK: Unwin Hyman.

British Columbia Environmental Assessment Office. (2014). *Procedures for mitigating impacts on environmental values: environmental mitigation procedures*. Victoria, BC: BC EAO.

Cronmiller, J., & Noble, B.F. (2018). The discontinuity of environmental effects monitoring in the Lower Athabasca region of Alberta, Canada: institutional challenges to long-term monitoring and cumulative effects management. *Environmental Reviews* 26, 169–80.

Environment and Climate Change Canada. (2018). *Code of practice for the management of air emissions from pulp and paper facilities*. Ottawa ON: Minister of Environment and Climate Change.

Fisheries Act R.S.C. (1985). c. F-14. Current to 2019-06-20 and last amended 2016-04-05.

Greenstone Gold Mines GP Inc. (2017). *Hardrock project final environmental impact statement/environmental assessment summary*. Guelph ON: Stantec Consulting.

Hegmann, G., et al. (1999). *Cumulative effects assessment practitioners guide*. Prepared by AXYS Environmental Consulting Ltd and the CEA Working Group for the Canadian Environmental Assessment Agency. Hull, QC: CEAA.

Hunsberger, C., Gibson, R., & Wismer, S. (2005). Citizen involvement sustainability-centred environmental assessment follow-up. *Environmental Impact Assessment Review* 25, 609–27.

Kilgour, B., et al. (2007). Aquatic environmental effects monitoring guidance for environmental assessment practitioners. *Environmental Monitoring and Assessment* 130(1–3), 423–36.

Lawe, L., Wells, J., & Mikisew Cree First Nations Industry Relations Corporation. (2005). Cumulative effects assessment and EIA follow-up: a proposed community-based monitoring program in the oil sands region, northeastern Alberta. *Impact Assessment and Project Appraisal* 25(3), 191–6.

Lott, E.O., & Jones, R.K. (2010). *Review of four major environmental effects monitoring programs in the oil sands region*. Oil Sands Research and Information Network, University of Alberta, School of Energy and the Environment, Edmonton, AB. OSRIN Report no. TR-6.

Morrison-Saunders, A., et al. (2001). Roles and stakes in environmental impact assessment follow-up. *Impact Assessment and Project Appraisal* 19(4), 289–96.

Morrison-Saunders, A., Baker, J., & Arts, J. (2003). Lessons from practice: towards successful follow-up. *Impact Assessment and Project Appraisal* 21(1), 43–56.

Munkittrick, K., et al. (2000). *Development of methods for effects-driven cumulative effects assessment using fish populations: Moose River Project*. Pensacola, FL: Society of Environmental Toxicology and Chemistry.

Murray, F., Needham, K., Gormley, K., Rouse, S., Coolen, J., Billett. D., . . . & Roberts, J. (2018). Data challenges and opportunities for environmental management of North Sea oil and gas decommissioning in an era of blue growth. *Marine Policy* 97, 130–8.

Noble, B. (2006). *Environmental impact assessment follow-up prescription and reporting framework: grasslands grazing experiment and reintroduction of plains bison, Grasslands National Park of Canada*. Val Marie, SK: Grasslands National Park.

Noble, B.F., & Birk, J. (2011). Comfort monitoring? Environmental assessment follow-up under community–industry negotiated environmental agreements. *Environmental Impact Assessment Review* 31, 17–24.

Noble, B., Martin, J., & Olagunju, A. (2016). *A review of the application of cumulative effects assessment in the context of section 22 environmental assessments conducted in the James Bay territory*. Report to the James Bay Advisory Committee on the Environment. Saskatoon SK: Aura Environmental Research and Consulting Ltd.

Ramos, T., Caeiro, S., & de Melo, J. (2004). Environmental indicator framework to design and assess environemntal monitoring programs. *Impact Assessment and Project Appraisal* 22(1), 47–62.

SENES Consultants Ltd. (2009). *YESAA five-year review: observations and conclusions*. Yellowknife, NT: SENES Consultants Ltd.

Squires, A., & Dubé, M.G. (2012). Development of an effects-based approach for watershed scale aquatic cumulative effects assessment. *Integrated Environmental Assessment and Management*. doi: 10.1002/ieam.1352.

Tinker, L., et al. (2005). Impact mitigation in environmental assessment: paper promises or the basis of consent conditions? *Impact Assessment and Project Appraisal* 23(4), 265–80.

Tomlinson, P., & Atkinson, S. (1987). Environmental audits: proposed terminology. *Environmental Monitoring and Assessment* 8(3), 187–98.

Wong, L., Noble, B.F., & Hanna, K. (2019). Water quality monitoring to support cumulative effects assessment and decision making in the Mackenzie Valley, Northwest Territories, Canada. *Integrated Environmental Assessment and Management* 15(6), 988–99.

Wood Buffalo Environmental Association. (2016). WBEA factsheet: traditional ecological knowledge (TEK). https://wbea.org/resources/reports-publications/program-fact-sheets.

Wood Buffalo Environmental Association. (2017). WBEA factsheet: terrestrial environmental effects monitoring. https://wbea.org/resources/reports-publications/program-fact-sheets.

10 Indigenous Consultation and Engagement

Indigenous Engagement

The engagement of Indigenous peoples (i.e., First Nations, Inuit, and Métis peoples) potentially impacted by development is foundational to effective EA and a cornerstone of sustainable resource development (Noble & Udofia, 2015). In Canada, Indigenous and treaty rights are entrenched in law under section 35 of the Constitution Act, 1982, requiring that federal, provincial, and territorial governments consult with rights-holders when legislative or regulatory actions, such as a project approval, may interfere with or infringe upon those rights (Olszynski, 2016). Even when rights are unproved, governments are obligated to consult with Indigenous peoples regarding potential impacts to their asserted rights. Under the federal Impact Assessment Act (2019), for example, one of the factors to be considered in EA is the impact on the rights of Indigenous peoples (sec. 16(2)). One of the purposes of the act is to promote communication and cooperation with Indigenous peoples and to ensure that impact assessment takes into account Indigenous knowledge. The Impact Assessment Act reflects, in part, Canada's commitment to the **United Nations Declaration on the Rights of Indigenous Peoples** (United Nations General Assembly, 2008) (Box 10.1).

Indigenous engagement is also a necessity for project proponents wanting to earn a **social licence** to operate (Noble & Udofia, 2015). The benefits of Indigenous engagement to project proponents are widely noted, including:

- improvements in project design;
- new knowledge and understanding about the potential impacts of a project;
- creating innovative and collaborative approaches to mitigate adverse impacts on the environment and communities; and
- increased legitimacy of project undertakings.

Isaac and Knox (2003) argue that the engagement of Indigenous peoples in EA not only increases the efficiency of the EA process, it's also a wise choice for developers. Ensuring opportunities for meaningful engagement, even if it adds to project timelines, is likely to be inexpensive in comparison to the potential costs associated with litigation because of insufficient or no engagement (Chrétien & Murphy, 2009).

Box 10.1 The United Nations Declaration on the Rights of Indigenous Peoples

The United Nations Declaration on the Rights of Indigenous Peoples (UNDRIP) is an international instrument adopted by the United Nations in 2007 to enshrine the rights that constitute the minimum standards for the survival, dignity, and well-being of Indigenous peoples. UNDRIP acknowledges several rights of Indigenous peoples to manage the natural resources on their land. Article 26 of UNDRIP, for example, notes that "Indigenous peoples have the right to own, use, develop and control the lands, territories and resources that they possess by reason of traditional ownership or other traditional occupation or use, as well as those which they have otherwise acquired." Article 32 notes that governments "shall consult and cooperate in good faith" in order to obtain free, prior, and informed consent from Indigenous peoples prior to the approval of any development project (United Nations General Assembly, 2008). Although UNDRIP is not a formally binding international treaty, it was initially endorsed by 144 nations. Four countries initially voted against UNDRIP, including Canada, the United States, New Zealand, and Australia. Australia and New Zealand reversed their positions in 2009. Canada issued a statement of support endorsing the UNDRIP principles in 2010 and committed in 2016 to fully implement the declaration. Canada is now evaluating changes to its laws and regulations considering this commitment. In 2018, the House of Commons also adopted Bill C-262, An Act to Ensure That the Laws of Canada Are in Harmony with the United Nations Declaration on the Rights of Indigenous Peoples. British Columbia is the first, and as of late 2019 the only, province in Canada to introduce legislation to enact UNDRIP—the British Columbia Declaration on the Rights of Indigenous Peoples Act sets out a process to align the province's laws with UNDRIP.

Duty to Consult

The **duty to consult** is a formal, legal obligation for governments to consult with Indigenous peoples. This legal obligation is separate from, and in addition to, any requirements for public participation and engagement under various federal and provincial EA systems. The duty to consult as practised by governments has been established through several court decisions, perhaps the most notable of which are *Haida Nation v. British Columbia* (2004), *Mikisew Cree First Nation v. Canada* (2005), and *Tsilhqot'in Nation v. British Columbia* (2014) (Bains & Ishkanian, 2016) (Box 10.2).

Consultation with Indigenous peoples must be carried out with the goal of addressing the concerns of the affected Indigenous group, but the extent of consultation is determined on a case-by-case basis according to the severity of the potential impact; the extent to which there is an asserted claim or treaty right; the status, merit, or strength of that claim; and whether the Indigenous or treaty right potentially affected is claimed but not yet established (Potes, Passelac-Ross, & Bankes, 2006).

> **Box 10.2   Duty to Consult and the Haida Case Ruling**
>
> In 2004, the Supreme Court of Canada released its decisions in *Haida Nation v. British Columbia (Minister of Forests) and Weyerhaeuser*. The Haida case involved a judicial review, pursuant to the British Columbia Judicial Review Procedure Act, of the minister's decision to replace and approve the transfer of a tree farm licence. In 1961, tree farm licences were issued to MacMillan Bloedel, a forest harvesting company, permitting the company to harvest in an area of Haida Gwaii, the Queen Charlotte Islands. The licences were replaced in 1981, 1995, and 2000; the tree farm licence was transferred to Weyerhaeuser in 1999. The Haida challenged these replacements and the transfer, arguing that they were made without the consent of the Haida and over their objections. The Haida's case was dismissed by the British Columbia Supreme Court, which noted that the law could not presume the existence of Indigenous rights based only on their assertion and concluded that the government had only a "moral duty" to consult. The decision was appealed, and the court of appeal found that the government had fiduciary obligations of good faith to the Haida with respect to their claims to Indigenous title and right and that both the province and Weyerhaeuser were aware of the Haida's claims to the area covered by the licence. The court concluded that both the Crown and Weyerhaeuser had a legally enforceable duty to consult with the Haida to address their concerns. Weyerhaeuser's appeal of the decision to the Supreme Court of Canada was allowed, but the government's appeal was dismissed. The Supreme Court of Canada ruled that the duty to consult rests with government and government must consult and accommodate when there is knowledge of the potential existence of an Indigenous right or title, regardless of whether that right or title has been legally established.
>
> Source: Bergner, 2005

The duty to consult applies even before the Indigenous claim, right, or title is established conclusively by the courts (Brackstone, 2002). There is no duty to reach an agreement during consultation; rather, the intent is to substantially address the concerns of affected Indigenous peoples. Consent during consultation is not a requirement. The duty to consult does not provide Indigenous peoples with the power to veto development (Newman, 2014); rather, it is about determining whether an adverse impact on established or asserted treaty rights is likely and, if so, it can be mitigated or a project modified or, in rare cases, abandoned to accommodate those rights.

Discharging the Duty to Consult through EA

The engagement of Indigenous peoples in EA is different from the duty to consult (Noble & Udofia, 2015), but in practice the two are not mutually exclusive (Craik, 2016). Governments hold the duty to consult with Indigenous peoples. This duty cannot wholly be delegated to other parties, such as a project proponent (Chrétien & Murphy, 2009). A government can and often will, however, download many of the substantive elements of its duty to consult onto project proponents as part of the proponent's EA participation and engagement processes (Chrétien & Murphy, 2009; Land, 2014) (Box 10.3). EA addresses the technical aspects of projects, which are important to understanding potential impacts on Indigenous peoples (Noble, 2016). In this sense, a project proponent can be well equipped to find ways to mitigate or avoid those effects. As Land (2014, p. 20) reports: "there is a significant practical overlap in the process of an EA on the one hand, and review of Aboriginal and treaty rights impacts on the other hand . . . a fulsome assessment of environmental impacts will inevitably go a great distance to assessing impacts of a project on Aboriginal and treaty rights as well."

In a report prepared for the MacDonald Laurier Institute, however, Noble (2016) asserts that underlying both Indigenous engagement in EA and the duty to consult is that the rights, interests, and values of Indigenous peoples will be better accounted for and accommodated in development decisions, but the function and objectives of EA engagement versus the duty to consult are quite different. Engagement of Indigenous peoples in EA is one means to find balance among competing values and to

Box 10.3 A Water Licence Retracted: Clarity of Roles and Responsibilities in Discharging the Duty to Consult

Governments will sometimes delegate certain procedural aspects of consultation to project proponents. The objective is that project proponents or their consultants can provide comprehensive information to affected communities and capture all aspects of their concerns about the proposed project and in doing so meet the government's own consultation requirements. Quite often, however, there can be confusion about the relative roles and responsibilities of government and industry—specifically in terms of a proponent's actions to build relationships with a community to earn a social licence versus a government's legal duty to consult. A recent example is a decision by the British Columbia Environmental Appeals Board regarding a challenge by the Fort Nelson First Nation over a water extraction licence issued to Nexen, an upstream oil and gas company, to support hydraulic fracturing in northern British Columbia. The Fort Nelson First Nation claimed that it was not made clear to them that their engagement with the project proponent also constituted the government's legal duty to consult and that engagement with the proponent certainly did not meet their expectations for engagement with the Crown. The Appeals Board found that the province failed to consult, in good faith, with the Fort Nelson First Nation, noting that meaningful consultation must be based on a clear framework and set of processes. Nexen's role in meeting the government's legal obligation to consult was not clearly communicated to the First Nation. The board explained that if government expects a project proponent to play a role in the consultation process, then it must make that role clear to the First Nation. The company's water extraction licence was retracted.

Source: Noble, 2016

facilitate reconciliation between development actions and environmental protection; the duty to consult is about reconciliation of the pre-existence of Indigenous peoples and Indigenous rights with the sovereignty of the Crown (Craik, 2016).

The intent of Indigenous peoples' engagement in the EA is to ensure that Indigenous values and knowledge inform, if not influence, project design and decisions, including how a project's impacts are understood and managed. In many cases, these values reflect the common interests of society to avoid potentially damaging impacts on environmental quality, which may, in certain cases, be traded off to ensure economic benefit (Craik, 2016). The values and interests addressed by the duty to consult, in contrast, are grounded in the rights held by a particular group and are based on legal claim or title or on the potential existence of a claim or right (Craik, 2016). EA can thus provide a venue for governments to discharge their duty to consult, but the focus of Indigenous engagement in EA, including a proponent's obligations to manage the environmental impacts of their project, may not always satisfy a government's duty to consult—even though it may satisfy good-practice EA principles.

Indigenous and Local Knowledge Systems

A credible EA process must be based on the best available knowledge. A combination of knowledge sources can provide a more comprehensive understanding of the potential impacts of a proposed undertaking and how best to manage them. Knowledge in the EA process is generally of two forms: scientific knowledge and **Indigenous and local knowledge** (ILK). Berkes, Colding, and Folke (2000) define ILK as "a cumulative body of knowledge, practice and belief, evolving by adaptive processes and handed down through generations by cultural transmission, about the relationship of living beings (including humans) with one another and with their environment." ILK encompasses language, systems of classification, resource use practices, social interactions, and rituals, among other things. ILK is thus a way of life for knowledge-holders rather than simply a collection of information that can be codified for use in impact assessment (Council of Canadian Academies, 2019).

Cash et al. (2006) characterize ILK as functioning at different spatial and temporal scales from those of Western scientific knowledge in that it is often based on observations from the land, personal and community experience, and relationships developed and shared over time at a local scale. It often captures long periods of observation of environmental change (Stevenson, 1996). ILK can thus play an important role in complementing and validating scientific data (e.g., information about fish and wildlife populations, movements, and health) and providing an improved understanding of the implications of change (e.g., interpreting the significance of development impacts and understanding how impacts have accumulated over time). However, the Council of Canadian Academies (2019) cautions that ILK is not merely an add-on to Western science or used only when there are gaps in scientific data or understanding.

Validation of information *within* knowledge systems is generally understood, but validation *across* knowledge systems is a major challenge (Tengö, Brondizio, Elmqvist, Malmer, & Spierenburg, 2014). It is thus more appropriate in EA to consider **bridging knowledge systems**. The Council of Canadian Academies

(2019) explains that scientific knowledge is often more easily disseminated to different actors than ILK, which is often held by fewer people and may be more context-specific (Huntington, 2013). That said, when scientific knowledge is used in EA to understand baselines or to predict or assess change, it is usually necessary to downscale it to local contexts—such as the project-specific setting—where data are sometimes limited and local dynamics must be interpreted to make the science applicable. Thus, in principle, ILK and science can be complementary, and the consideration of multiple forms of knowledge can contribute to better decision-making (Dietz, 2017). This does not mean that different knowledge systems always align (Box 10.4). Although this may cause conflict at times, it can also stimulate a discussion that challenges the status quo and changes how resources are managed (Council of Canadian Academies, 2019). The objective is not to reduce science and ILK into a single, unified source, or to prove (or disprove) one with the other, but to carefully consider knowledge in its own context (Berkes, 2018) and in terms of its individual and collective contribution to more informed decisions about the nature, risks, and distribution of impacts and benefits associated with a proposed development action.

Box 10.4 Indigenous and Local Knowledge in Keeyask Hydroelectric Generation Project EIS

The Keeyask Generation Project was recently approved for development on the lower Nelson River, northern Manitoba. Keeyask is a 696-megawatt hydroelectric facility, slated for development, about 725 kilometres northeast of Winnipeg and less than 60 kilometres from the communities of Split Lake and Gillam. Keeyask was subject to EA under the Environment Act (Manitoba) and the Canadian Environmental Assessment Act. Manitoba Hydro submitted its EIS in 2012, with a project completion date proposed for 2022. The Keeyask project is a collaboration between Manitoba Hydro, a provincial Crown energy corporation, and four Treaty 5 First Nations—Tataskweyak Cree Nation, War Lake First Nation, York Factory First Nation, and Fox Lake First Nation—collectively referred to as the Keeyask Hydropower Limited Partnership.

The Keeyask EIS included ILK as a separate, independent analysis, with each First Nation partner leading their own community consultations and preparing their own assessments. Each First Nation partner, with support provided by Manitoba Hydro, engaged independent consultants to assist with their contribution to the project's assessment, including baseline and impact analysis. In its public review of the Keeyask EIS, the Manitoba Clean Environment Commission reported: "A new twist to the Keeyask process was that the First Nations produced their own environmental assessments, based on their own Cree worldview—not on western science" (2014, p. xii). In addition to the usual technical analyses, the Keeyask EIS included three self-contained reports presenting the assessment results of the First Nation partners.

continued

Many EAs claim to integrate ILK, but ILK is often treated as another form of technical assessment or applied outside of its cultural context. The Keeyask EA is an interesting example in which ILK was presented through an independent assessment process that, at times, conflicted with the science presented elsewhere in the EIS. For example, Manitoba Hydro reported that elevated levels of suspended sediment in the project's hydraulic zone of influence were unlikely to have a measurable effect on fish and wildlife. However, the First Nations' ILK-based assessments reported higher levels and longer-term effects from suspended sediment. Manitoba Hydro's technical analysis identified no adverse effects of the project on fish populations; ILK conclusions indicated a larger spatial and temporal extent of effects and a decline in fish populations.

Although the project and the quality of the EA were highly contested on multiple fronts, Keeyask is an example of how an EIS can present ILK and science without forced integration. The provincial panel commissioned to review the assessment, however, noted that the approach did present a challenge to the panel, since only one of five panel members was Indigenous (Manitoba Clean Environment Commission, 2014).

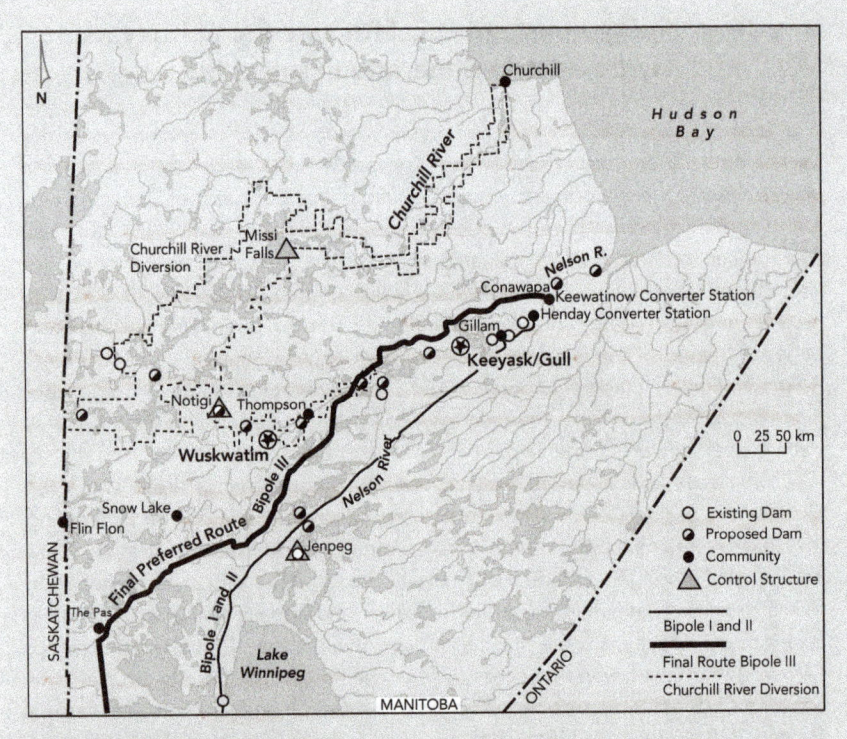

Hydroelectric developments (major generating stations and transmission lines) in northern Manitoba, Canada.

Source: Based on Noble, 2017

Enduring Challenges to Indigenous Engagement

Historically, the participation of Indigenous peoples in EA is often triggered by conflict because of the failure of proponents or governments to provide meaningful opportunities for engagement. The result has been increasing litigation, and often distrust of the EA process, because of the limited influence of Indigenous peoples on regulatory decisions about project proposals that directly affect their traditional lands and livelihoods (Noble, 2016). Voicing concerns about Pacific Northwest's proposed liquefied natural gas terminal north of Prince Rupert, British Columbia, for example, the Lax Kw'alaams First Nation explained they are "open to development ... but not the way the project is currently constituted," identifying concerns raised about the potential environmental impacts of the project that had not been adequately resolved. For many project proponents, in contrast, the primary concerns are usually about ensuring a cost-efficient review process and timely decisions about their project application (Noble, 2016).

The Centre for Indigenous Environmental Resources suggests that EA "has not met the needs or expectations of Aboriginal peoples for an inclusive process that respects their unique place within the legal and political fabric of Canada" (CIER, 2009, p. 3). Land describes the current situation as an "escalating drift ... towards more litigation arising from conflicts over Aboriginal consultation ... and this will inevitably lead to more project delays and more economic risks and losses for the Crown, industry and Aboriginal groups" (2014, p. 22). This has certainly been reflected by a growing number of court challenges (Assembly of First Nations, 2011; Udofia, Noble, & Poelzer, 2017), Indigenous protests, and outright project rejection. In the case of the Kemess North copper-gold mine, north-central British Columbia, the First Nation felt marginalized during the proponent's EA and participated in the federal review panel hearings only by way of protest as a means of expressing their concerns about the impacts of the project on their cultural and spiritual values. The federal review panel listened and recommended that the project be rejected. In the Northwest Territories, Ur-Energy Screech Lake's uranium exploration project was one of several projects rejected by the Mackenzie Valley Environmental Impact Review Board (MVEIRB) after hearing concerns from the affected Indigenous communities about impacts to landscapes that had cultural and spiritual significance (Box 10.5).

Opportunities and provisions for Indigenous engagement in EA have improved substantially over the past decade; however, there remain many challenges to meaningful engagement. The following list may not be comprehensive of all challenges and contexts, but it does capture some of the enduring problems and concerns.

Late Timing of Engagement

The Canadian federal Impact Assessment Act (2019) requires early engagement of Indigenous peoples in the pre-project planning phase. However, traditionally, the timing of Indigenous engagement is late in the project planning and development cycle. As a result, potentially adverse impacts on Indigenous lands and resources are either unaccounted for or not properly compensated (Noble & Udofia, 2015). Under many EA systems, participation is not *required* until late in the project planning

Box 10.5   Ur-Energy Inc. Screech Lake Uranium Exploration Project

In 2006, Ur-Energy Inc. applied for a uranium exploration licence in the Upper Thelon watershed, Northwest Territories. The project would involve exploration activity and the construction of a temporary work camp. The exploration activity would occur 300 kilometres east of Łutsël K'e Dene First Nation in the Akaitcho. The Thelon River, a designated Canadian Heritage River, runs through the watershed. The region is also of significant spiritual and cultural importance to the Łutsël K'e (MVEIRB, 2007). The project was referred to the Mackenzie Valley Environmental Impact Review Board (MVEIRB) for an EA under the Mackenzie Valley Resource Management Act. In 2007, during a public hearing in Łutsël K'e, numerous social and cultural concerns were raised about the project, including the importance of the region to Indigenous peoples—namely, because of its spiritual significance and importance for wildlife. In its report, the MVEIRB noted that Łutsël K'e people "described their distress at the prospect of industrial development of an area they wish to pass on to their children as they inherited it from previous generations" (MVEIRB, 2007, p. 1). Concerns were also raised during the hearings about the potential for exploration activity to pave the way for future development in the Akaitcho region.

The project proponent indicated that exploration activity would be localized, would last for only a short time period, and was unlikely to pose any significant adverse effects on traditional land use or to wildlife (MVEIRB, 2007). In the MVEIRB's decision statement, however, it concluded that the project, when considering other present land uses and the prospects of future development in the Upper Thelon Basin, "will cause adverse cultural impacts of a cumulative nature to areas of very high spiritual importance to aboriginal peoples" (MVEIRB, 2007, p. ii). The MVEIRB recommended that the project be rejected, stating:

> The Review Board considered the evidence of cultural impacts from the people of Łutsël K'e and other aboriginal groups. This included traditional knowledge shared by the Elders. In the view of the Review Board, the Upper Thelon area is of high spiritual and cultural importance to the Akaitcho and other aboriginal peoples. They see industrial development, including this proposed development and others, as a desecration of a spiritual area of intrinsic value. The Review Board is of the view that although the proposed development is physically small, the potential cultural impacts are not. (MVEIRB, 2007, p. 1)

The MVEIRB's recommendation to reject the project was accepted by the minister of Indian and Northern Affairs. Ehrlich (2010) describes the case as an illustration of EA going beyond "bones and stones" to incorporate non-tangible cultural values and concerns.

Source: Based on Noble, 2016

process or even not until after a draft assessment report is complete, meaning that many of the most important decisions about project design, including the identification and mitigation of potential impacts on Indigenous lands and resources, have already been made. The result is an adversarial EA process characterized by distrust.

In the case of Exxon Mobil's Sable offshore gas project, public hearings conducted by the federal regulator, the National Energy Board, were characterized as discouraging participation, fostering an environment where not all evidence was given equal consideration, and limiting the opportunity for open dialogue about potential solutions to the project issues raised (Fitzpatrick & Sinclair, 2003). Seeking Indigenous engagement late in the process can also be costly to the proponent, as learned by Platinex, a junior mining exploration company in northern Ontario. Planitex was issued drilling rights by the Ontario Ministry of Northern Development on land to which the Kitchenuhmaykoosib Inninuwug First Nation claim treaty rights, without prior consultation with the First Nation. No exploration agreement was reached between Planitex and the Kitchenuhmaykoosib Inninuwug First Nation. The result was litigation, project delays, road blockades by the First Nation, and the jailing of several community members and the First Nation Chief (Noble & Udofia, 2015).

Misaligned Expectations about the Scope and Intent of EA

EA is sometimes approached by Indigenous communities as a means to address right-based issues (Noble & Udofia, 2015), with expectations about engagement processes and outcomes shaped by treaty rights. As a result, some of the issues brought to the table during EA, such as whether certain resource sectors or types of development should even occur or the types of land uses deemed appropriate, are beyond the scope of project-based reviews and not within the mandate of project proponents to address (Noble, 2017). Rightfully, Indigenous peoples expect that potential impacts to recognized rights are adequately considered during the project review process, but many of the issues raised are not EA issues per se; rather, they are land-use planning, policy, legal, and even constitutionally based issues concerning land title and the rights of Indigenous peoples. Noble and Udofia (2015) explain that these are not issues that EA is well equipped to address and resolving them is often beyond the scope of a single project, such as a mine operation or wind generation facility.

Limited Financial and Human Resource Capacity

The size and technical complexity of EAs, and the complexity of the regulatory process, can pose significant challenges to Indigenous engagement. Many Indigenous communities, especially those located in remote locations, do not have the human resources or financial or technical capacity to engage in EA. This includes the resources needed to carry out **traditional use studies** to demonstrate the potential for impact on land uses, to review and comment on project applications (e.g., technical design, mitigation measures), and to intervene in regulatory or public hearing processes (Noble & Udofia, 2015).

In the case of De Beers Canada's proposed Victor diamond mine, a remote fly-in fly-out mine located in the James Bay Lowlands of northern Ontario, the First Nation lacked the resources to secure the technical and legal expertise necessary to participate in the EA process, including data collection, analysis, and regulatory reviews (Whitelaw, McCarthy, & Tsuji, 2009). Similar capacity constraints were revealed in Spectra Energy's Westcoast Connector Gas Transmission Project, 2014—a proposal to transport LNG from northeastern British Columbia to the northwest coast. Spectra's EA report indicated that of the 24 potentially affected First Nations located along the pipeline corridor, 17 said that that they had insufficient technical and financial resources to participate in the project review (Spectra Energy, 2014).

Participation Fatigue in Resource-Intense Development Regions

A much less discussed challenge is participation fatigue. Noble and Udofia (2015) report that the capacity constraints to engage in EA are exacerbated in regions subject to major resource developments, such mining or energy resource–rich regions, where increasing numbers of EA applications and project developments mean ongoing demands for consultation as proponents and governments meet their consultation obligations. In the Inuvialuit Settlement Region of Canada's western Arctic, for example, concerns by the Inuvialuit about the number of energy exploration projects and demands on community time and resources for consultation concerning energy development projects highlight the importance of building community capacity to engage in EA—and the need to unload the burdens of participation in multi-project contexts to more strategic assessment or regional land-use planning processes (Fidler & Noble, 2013). In the Mackenzie Valley of the Northwest Territories, the Mackenzie Valley Environmental Impact Review Board (2008) similarly reports that Indigenous communities are struggling to participate in EA under increasing workloads and responsibilities to coordinate engagement processes with land users, elders, and community members.

Toward Meaningful Indigenous Engagement in EA

Meaningful engagement means that those potentially affected by development, or who have a vested interest in development, contribute to the planning, assessment, and decision process. Meaningful engagement means opportunities for the exchange of information, opinions, interests, and values and that those responsible for engagement—namely, project proponents and governments—are open to modifying project designs and to working with communities to change project plans or even abandon proposals (Noble, 2016). There are several foundational principles to ensuring meaningful Indigenous engagement in EA; of particular importance are the following practices.

Governments Take Leadership to Address Strategic Issues Pre-EA

When government agencies engage communities early in the pre-EA stages, there is an opportunity for government and communities to identify any issues associated with development that need to be addressed, such as rights-based issues and the completion of traditional use studies that cannot be meaningfully addressed through a project-focused EA. Commencing traditional use studies only when approached by a project proponent, for example, results in participation that is focused more on information-gathering than genuine engagement (Noble & Udofia, 2015). A well-prepared community, with land-use values and priorities already documented, can go a long way in facilitating meaningful engagement.

Addressing such issues *before* the EA process commences can help to ensure a more efficient, meaningful, and focused EA that addresses the project's potential impacts and devises the best ways to manage them given the expectations of the affected community. This means that governments also take responsibility for shaping expectations about engagement—including clarifying the roles of government and the proponent in consultation processes. Government must consult potentially affected Indigenous communities *before* land is opened for development (Cooney, 2013) and before EA applications are accepted from project proponents. The federal government's Major Projects Management Office (2012) explains: "[e]xperience has shown that engagement with Aboriginal groups early in the planning and design phases of a proposed project can benefit all concerned. Conversely, there have been instances where failure to participate in a process of early engagement with Aboriginal people has led to avoidable project delays and increased costs to proponents."

Proponents Engage Communities in the EA Planning Process

Indigenous engagement is often initiated *after* a project proponent initiates its EA process or once the draft EIS is complete. Good practice means that Indigenous communities are engaged early in the EA planning process and in jointly setting the scoping requirements or terms of reference for the EA. The EA terms of reference establish the objectives and expectations for an assessment, including specific engagement strategies. The terms of reference are often drafted by the project proponent and submitted to government for formal public consultation. However, when terms of reference are developed in partnership with potentially affected Indigenous communities, expectations are established before the start of the EA process and Indigenous values can inform EA from the outset. This is not common practice, but it is not unprecedented. In 2004, Polaris Minerals submitted an EA to develop the Orca project sand and gravel mine located in 'Namgis First Nation territory, northwest Vancouver Island, British Columbia. Polaris engaged the 'Namgis in the creation of the EA terms of reference, and the 'Namgis had a say in the choice of EA consultants, thus ensuring an EA process that was acceptable to the community. Plate, Foy, and Krehbiel (2009) report that the Orca project "set a standard for meaningful participation of a First Nation in an environmental assessment . . . the completion of the EA process was swift and mutually supported."

Communities Have Access to Capacity-Building and Financial Support

Engagement in EA can be meaningful to both the Indigenous community and the project proponent when the community is prepared to engage and has the capacity to engage. Indigenous communities must have the resources to engage their membership during the EA process and engage with the dominant scientific discourse of EA (O'Faircheallaigh, 2007). Unless financial and technical resources are provided, such participation is not likely to occur (Udofia, Noble, & Poelzer, 2017). Government agencies, and project proponents, sometimes provide Indigenous communities with participant funding to engage in EA—at different stages of the process. The National Energy Board's program was intended to support **intervenors** by offsetting the costs of their participation in a formal project hearing. For the Trans Mountain Pipeline expansion project public hearing, a pipeline-twinning project from Strathcona County, Alberta, to Burnaby, British Columbia, for example, Noble (2016) reports that a total of $24 million in intervenor funding was requested by 95 applicants. A total of $3 million was awarded to 71 applicants, and about 80 per cent was awarded to Indigenous groups along the pipeline route to attend project hearings and present their concerns about potential impacts. The Fair Mining Collaborative (2015), a charitable foundation that provides technical and practical assistance to First Nations and communities on issues related to mining, reports that government-provided participant funding is not sufficient to ensure meaningful participation. They argue that support is needed to ensure that Indigenous communities can independently assess a project's potential impacts on their lands and to engage project planning, impact management, and monitoring activities.

Communities Have Access to Information

Access to information about a proposed project, including information about its location, design, baseline conditions and impacts, and any monitoring results, is necessary for informed engagement. This is sometimes provided through a centralized online registry, making EA information available to the public, Indigenous groups, and other interested parties. A 2009 report of the Commissioner of the Environment and Sustainable Development found that for federal EAs, project files were complete for all the comprehensive studies and panel reviews contained in a sample of studies and reviews. Requirements also exist under the Impact Assessment Act (2019) for the establishment of a Canadian Impact Assessment Registry that would contain such material as public notices for participation in an assessment, a record of the factors considered in setting the scope of an assessment, assessment reports and any information taken into consideration during the assessment, public comments, and results of any follow-up programs, among others. Hanna and Noble (2011) caution that it is one thing to have information available in the form of a registry but it is another to have it available in a language, format, and style that are easily accessible and understood by diverse audiences (Box 10.6).

Box 10.6 Researchers' Experiences Working with Environmental Assessment Registries

Findlay (2010): In a sample of 30 road EAs for which decisions had been posted, obtaining documents was neither easy nor often successful. In 23 instances, requests for reports were met with no reply or "endless redirections." In only one instance was the requested material available through the Internet site, and in the remainder the wait was one to two weeks. Reports were not uncommonly copyrighted by consultants, and thus several approvals were required to obtain material. Older EAs were often not available in electronic form.

Sinclair and Diduck (2009): Cognitive inaccessibility is a persistent problem in EA registries. EA documentation is characterized by "overly technical language and general lack of readability"—an obstacle for most members of the public.

Ball, Noble, and Dubé (2013): When sampling federal and provincial (Saskatchewan and Alberta) EISs from the South Saskatchewan watershed, approximately 25 per cent of the EISs initially selected for review could not be accessed because of the incompleteness of public registries, limited logistical support from some responsible authorities to locate the documents, and the reluctance of some proponents to share documents. It required three months of active soliciting to collect the sample of "publically available" EISs. Among the EISs that were reviewed, many referred to technical studies and to data that were not included in the registry.

Engagement Continues after the EA Approval

Indigenous engagement, and financial support for engagement, is typically provided only up to the point of project approval (Noble & Udofia, 2015). Meaningful engagement means that Indigenous communities are involved beyond project reviews and play a meaningful role post–EA approval. Relationships established during the EA process, and recommendations emerging from EA authorizations, can result in opportunities for meaningful participation in a project's management beyond the approval stage (Plate et al., 2009). Noble and Udofia (2015), however, report that most government-led participant funding programs do not support Indigenous community participation beyond the regulatory decision-making process—even though continued support is essential for meaningful participation (O'Faircheallaigh, 2007).

The ongoing participation of Indigenous communities in environmental monitoring and follow-up programs is increasingly required as a condition of EA approval. In the case of Northcliff Resources Ltd Sisson tungsten-molybdenum project, an open-pit mining operation near Napadogan, approximately 60 kilometres northwest of Fredericton, the project was approved by the government of New Brunswick subject to 40 conditions. One of these conditions was the establishment of an environmental and socio-economic effects monitoring program and that the proponent provide the First Nations with capacity funding to participate in the program's development, planning, and implementation (New Brunswick Department of Environment and Local Government, 2016).

Opportunities for Indigenous-Led EA

Indigenous engagement can also mean a much more substantive role in EA: assuming leadership of the EA process. **Indigenous-led impact assessment** is defined by Gibson, Hoogeveen, MacDonald, and The Firelight Group (2018) as:

A process that is completed prior to any approvals or consent being provided for a proposed project, which is designed and conducted with meaningful input and an adequate degree of control by Indigenous parties—on their own terms and with their approval. The Indigenous parties are involved in the scoping, data collection, assessment, management planning, and decision-making about a project.

There are examples of Indigenous communities leading EA processes—although most often these examples are in northern Canada where EA systems are established under comprehensive land-claims or self-government agreements, such as the review by Tłı̨chǫ of Fortune Minerals' NICO poly-metallic mine project in the Northwest Territories or the review by Glencore and Inuit of the Slvumut project, a proposed expansion to the existing Raglan nickel mine, Quebec. Gibson et al. (2018) explain that Indigenous-led impact assessments can vary from existing legislated processes, partly because they are based on very different goals and aspirations; however, as illustrated by the Squamish Nation's independent EA of the Woodfibre LNG project, this is not always the case (see Environmental Assessment in Action: Squamish Nation EA of the Woodfibre Liquefied Natural Gas Facility).

Environmental Assessment in Action

Squamish Nation EA of the Woodfibre Liquefied Natural Gas Facility

In 2013, Woodfibre LNG Limited, a subsidiary of Royal Golden Eagle, submitted an EA application to the province of British Columbia to develop a liquefied natural gas (LNG) export facility approximately seven kilometres from Squamish. The project would include a natural gas liquefaction facility and an LNG transfer facility from which product would be shipped to international markets. The facility would have a storage capacity of 250,000 cubic metres and would produce up to 2.1 million tonnes per year of LNG, with an expected project lifespan of approximately 25 years (Woodfibre LNG Ltd, 2015).

The proposed Woodfibre site is on the traditional territory of the Squamish Nation. The Squamish are Coast Salish people with approximately 3600 members, of whom more than 60 per cent live on Squamish First Nation urban reserves in the Vancouver area and in the municipality of Squamish. The Squamish initiated treaty negotiations with the province in 1993, but no framework agreement has yet been established. In 2001, the Squamish Nation developed the Xay Temíxw land-use plan, identifying forest stewardship, sensitive areas, restoration areas, and wild spirit places land-use zones. The Squamish traditional territory encompasses 6732 square kilometres.

The Woodfibre LNG project was subject to EA under the British Columbia Environmental Assessment Act and the Canadian Environmental Assessment Act, 2012. Woodfibre LNG Ltd started engaging with Indigenous communities in 2013, including the Tsleil-Waututh First Nation, the Musqueam First Nation, and the Squamish First Nation. The Squamish Nation passed a resolution to conduct its own Squamish Nation–led EA and communicated to the proponent that the Squamish Nation would not negotiate economic benefits until all potential environmental and cultural impacts had been addressed (Squamish Nation, 2015).

Starting in May 2013, the Squamish Nation–led EA operated independently of, and parallel to, the provincial and federal EA processes and was the first independent EA led by a First Nation in British Columbia (Squamish Nation, 2015). The Squamish Nation–led EA included a public education campaign about the project, meetings with communities to hear concerns about the project, and use of an independent consultant to review the project's technical aspects and specific Squamish Nation concerns. The EA report is described by the Squamish as a "science and fact-based . . . rigorous investigation of all aspects of the proposal" (Squamish Nation, 2015). There was no obligation for Woodfibre to participate in the Squamish Nation EA process—it did so voluntarily.

There were 25 project conditions identified in the Squamish EA report, 13 of which applied to the proponent and the remainder to the province and to FortisBC—which will construct the pipeline and compressor station for the Woodfibre facility. Most of the conditions were related to project design and operations, including the need to: further assess the proposed seawater cooling method and impacts on marine environments; reroute project activities to avoid disturbances to the Skwelwil'em Wildlife Management Area; avoid fuelling tankers with bunker fuel in Squamish territory; work with the Squamish Nation to develop an Emergency Response Plan in the event of a spill the Squamish Valley; and relocate the project's compressor station that was initially proposed near Squamish Nation members. The Squamish Nation demanded that the proponent sign a legally binding Squamish Nation Environmental Certificate to ensure that all conditions would be met.

The Squamish Nation Council approved the EA for the Woodfibre LNG project in October 2015, subject to the conditions identified. Woodfibre subsequently requested a suspension of the regulatory EA process in order to address the Squamish Nation's EA report. Woodfibre accepted the Squamish Nation's EA process and conditions and filed an addendum to its project application, including, among other things, alternative design options to avoid disturbances in the Squamish Estuary and Skwelwil'em Squamish Estuary Wildlife Management Area and relocation of the proposed compressor station. The project received approval by the province in late 2015 and received federal approval in early 2016.

The Squamish viewed the regulatory EA process as inadequate—not addressing Indigenous rights and not providing solutions to manage the impacts that it identified. The Squamish Nation reported that its own Nation-led EA not only identified important environmental issues but also determined how they should be addressed (Squamish Nation, 2015).

Source: Noble, 2017

Key Terms

bridging knowledge systems
duty to consult
Indigenous and local knowledge
Indigenous-led impact assessment
intervenors
social licence
traditional use studies
United Nations Declaration on the Rights of Indigenous Peoples

Review Questions and Exercises

1. What is the United Nations Declaration on the Rights of Indigenous Peoples, and how does it differ from the duty to consult?
2. What is the role of early engagement in a proponent's social licence to operate?
3. Discuss some of the advantages and disadvantages of government discharging its duty to consult through the EA participation and engagement initiatives of project proponents.
4. What are the key differences and complementarities between Indigenous and Western science knowledge systems?

References

Assembly of First Nations. (2011). *Assembly of First Nations Submission to the House of Commons Standing Committee of the Environment and Sustainable Development, Canadian Environmental Assessment Act Seven-Year Review*.

Bains, R., & Ishkanian, K. (2016). *The duty to consult with Aboriginal peoples: a patchwork of Canadian policies*. Vancouver, BC: Fraser Institute. https://www.fraserinstitute.org/sites/default/files/duty-to-consult-with-aboriginal-peoples-a-patchwork-of-canadian-policies.pdf.

Ball, M., Noble, B.F., & Dubé, M. (2013). Valued ecosystem components for watershed cumulative effects: an analysis of environmental impact statements in the South Saskatchewan River watershed, Canada. *Integrated Environmental Assessment and Management* 9(3), 469–79.

Bergner, K. (2005). *The Crown's duty to consult and accommodate*. Paper presented at the Canadian Institute's Fifth Annual Advanced Administrative Law and Practice, Ottawa, ON.

Berkes, F. (2018). *Sacred ecology*. 4th ed. New York, NY: Routledge.

Berkes, F., Colding, J., & Folke, C. (2000). Rediscovery of traditional ecological knowledge as adaptive management. *Ecological Applications* 10(5), 1251–62.

Brackstone, P. (2002). *Duty to consult with First Nations*. Victoria, BC: Environmental Law Centre.

Cash, D., Adger, W., Berkes, F., Garden, P., Lebel, L., Olsson, P., . . . & Young, O. (2006). Scale and cross-scale dynamics: governance and information in a multilevel world. *Ecology and Society* 11(2), 8.

CIER (Centre for Indigenous Environmental Resources). (2009). *Meaningful involvement of Aboriginal peoples in environmental assessment*. Winnipeg, MB: CIER.

Chrétien, A., & Murphy, B. (2009). *Duty to consult, environmental impacts, and Metis Indigenous knowledge*. Ottawa, ON: Institute on Governance. http://iog.ca/wp-content/up-loads/2013/01/April2009_DutytoConsult-Chretien_Murphy.pdf.

Cooney, J. (2013). *Mining, economic development and Indigenous peoples: getting the governance equation right*. Montreal, QC: Institute for the Study of International Development.

Council of Canadian Academies. (2019). *Greater than the sum of its parts: toward integrated natural resource management in Canada*. Ottawa, ON: The Expert Panel on the State of Knowledge and Practice of Integrated Approaches to Natural Resource Management in Canada.

Craik, A. (2016). Process and reconciliation: integrating the duty to consult with environmental assessment. Osgoode Legal Studies Research Paper Series, paper 122. http://digitalcommons.osgoode.yorku.ca/olsrps.

Dietz, T. (2017). Science, values, and conflict in the national parks. In S.R. Beissinger, D.D. Ackerly, H. Doremus, & G.E. Machlis (eds), *Science, conservation, and national parks*. Chicago, IL: University of Chicago Press.

Ehrlich, A. (2010). Cumulative cultural effects and reasonably foreseeable future developments in the Upper Thelon Basin, Canada. *Impact Assessment and Project Appraisal* 28(4), 279–86.

Fair Mining Collaborative. (2015). Fair mining practices: a new code for BC. Section 6—Environmental assessment for mining activities. Comox, BC: Fair Mining Collaborative. http://www.fairmining.ca/code/environmental-assessment-for-mining-activities-2.

Fidler, C., & Noble, B. (2013). Advancing regional strategic environmental assessment in Canada's western Arctic: implementation opportunities and challenges. *Journal of Environmental Assessment Policy and Management* 15(1), 1–27.

Findlay, S. (2010). The CEAA registry as a tool for evaluating CEAA effectiveness. Presentation to the Ontario Association for Impact Assessment annual general meeting and conference. Ottawa, ON: OAIA.

Fitzpatrick, P., & Sinclair, J. (2003). Learning through public involvement in environmental assessment hearings. *Journal of Environmental Management* 67(2), 61–174.

Gibson, G., Hoogeveen, D., MacDonald, A., & The Firelight Group. (2018). *Impact assessment in the Arctic: emerging practices of Indigenous-led review*. Report prepared for Gwich'in Council International (GCI). Vancouver, BC: The Firelight Group.

Hanna, K., & Noble, B.F. (2011). The Canadian environmental assessment registry: promise and reality. *UVP-report* 25(44), 222–5.

Huntington, H.P. (2013). *Traditional knowledge and resource development*. Whitehorse, YT: Resources and Sustainable Development in the Arctic.

Isaac, T., & Knox, A. (2003). The Crown's duty to consult Aboriginal people. *Alberta Law Review* 41, 75.

Land, L. (2014). *Creating the perfect storm for conflicts over Aboriginal rights: critical new developments in the law of Aboriginal consultation*. Toronto, ON: The Commons Institute.

MVEIRB (Mackenzie Valley Environmental Impact Review Board). (2007). *Report of environmental assessment and reasons for decision on Ur-Energy Inc. Screech Lake uranium exploration project (EA 607-003)*. Whitehorse, YT: MVEIRB.

MVEIRB (Mackenzie Valley Environmental Impact Review Board). (2008). *Environmental impact assessment (EIA) practitioner's workshop report*. Whitehorse, YT: MVEIRB. www.reviewboard.ca.

Major Projects Management Office. (2012). *Early Aboriginal engagement: a guide for proponents of major resource projects*. Ottawa, ON, Major Projects Management Office.

Manitoba Clean Environment Commission. (2014). *Report on public hearing: Keeyask generation project*. Winnipeg, MB: Manitoba Clean Environment Commission.

New Brunswick Department of Environment and Local Government. (2016). *Summary of public and First Nations participation, environmental impact assessment. Proposal by Sisson Mines Ltd to construct and operate an open-pit tungsten and molybdenum mine near Napadogan, New Brunswick*. Fredericton, NB: Government of New Brunswick.

Newman, D.G. (2014). *Revisiting the duty to consult Aboriginal peoples.* Saskatoon, SK: Purich Publishing.

Noble, B.F. (2016). *Learning to listen: snapshots of Aboriginal participation in environmental assessment.* Ottawa ON: Macdonald-Laurier Institute

Noble, B.F. (2017). *Getting the big picture: how regional assessment can pave the way for more inclusive and effective environmental assessment.* Ottawa, ON: Macdonald-Laurier Institute.

Noble, B.F., & Udofia, A. (2015). *Protectors of the land: toward an EA process that works for Aboriginal communities and developers.* Ottawa ON: Macdonald-Laurier Institute.

O'Faircheallaigh, C. (2007). Environmental agreements, EIA follow-up and Aboriginal participation in environmental management: the Canadian experience. *Environmental Impact Assessment Review* 27(4), 319–42.

Olszynski, M. (2016). The duty to consult and accommodate: an overview and discussion. In *Seizing six opportunities for more clarity in the duty to consult and accommodate process.* Ottawa, ON: Canadian Chamber of Commerce.

Plate, E., Foy, M., & Krehbiel, R. (2009). *Best practices for First Nation involvement in environmental assessment reviews of development projects in British Columbia.* West Vancouver, BC: New Relationship Trust.

Potes, V., Passelac-Ross, M., & Bankes, N. (2006). Oil and gas development and the Crown's duty to consult: a critical analysis of Alberta's consultation policy and practice. Paper no. 14 of the Alberta Energy Futures Project. Calgary, AB: Institute for Sustainable Energy, Environment and Economy, University of Calgary.

Sinclair, J., & Diduck, A. (2009). Public participation in Canadian environmental assessment: enduring challenges and future direction. In K. Hanna (ed.), *Environmental assessment: practice and participation*, 2nd edn (pp. 58–82). Don Mills, ON: Oxford University Press.

Spectra Energy. (2014). Environmental assessment certificate application for the Westcoast Connector gas transmission project. Prepared under the British Columbia Environmental Assessment Act for Westcoast Connector Gas Transmission Ltd. Prepared by TERA Environmental Consultants, Victoria, BC.

Squamish Nation. (2015). PGL's environmental report on Woodfibre LNG proposal. *Squamish Nation Update*, Issue 3. Squamish Nation. http://www.squamish.net/wp-content/up-loads/2015/07/SN-WoodfibreUpdate-Summary-03.pdf.

Stevenson, M.G. (1996). Indigenous knowledge in environmental assessment. *Arctic* 49(3), 278–91.

Tengö, M., Brondizio, E., Elmqvist, T., Malmer, P., & Spierenburg, M. (2014). Connecting diverse knowledge systems for enhanced ecosystem governance—the multiple evidence base approach. *Ambio.* doi: 10.1007/s13280-014-0501-3.

Udofia, A., Noble, B., & Poelzer, G. (2017). Meaningful and efficient? Exploring the challenges to Aboriginal participation in environmental assessment. *Environmental Impact Assessment Review* 65, 164–74

United Nations General Assembly. (2008). United Nations Declaration on the Rights of Indigenous Peoples. http://www.un.org/esa/socdev/unpfii/documents/DRIPS_en.pdf.

Whitelaw, G., McCarthy, A., & Tsuji, J. (2009). The Victor diamond mine environmental assessment process: a critical First Nation perspective. *Impact Assessment and Project Appraisal* 27(3), 205–15.

Woodfibre LNG Ltd. (2015). Application for an environmental assessment certificate. Government of British Columbia. https://a100.gov.bc.ca/appsdata/epic/html/deploy/epic_proj-ect_home_408.html.

11 Cumulative Effects Assessment

Cumulative Effects

Describing the loss of coastal wetlands along the east coast of the United States between 1950 and 1970, Odum (1982, p. 728) explains:

> No one purposely planned to destroy almost 50% of the existing marshland along the coasts of Connecticut and Massachusetts. . . . However, through hundreds of little decisions and the conversion of hundreds of small tracts of marshland, a major decision in favour of extensive wetlands conversion was made without ever addressing the issue directly.

It is not possible to determine the true significance of a project's effects without the consideration of cumulative environmental effects. Each additional disturbance or impact, regardless of its magnitude, can represent a high marginal cost to the environment. Cumulative effects are often described as "progressive nibbling," "death by a thousand cuts," or the "tyranny of small decisions." In other words, cumulative effects are the culmination of effects—many of which can be individually small and seemingly insignificant, such as seismic lines, pipelines, water withdrawals, or the incremental filling of wetlands. Such characterizations are based on the notion that a significant adverse effect can result over space or over time because of the culmination of seemingly small and insignificant actions. For each action, the effects are deemed marginal or relatively insignificant when compared to other types or scales of change or disturbances. But over time, such seemingly insignificant effects can result in significant cumulative environmental change (Gunn & Noble, 2012).

Assessing Cumulative Environmental Effects

The foundational principles of CEA were established in the 1980s through the pioneering work of Gordon Beanlands and Peter Duinker (1983) and expanded throughout the 1990s by scholars from diverse fields. Scholars have advanced cumulative effects principles, frameworks, and applications in wildlife management (Gunn, Russell, & Greig, 2014), aquatic systems (Squires & Dubé, 2012), and land-use planning (Noble, 2008), to name a few. Multiple supporting tools and approaches have also been developed, including scenario-based modelling (Francis & Hamm, 2011) and risk-based methods (Dubé & Munkittrick, 2001) (Box 11.1).

Box 11.1 Cumulative Effects on Social Systems—the Orphan of CEA

Notwithstanding advances in CEA practice, the focus of attention has been on biophysical systems. This is not surprising, since the biophysical environment has been the dominant focus of impact assessment since it was first introduced in the early 1970s. Hackett, Liu, and Noble (2018), however, explain that development projects can generate many important positive impacts for communities but they are also responsible for many significant adverse impacts—including the cumulative loss of traditional lands and livelihoods and cultural practices. Accounting for the historical build-up of effects on social systems is an essential component of sound resource development, and good CEA, and a prerequisite to understanding the significance of a project's actions on the current and future well-being of communities. Hackett, Liu, and Noble argue that focusing on legacy effects can improve our understanding of why adverse cumulative impacts to communities may arise because of newly proposed projects and how best to manage them.

The basic concept of "cumulative effects" was introduced in Chapter 6. Attention focused on the importance of identifying a project's *additional* effects to valued components in the project's local to regional environment. This is the standard approach under project EA systems, but it falls significantly short of assessing, understanding, and effectively managing cumulative effects. CEA is the systematic process of identifying, analyzing, and evaluating cumulative effects—that is, identifying effects and pathways in order to avoid, wherever possible, the potential triggers or sources that lead to cumulative environmental change (Spaling & Smit, 1994). Good CEA is thus focused on the condition of environmental receptors and whether the *total effects* via all stressors are acceptable. This requires what Greig (2019) refers to as a **valued component–centred** approach versus a **project- or activity-centred** approach (Figure 11.1). The project- or activity-centred approach focuses on the effects of a project or activity on each of the valued components of concern. Multiple projects or activities may be affecting the same valued components, but the assessment is conducted one project at a time. It is not possible to understand cumulative effects. The valued component–centred approach focuses on the valued component of concern and considers the impacts of all sources of disturbance or stress—whether caused by individual projects subject to EA, projects exempt from EA, other types of land uses and human activities, or natural drivers. This approach provides a true picture of cumulative effects, though it can be challenging (and at times impossible) to attribute a cumulative effect to individual sources.

Challenges to Project EA

A consistent theme in EA is the challenges to assessing and managing cumulative effects under regulatory, project EA. Some would even argue that although project EA can identify a project's additional stress to the environment (e.g., additional

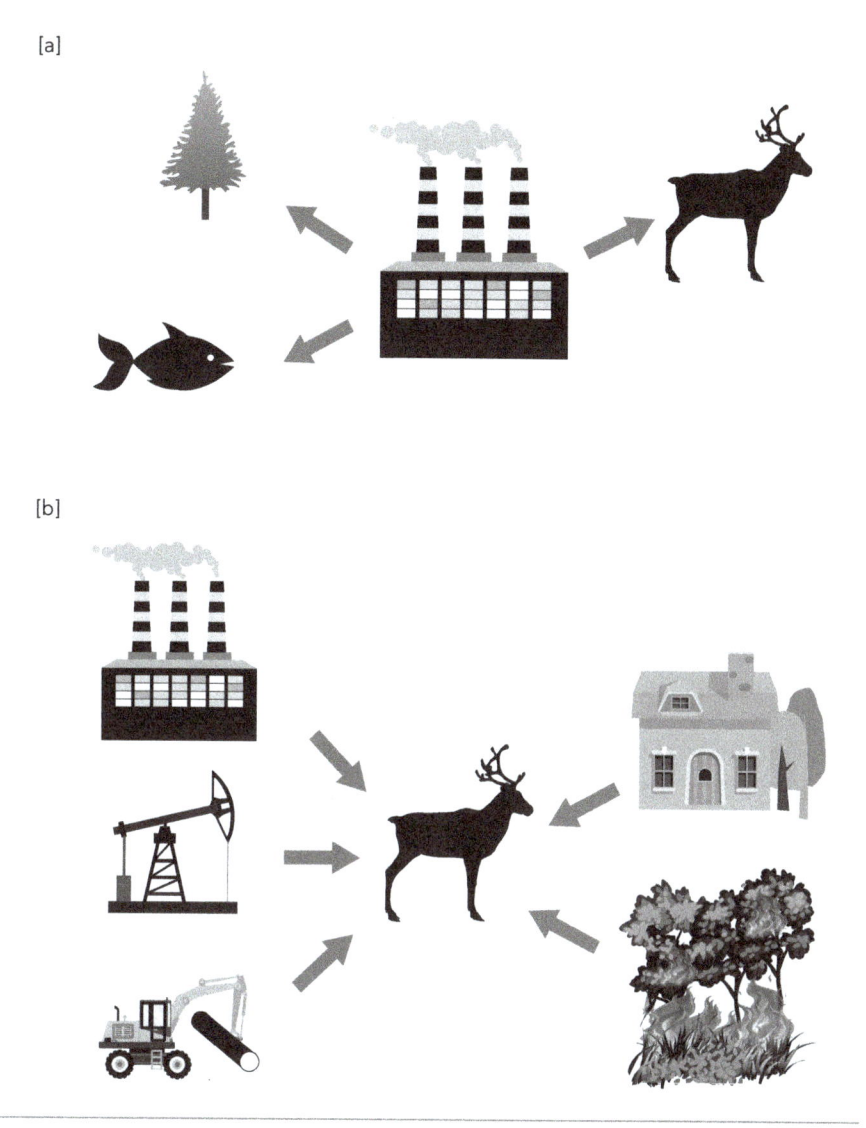

Figure 11.1 Project- or activity-centred [a] and valued component–centred [b] approaches to effects

discharge to a lake system, additional linear disturbance), it's simply not possible to assess cumulative effects to environmental systems within the scope, scale, and objectives of a single project. Parkins (2011) suggests that "thinking cumulatively and regionally" simply "does not emerge naturally from a project-based perspective." In a review of the state of cumulative effects assessment in Canada, Duinker and Greig (2006) conclude that "continuing the kinds and qualities of CEA currently undertaken may be doing more harm than good." These "kinds and qualities" that Duinker and Greig are referring to are CEAs conducted solely within the scope of a single development project, where the focus is on minimizing the project's additional impact contribution to an acceptable level and coming to a conclusion of "insignificant impacts," versus considering the total impacts of all activities on the valued components of concern in the project's regional environment.

Duinker and Greig (2006), Canter and Ross (2010), Seitz, Westbrook, and Noble (2011), and Noble, Liu, and Hackett (2017), among others, point to several enduring challenges and concerns with the current practice of CEA. The first problem concerns the context of CEA as currently required in most jurisdictions, including in Canada under the Impact Assessment Act (2019)—situated within a project-based review. Cumulative environmental effects concern the total effects of activities on an environmental component or environmental system. Project EA, in contrast, is concerned with project-induced stress and making sure that the impacts of a project are acceptably small rather than understanding the total effects of all stressors, project and non-project, on any single component or system. Parkins (2011) reports that treating cumulative effects simply as additive impacts from multiple projects on indicators, such as water use or pollution, does not facilitate broader discussions about regional limits to development or change and the ways in which specific projects and impacts are aligned or misaligned with regional development goals and objectives.

Second, and closely related, is that project EA is concerned primarily with minimizing project stress to a level of acceptability. The objective of proponents, and rightfully so, is to ensure that their project meets regulatory approval—this usually means minimizing any efforts regarding CEA and paying little attention to understanding regional environmental conditions, the impacts of other actors or drivers, and longer-term sustainability. The result is often findings of "non-significance" when in reality the project is contributing to incremental, if not synergistic, cumulative environmental change (Box 11.2).

A third concern relates to thresholds. Understanding the cumulative effects of human activity, and the implications of such effects, requires some understanding of thresholds and carrying capacities. The challenge, however, is that thresholds are not easily determined, particularly within the spatial and temporal confines of a project assessment. There is often reluctance to set thresholds or to limit development when our understanding of natural variability and adaptability within the system is poor. However, in general, for any assessment it is useful to have a management target or benchmark against which to assess condition change (either effects-based change or stressor-based change); otherwise, it is difficult to determine when to take action and what action to take when undesirable change occurs. When thresholds are addressed, they are usually defined within the context of the project, such as discharge limits, as opposed to the total effects on an environmental component.

Fourth, CEA and management are ultimately about the future and require looking far enough into the future to capture the full array of human activities and natural changes that may affect the sustainability and functioning of the environmental components or systems of concern. This is a highly uncertain environment and one that is about possible futures and outcomes—a view that stands in sharp contrast

Box 11.2   Project EA and Individually Insignificant Actions

Oil and gas development in southern Saskatchewan: In southwest Saskatchewan, a 1940 km^2 ecologically rich land base, consisting of active sand dunes, rare and endangered species, and plants of Indigenous cultural importance is subject to the pressures of approximately 1500 natural gas wells, cattle grazing, and more than 3000 kilometres of access roads and trails. The landscape is significantly fragmented, and biodiversity, in a once-native grassland ecosystem, is at risk. Cattle grazing and roads and trails in the region have not been subject to EA. Of the 1500 wells in the area, only five proposals were subject to assessment—none of which was deemed to have significant environmental effects (GSH SAC, 2007). Nasen, Noble, and Johnstone (2011), however, found that the ecological footprint of petroleum and natural gas wells in southwest Saskatchewan grasslands has an effect on soils and range health up to 25 metres from the well head—well beyond the physical footprint of the infrastructure and with a duration of at least 50 years.

Water discharge and withdrawals in the Athabasca region, Alberta: The Athabasca River basin, Alberta, is exposed to a wide range of land-use activities, including agriculture, forestry, pulp and paper operations, and petroleum extraction. Roads, power lines, pipelines, and other disturbances have fragmented forests, and the amount of old-growth forest has been significantly reduced. Between 1966–76 and 1996–2006, the number of pulp mills discharging into the Athabasca basin increased from one to five; total farm area increased from 47.2 million acres to 52.1 million acres; the number of operating oil sands leases increased from two to 3360; water withdrawals increased from approximately 12 million m^3/yr to 595 million m^3/yr, of which more than 70 per cent can be attributed to oil sands operations. Between these two time periods, the cumulative annual flow in the Athabasca River decreased by more than 500 m^3/s, and temperature increased by 1.4°C; conductivity, turbidity, and phosphorous levels also increased (Squires, Westbrook, & Dubé, 2010). Many of these disturbances, such as urban growth and agricultural expansion, have not been subject to assessment. For others, the effects of each project have been deemed unlikely to cause significant adverse environmental effects.

Hydroelectricity infrastructure development in northern Manitoba: Hydroelectric development in northern Manitoba has been ongoing for more than 50 years. In 2003, Manitoba Hydro, a provincial Crown corporation and the primary developer and operator of hydroelectricity facilities in the province, submitted an EIS for the proposed Bipole III Transmission Line project—an approximately 1400-kilometre transmission line from northern Manitoba south to Winnipeg. The transmission line would traverse boreal forest and caribou habitat in the north and agricultural land in the south, including several river and stream

continued

crossings along the route. Coupled with the existing Bipole I and II transmission lines, Bipole III will improve the reliability of electricity supply to Manitoba and reduce the risk of supply interruptions due to severe weather events. The project was subject to EA under the Environment Act of Manitoba and the proponent directed to consider the potential cumulative effects of the project. An analysis of the proponent's CEA was undertaken by Gunn and Noble (2012) on behalf of the Consumers' Association of Canada and found that: the baseline against which cumulative effects were assessed largely ignored the effects of past actions and changing conditions over time; the temporal scope of the assessment was shorter than the lifetime of the project and adopted only a five-year horizon for assessing cumulative effects to caribou; the focus was on whether the regional environment can absorb the additional stress versus cumulative effects; much of the effects analysis was restricted to the transmission line right-of-way, ignoring the effects of the Bipole I and II projects; and the magnitude of the project's impacts were interpreted relative to the effects of other actions The Manitoba Clean Environment Commission, an arm's-length provincial agency mandated to provide advice to the minister and hold public hearings, reported that it was "simply inconceivable—given the 50-plus-year history of Manitoba Hydro development in northern Manitoba and given that at least 35 Manitoba Hydro projects had been constructed in the north in that time—that there are few, if any, cumulative effects identified in this EIS" (Manitoba Clean Environment Commission, 2013, p. 112). That said, the project was approved under the Environment Act (1987) and issued a licence for development.

Part of what leads to scenarios like these is that cumulative effects are often ignored or diminished in project assessment, sometimes deliberately and sometimes because the project in question is considered too small to warrant attention. Quite often, individual developments are evaluated independently of other activities and thus deemed "unlikely" to cause significant adverse environmental effects. In other cases, the magnitude of a project's impacts is sometimes erroneously "measured against" or "compared to" the effects of other projects versus focusing on the overall effects on environmental conditions. When the significance of a project's effects, no matter how small the effect, is evaluated from the perspective of the additional stress placed on an environmental component or system that is already stressed by other sources, it is far more likely to be deemed unacceptable, particularly in regions of concentrated development where environmental thresholds may already be exceeded (Gunn & Noble, 2012).

to the shorter-term perspective of project approval and predicting the "most likely," versus the most desirable, effects of development. Assessing cumulative effects means exploring alternative futures or possibilities; this includes the consideration of hypothetical development scenarios—a requirement that extends beyond the capacity of a single project proponent.

Table 11.1 Characteristics of Status Quo Thinking versus Requirements for CEA

	Status quo	Required for CEA
Assumptions	abundance	limits
Receptors	single media	environmental systems
Spatial context	project	multiple scales
Temporal context	present	past, present, future
Scope	regulated activities	all disturbances
Assessment	stressors *or* effects	stressors *and* effects
Futures	predicted impacts	probable outcomes
Management	mitigation	avoidance
Monitoring	regulatory compliance	thresholds and capacity
Responsibility	individual proponents	multi-stakeholder
Performance	increased efficiency	increased efficacy

Fifth, our assumptions about cumulative effects, as evidenced by project EA practice, are not always consistent with the nature of how environmental systems function. Assessing and managing cumulative effects requires a set of assumptions different from what is normally adopted when managing the stress of a single project action (Table 11.1). As Ross (1994, p. 6) points out, "the environmental effects of concern to thinking people are . . . not the effects of a particular project; they are the cumulative effects of everything." In particular, there is a need to think about limits of environmental systems in terms of the types, amounts, and rates of development that can be accommodated. Identifying such limits is not within the scope of individual project reviews, but knowing these limits is important to informing project decisions and whether a project is acceptable given potential cumulative effects.

Finally, managing cumulative effects requires collaboration. Project EA is focused on regulating and managing the impacts of the project at hand—not managing the impacts of other projects that have already been approved or that may be approved in the future. For example, should it be discovered during an EA that wildlife habitat is already fragmented beyond an acceptable limit, or will approach or exceed an acceptable limit should the project be approved, it is difficult (legally, politically), though not impossible, to require changes in the management actions of other projects in the area that have already been approved to accommodate the additional development. The argument is often presented in project EA that it is not the responsibility of a project proponent to understand and manage the cumulative effects of other activities. This may be a valid argument given the role of project EA; however, it also reflects the need for a collaborative approach to cumulative effects management that extends beyond the responsibilities of any single EA application (Box 11.3).

Box 11.3 Eastmain 1A Project, Quebec, and the Need for Collaborative Approaches to Managing Cumulative Effects

The Eastmain 1A project, Quebec, involves powerhouse development and diversion of part of the Rupert River to produce hydroelectricity. Hydro-Québec, the project proponent, submitted an EIS in 2004 for approval under the Environmental Quality Act, the James Bay and Northern Quebec Agreement, and the Canadian Environmental Assessment Act. Both the federal panel report and the provincial panel report, issued by the Provincial Review Committee to the Administrator of Chapter 22 of the James Bay and Northern Quebec Agreement (COMEX), speak specifically to concerns about the project's cumulative effects. The federal panel, for example, notes: "It feels that the cumulative impact is real, but difficult to quantify, making it impossible to specify the relative significance of the incremental cumulative impact and the causes of change attributable to the project" (Federal Review Panel, 2006, p. xxi). The federal panel report also notes the narrow extent of the CEA regarding the Cree communities examined and suggests that a study of other potentially affected communities is needed. The federal panel report then goes on to state that "because the effects extend beyond the borders of the province and affect many jurisdictions . . . a formal analysis of these impacts cannot be conducted without establishing a large-scale research and follow-up program, a measure which the Panel recommends" (Federal Review Panel, 2006, p. xxii).

The COMEX report also speaks to the adverse cumulative effects of the project, noting, for example, that "the cumulative effects of the project being studied will complement an already large and complex 'baggage', and this in a territory that has experienced rapid and intense development over the past 30 years" (COMEX, 2006, p. 407). Several conditions are included in the provincial authorization that speak to the need for follow-up programs for monitoring baseline conditions and verifying the effectiveness of proposed impact mitigation measures. However, any conditions regarding cumulative effects are deferred to a future assessment or study. For example, the panel notes that "The evaluation of the cumulative impacts of the hydroelectric projects of James Bay and Hudson Bay, by reason of their scope, concerns several jurisdictions and goes beyond the responsibility of one single proponent. The analysis of these impacts cannot be done without setting up a large-scale research and follow-up program carried out by a consortium . . . responsible for this issue which devolves only partially on the proponent." The report goes on to note that the cumulative effects of this latest project can only be determined through a wide-scale and long-term follow-up of social impacts on Cree society, that "a single proponent should not be responsible for these cumulative impacts, and that their analysis should be the responsibility of all the jurisdictions involved" (Federal Review Panel, 2006, p. 416).

Source: Noble, Martin, & Olagunju, 2016

Regional Assessment

The above challenges are not to say that project EAs should not consider cumulative effects; rather, something more is needed to address and manage cumulative effects in an effective manner. **Regional assessment** presents an opportunity to expand the CEA, and the implementation of management solutions, beyond the evaluation of site-specific direct and indirect project impacts to address broader regional environmental impacts, issues, and concerns. Cumulative effects are the product of multiple, interacting development actions, natural disturbances, and the multiplicity of development decisions in a region that accumulate across space and over time. Regions may be physical units, such as watersheds, ecological units, such as ecosystems, or jurisdictional units, such as planning areas. Proponents of regional assessment initiatives for addressing cumulative effects are usually partnerships involving both public and private sectors; triggers are typically non-regulatory but often emerge from recommendations from regulatory-based processes; and the focus of attention is on planning to ensure sustainable land use and resource development over extended periods of time—usually decades (Gunn & Noble, 2015).

Under the current federal Impact Assessment Act (2019), the emphasis of CEA remains project-focused; however, the act does provide for the assessment of cumulative effects of existing or future activities in a specific region through regional assessment. One of the stated purposes of the act is "to encourage the assessment of the cumulative effects of physical activities in a region" (paragraph 6(1)(m)), and sections 92 and 93 of the act provide the minister of environment and climate change with discretionary powers to establish a committee—or authorize the Impact Assessment Agency of Canada—to conduct a regional assessment of the effects of existing or future physical activities carried out in a region—whether led by the federal government when on federal lands or in collaboration with other jurisdictions when in part on federal lands or not on federal lands. This is not an entirely new provision. A stated purpose of the previous act, the Canadian Environmental Assessment Act, 2012 (paragraph 4(1)(i)) was similarly "to encourage the study of cumulative effects of physical activities in a region," with provisions for regional assessment under subsections 73(1) and 74(1)—referred to then as "regional studies." A regional assessment was recently completed for offshore oil and gas exploratory drilling, east of Newfoundland and Labrador, initiated under the former Canadian Environmental Assessment Act, 2012 (Box 11.4).

Regional assessments can be traced to the formative years of EA—such as the Churchill River Basin study, which examined the implications of a program of water development projects on the Churchill River system. It is only in more recent years, however, owing in part to the constraints of project EA in dealing with cumulative effects, that regional assessments have emerged with an explicit cumulative effects focus. A regional approach to CEA was promoted by the Alberta Society of Professional Biologists in the mid-1990s (Kennedy, 1995), and several regional initiatives for cumulative effects monitoring emerged throughout the late 1990s and 2000s—especially in the Alberta oil sands region, including the Regional Aquatics Monitoring Program (1997–2012), the Northern Rivers Ecosystem Initiative (1998–2003), and the Cumulative Environmental Management Association (2000–16)

Box 11.4 Regional Assessment of Offshore Oil and Gas Exploratory Drilling East of Newfoundland and Labrador

In April 2019, the governments of Canada and Newfoundland and Labrador signed an agreement to conduct a regional assessment of offshore oil and gas exploratory drilling east of Newfoundland and Labrador. The assessment will focus on the effects of existing and anticipated offshore oil and gas exploratory drilling, with the intent to improve the efficiency of project EA processes as applied to oil and gas exploration drilling and reduce duplication in processes and information. Included among the factors to be considered in the assessment are the changes to the environment or to health, social, or economic conditions and the positive and negative consequences of these changes that are likely to be caused by offshore exploratory drilling, including: i) the effects of malfunctions or accidents that may occur in connection with exploratory drilling; ii) any cumulative effects that are likely to result from offshore exploratory drilling in combination with other physical activities that have been or will be carried out; and iii) the result of any interaction between those effects. Four oil projects currently operate in offshore Newfoundland and Labrador (Hibernia, White Rose, Hebron, Terra Nova), with an estimated resource potential of more than six billion barrels of oil and 60 trillion cubic feet of natural gas. The regional assessment report is available on the Canadian Impact Assessment Registry.

Source: Government of Canada & Government of Newfoundland and Labrador, 2019

Environmental Assessment in Action

Elk Valley Cumulative Effects Management Framework

The Elk Valley is nestled in the Rocky Mountains in the southeastern region of British Columbia and home to the communities of Elkford, Sparwood, Hosmer, Fernie, Morrissey, and Elko. The Elk River, approximately 220 kilometres long, cuts through the Elk Valley. Its total drainage basin is 4450 square kilometres, providing important habitat for cutthroat trout—a popular freshwater sport fish and an important indicator species of ecosystem health. The valley is also home to grizzly bears, bighorn sheep, American dipper elks, northern goshawks, and bald eagles and is the traditional territory of the Ktunaxa Nation. The Elk Valley plays a significant role in the province's coal mining industry and contains the largest producing coalfield in British Columbia. There are five surface metallurgical coal mines operating in the valley, all owned by Teck Coal.

In 2012, triggered by an EA approval condition attached to Teck's proposed Line Creek Operations Phase II expansion project, coupled with increasing concerns about the potential cumulative impacts of human development in the Elk Valley, including tourism, angling, roads and trails, railway operations, and coal mining, Teck Coal and the Ktunaxa Nation Council initiated a multi-stakeholder process to develop and implement a cumulative effects management framework (CEMF) for the Elk Valley. Other stakeholders involved in the initiative included various provincial government agencies (e.g., Ministry of Environment, Ministry of Forests, Lands and Natural Resource Operations, Ministry of Aboriginal Relations and Reconciliation, Ministry of Energy and Mines), municipal and regional government

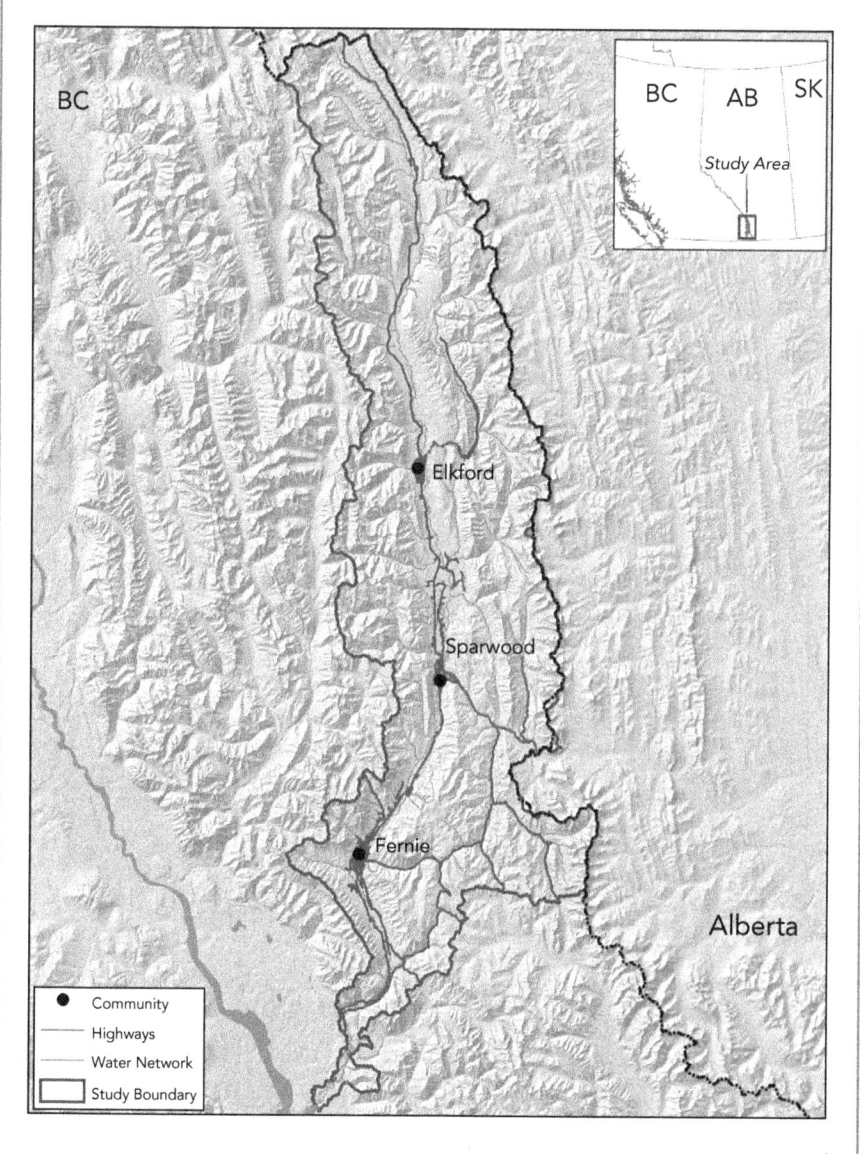

continued

representatives, other resource developers, and environmental non-government organizations—including the Elk River Alliance.

The CEMF is an iterative process comprised of four stages, shown below:

- *Context-setting*, which includes establishing the spatial and temporal boundaries for the CEMF and identifying the values and VCs that will be the focus of assessment.
- *Retrospective assessment*, which examines past and present conditions and trends in VCs using indicators of spatial and temporal change, such as riparian disturbance, linear feature density, equivalent clear-cut area, and selenium concentration hazard ratings, and setting benchmarks for those indicators.
- *Prospective assessment*, focused on simulating future conditions and exploring how different scenarios of disturbance patterns, rates, and intensities, including climate scenarios and mitigations, might affect the VCs and their indicators.
- *Management*, which includes management and monitoring recommendations and informing regulatory decisions such as permitting and project reviews.

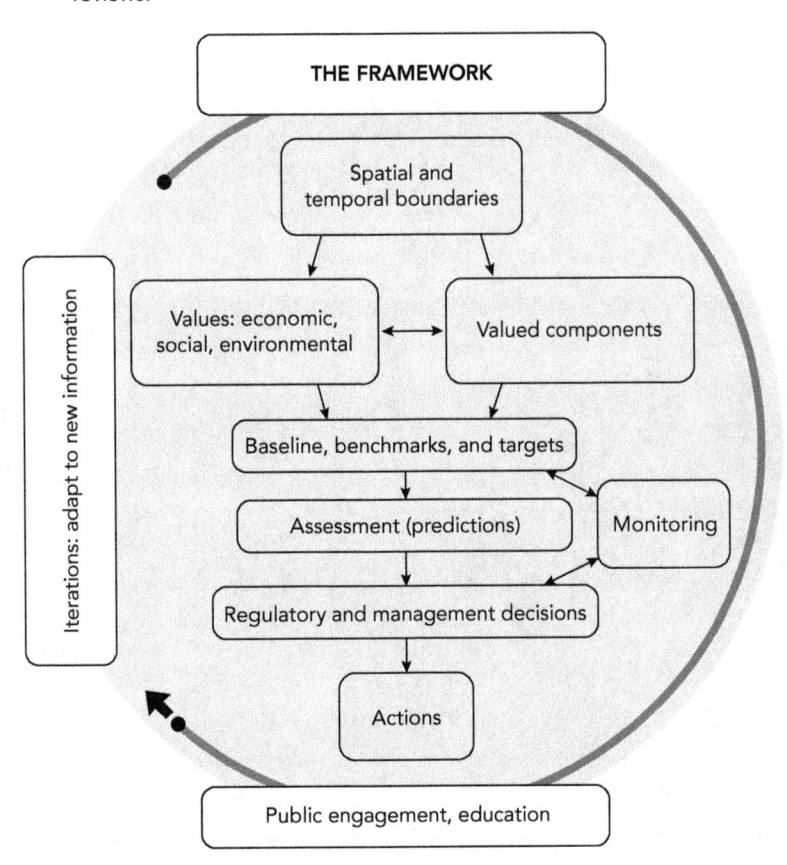

THE FRAMEWORK

- Spatial and temporal boundaries
- Values: economic, social, environmental ↔ Valued components
- Baseline, benchmarks, and targets
- Assessment (predictions)
- Monitoring
- Regulatory and management decisions
- Actions
- Public engagement, education

Iterations: adapt to new information

ALCES was used to explore the cumulative effects of land uses and natural disturbances and to simulate future scenarios to 2065. The CEMF focused on seven scenarios: i) a reference scenario, characterized by business as usual and simulating past trends into the future; ii) a minimum scenario, focused on reduced rates of forest harvesting, mining, and municipal growth; iii) a maximum scenario, simulating increased resource development and intensifying land uses; iv) a high natural disturbance scenario, exploring increased rates of fire and pest outbreaks in forests resulting from climate change; v) moderate mitigation, assuming modest efforts or standard practices to reduce or manage impacts associated with either the minimum or maximum scenarios; vi) intensive mitigation, based on "better" practices and more ambitious impact management actions; and vii) two future climate scenarios.

Results from ALCES simulations showed that past land uses, especially the extensive road network associated with forestry, mining, gas exploration, and recreation, have had major impacts on the region that will likely persist into the future. Widespread disturbance, such as forestry, coupled with climate change, were found to pose high risk to supporting habitat for grizzly bears and other VCs, including riparian habitat and cutthroat trout. The future development of existing mineral leases was found to have potentially significant cumulative impacts on the Elk Valley at smaller spatial scales. Impact management efforts focused on reducing road density, including removal of non-essential stream crossings and culverts, were found to be the most important actions for mitigating impacts and likely to generate substantial benefits for many important species and habitats. This suggests that in the future, when new roads are proposed as part of a development

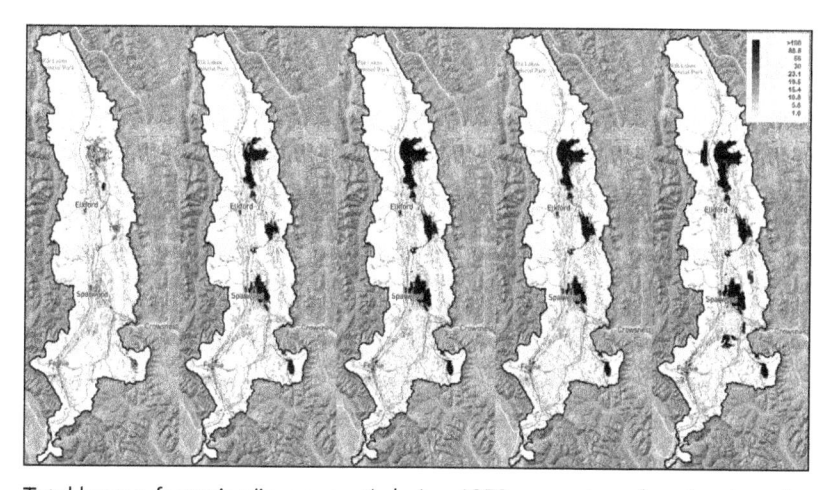

Total human footprint (in per cent) during 1950, current, and under the reference, minimum, and maximum scenarios in 2065 (from left to right). The footprint includes only the unbuffered, direct physical footprint.

continued

> proposal, it may be necessary to consider the closure or reclamation of an existing road (or abandoned road) as a condition of approval.
>
> The CEMF identified several important impact management and monitoring priorities for the Elk Valley and recommended coordination of development policies at the broader regional scale—beyond the boundaries of the Elk Valley—to support land-use zoning to better manage human development. The Elk Valley CEMF is illustrative of a regional and strategic approach to CEA and an example of a collaborative and transparent approach to addressing cumulative effects. Given the success of the initiative, and the opportunity to link the Elk Valley CEMF to regulatory decisions, leadership was transitioned from the initial working group to the provincial government, and it became a key component of a province-wide cumulative effects management and decision support framework (FLNROD, 2019).
>
> Sources: Based on the Elk Valley Cumulative Effects Working Group, 2018; meetings of the EVCEMF Working Group

(Cronmiller & Noble, 2018). A more recent example of regional CEA is the Elk Valley Cumulative Effects Management Framework, British Columbia (see Environmental Assessment in Action: Elk Valley Cumulative Effects Management Framework).

Effects- and Stressor-Based Approaches

Cumulative effects assessments for regions are focused on ascertaining the condition of environmental receptors and whether the total effects via all stressors in a regional environment are acceptable, including the potential additional stress caused by future development and disturbance. Although various approaches to and interpretations of CEA exist, regional CEA adopts two complementary approaches: **effects-based** CEA and **stressor-based** CEA. The concepts of "effects-based" and "stress-based" were introduced in Chapter 9, in the context of monitoring programs. In terms of CEA, effects- and stressor-based represent two important models for assessing effects.

Effects-based CEA is focused on assessing existing environmental conditions relative to a reference condition and is typically retrospective in design—*what has happened*. Examples include environmental effects monitoring programs (see Environment Canada, 2010; 2011) and ecological modelling and baseline studies (see Culp, Cash, & Wrona, 2000; Munkittrick et al., 2000; Dubé et al., 2006; RAMP, 2010). The strength of effects-based approaches is in measuring the accumulated environmental state of a system and identifying whether performance indicators are at or below an acceptable level (Dubé & Munkittrick, 2001). Doing so can inform the identification of thresholds and help to inform risk assessment processes that, in principle, support decision-making about the impacts of development.

Emphasis is on understanding the total effects on a particular valued component from all sources of stress (point, non-point, direct, indirect) and comparing these effects to some reference condition in order to determine an actual measure of cumulative change, irrespective of the number and nature of the impacts causing that change. Under this model, the focus of cumulative effects allows for questions of a broader nature related to ecological thresholds and synergistic effects. The underlying premise of this approach is that cause–effect relationships can be established through long-term monitoring, which can then be used to predict cumulative impacts.

Stressor-based CEA is prospective in design—*what might or could happen.* The focus is typically on quantifying past and present patterns and trends in the distribution of human disturbance in a region (e.g., industrial footprints, road densities, habitat fragmentation metrics) and then projecting disturbances into the future under different scenarios of land use and development. Attention is placed on predicting the cumulative stress associated with particular agents of change—such as different projects or types of disturbances. This involves an analysis of the distribution and rates of change in disturbance in the baseline and predictive modelling of future disturbance patterns.

There is often a strong statistical association between disturbances and ecological responses and an assumption that stressors are a good proxy for threats to the sustainability of valued components. In cases where data on indicators (e.g., water quality) are limited, there is the potential to use surrogate indicators, such as land-use and land-cover metrics (e.g., riparian zone habitat, stream crossing density, percent impervious surfaces), which can act as indicators of responses by affected systems to environmental change and can be used in regression and correlation analyses to provide an indication of cause–effect relationships (Seitz, Westbrook, & Noble, 2011). Such stressor-based indicators can often be monitored efficiently and cheaply with the assistance of remote sensing and Geographic Information Systems tools. The utility of the indicators (i.e., landscape metrics) will depend on the response variable, the strength of the relationship between the indicator and response, and the relative importance of other controlling variables (e.g., topographic relief, bedrock, or climate variability) (Gergel, Turner, Miller, Melack, & Stanley, 2002). The relationship between the stressor (e.g., road density) and the effect (e.g., water quality) is an assumed or statistical association rather than cause–effect per se. Such analysis is thus indicative of cumulative effects, but not explanatory of cause–effect.

Both the effects-based and the stressor-based approaches are useful, but each offers a different type of understanding of cumulative effects—the first from the perspective of change in the receiving environment, the second from the perspective of change in human disturbance or stress to the environment. If the role of CEA is solely to understand the accumulated state and set thresholds through monitoring, then further development of effects-based models is required. If the role of CEA is to guide decisions about the potential implications of proposed land and resource use, then further development of stressor-based models is required. Good CEA requires both effects- and stressor-based approaches. As Duinker and Greig (2006) report, dwelling on the past is useful but only in the sense of possible learning about interactions, knowledge that can be used to sharpen predictive analysis for the future. At the same time, focusing solely on the future is useful only if we are able to understand the implications of future environmental change, which is often based on learning from the past and understanding thresholds or limits of change.

Basic Science Components of a CEA Framework

There are a variety of frameworks that present steps or phases for CEA. Good CEA can be distilled to five necessary components: scoping, retrospective analysis, prospective analysis, monitoring of cumulative effects, and the management of cumulative effects (Figure 11.2). In the absence of any one of these components, CEA is incomplete.

Issues Scoping

Scoping establishes all that will be included and all that will be excluded when evaluating cumulative effects and subsequent impacts to valued components or environmental systems. Scoping was discussed in Chapter 5 in the context of project EA and plays a similar role in CEA. Of particular importance in the context of CEA is setting the spatial and temporal boundaries for assessment—these boundaries are typically much more ambitious for CEA than for project-based reviews.

Spatial boundaries

The spatial boundaries for CEA can vary considerably but are defined by a combination of factors, including: i) the specific land uses or industrial activity of interest; ii) planning or management jurisdictions; and iii) the characteristics or distribution of the valued components or indicators of concern. Cumulative effects occur over multiples scales and over time. It is generally acknowledged that in order to assess cumulative effects effectively, the spatial boundaries of assessment must be extended well beyond those typically scoped for a single project. This requires consideration of at least four types of scale.

Spatial scale refers to the actual geographic extent of the assessment and is typically based on one or both of natural boundaries, such as watersheds, or

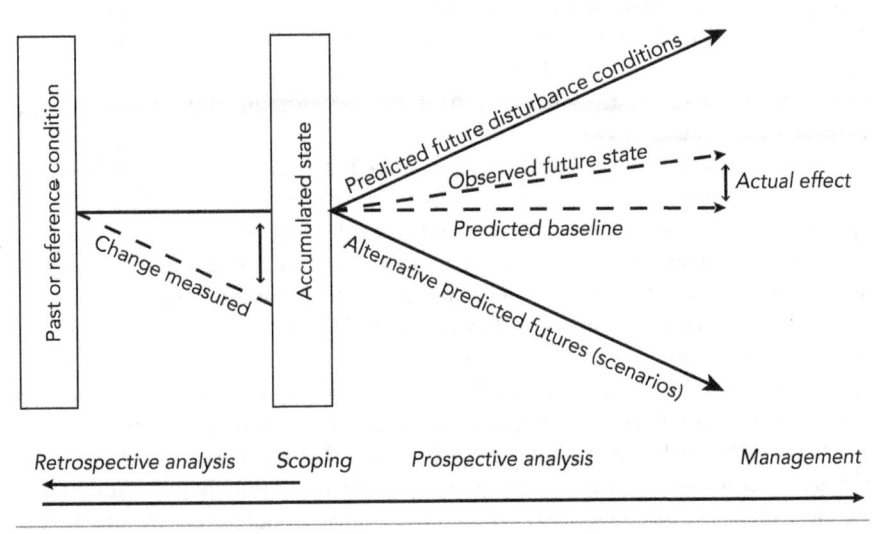

Figure 11.2 Components of a cumulative effects model
Source: Adapted from Dubé et al., 2013

administrative boundaries, such as townships or landownership. This is usually the most common interpretation of "scale" in CEA; however, it is certainly not the most functional with regard to the actual analysis of cumulative effects. **Analysis scale** is used to examine valued components and impacts across space and is represented by such ideas as data resolution, detail, and granularity. **Phenomenon scale** is perhaps the most important type of scale in CEA, since it refers to the spatial units within which various processes operate or function. Thus, in any single assessment, different spatial boundaries may be appropriate for different cumulative effects and for different indicators of interest. The boundaries selected for cumulative effects on air quality, for example, might be quite different from those chosen for cumulative effects on soil quality or sedentary versus migratory wildlife. **Administrative scale** reflects the realities imposed by jurisdictions, land-use plans, or regulatory processes. In principle, CEA should focus on ecological units, such as watersheds or ecoregions. In practice, administrative scale plays an important role in the success of CEA programs. Ambitious ecological boundaries often need to be tempered by institutional arrangements and the administrative authority to implement CEA, including mitigation and monitoring programs.

The spatial scale for CEA must also be practical. In the case of river systems, for example, some authors have argued that to account for cumulative change in the biophysical properties of a river system, any CEA should consider the entire river system from headwaters to mouth (Squires, Westbrook, & Dubé, 2010). In principle, this is the most desirable approach. However, such a scale may not be achievable due to such constraints as data availability, human and financial resource capacity, and the multi-jurisdictional authority over land and water use decisions that may exist in a large river basin. A trade-off approach is to focus on management units or on regions differentiated by land-use or disturbance intensities. For example, in the context of river systems, Seitz, Westbrook, and Noble (2011) suggest the use of a river reach as a more practical yet meaningful spatial scale. This may be an area subject to the most intense or diverse land use. Seitz, Westbrook and Noble argue that if a CEA fails when applied to a manageable region with the most landscape disturbance or if the cumulative effects model or monitoring program is not responsive to the range of stressors that exist in the most developed reach, then it will make no difference if the rest of the river continuum is considered in the assessment.

While there is no best method for determining the spatial boundaries for any particular CEA, literature offers a number of guiding principles to assist the practitioner:

- *Adequate scope.* Boundaries must be large enough to include relationships between the dominant disturbances in a region and the valued components or indicators of concern.
- *Natural boundaries.* Natural boundaries such as watersheds, airsheds, or ecosystems are perhaps the best reflection of the natural components of a system and should be respected where possible.
- *Maximum zones of detectable influence.* Impacts related to specific disturbances activities typically decrease with increasing distances from the

disturbance source; thus, boundaries should be established where impacts are no longer detectable.
- *Multi-scaled approach.* Multiple spatial scales, such as local and regional boundaries, should be assessed to allow for a more in-depth understanding of the scales at which environmental processes and impacts operate.
- *Flexibility.* CEA boundaries must be flexible enough to accommodate changing natural and human-induced environmental conditions.

Temporal boundaries
The delineation of temporal scale for CEA should consider how interactions from past, present, and future developments or disturbances might interact with, and affect, the specific indicators or system properties of concern. Future boundaries are constrained only by modelling capabilities or, in the case of scenario-based approaches, by one's imagination and willingness to tolerate uncertainties (see Chapter 6). Delineating past boundaries can be much more challenging and controversial. A common (though incorrect) approach in project EA is to define only the current environment as the basis for assessment (Dubé, 2003). But because the existing environment is a result of the influence of past actions, this approach assigns the effects of past and present human actions and disturbances to the current condition rather than to contributions to cumulative change (McCold & Saulsbury, 1996). Therivel and Ross (2007) explain that CEA temporal boundaries must reflect how past actions and incremental changes have influenced the present and the resulting long- and short-term effects as well as future effects.

There is some argument that the historical reference point or benchmark for cumulative effects should be pre-disturbance conditions. This is ideal but rarely possible. McCold and Saulsbury (1996) suggest that a more practical temporal reference for cumulative effects is a time when the valued component or indicator of concern was "more abundant" or "less affected" by human action. In the context of the Athabasca oil sands region, Alberta, for example, a heavily modified landscape, Seitz, Westbrook, and Noble (2011) suggest that an appropriate historical reference for CEA in the region would be the late 1960s—after which the Athabasca region experienced unprecedented landscape transformation and industrial development. Other options might be to consider the time when a certain land-use designation was made (for example, the establishment of a park or the lease of land for development) or the time when effects similar to those of concern first occurred. These are subjective reference points but practical ones in terms of data availability and given the onset of the major human activities that have led to current concerns about cumulative effects.

Retrospective Analysis
Retrospective analysis involves looking to the past to identify and determine key condition changes and drivers of change over time and understanding the present **accumulated state**. This typically requires both stress- and effects-based analysis of indicators and condition change, capturing multiple stressors and responses (i.e., cumulative contributions to a cumulative response) and establishing benchmarks or limits for both ecological response to stress and for the level of the stress itself (Dubé

et al., 2013). The accumulated state of a valued component or system must be understood over space and time and relative to a reference state such that current conditions are not set as the "new normal." Changes or trends in the condition of valued components and their indicators, or system properties, are then related to changes or trends in stressor or levels of development or disturbance on the landscape—usually through some form of correlative analysis or **dynamic systems-based modelling**.

Examples of disturbance measures of interest in cumulative effects analysis may include the density of linear features per unit area on the landscape (e.g., road or trail density—km/km^2), percentage disturbed landscape (e.g., cleared area), edge density or perimeter area ratio, the rate of land conversion (e.g., rate and area of change from forested to non-forested), the number or density of river crossings (e.g., number of crossings per river kilometre in a river reach), the density of impervious or hard surfaces in a watershed (e.g., road surfaces and parking lots have been linked to contaminant transfer and measurable responses in water quality [see Brydon et al., 2009]), and broader natural processes of change such as flood or fire frequency (Gunn & Noble, 2012). The objective is to identify measures of the drivers of change in the region, characterize indicator responses, and identify areas of concern based on spatial and/or temporal change that are outside of reference conditions or benchmarks. The relationships established between multiple stressors and responses form the basis to support predictive scenario-modelling.

Prospective Analysis

Accumulated state assessments provide the existing condition of the watershed and the trajectory for change that has occurred from past conditions and identifies the indicators that have changed and, as such, may be of higher risk or sensitivity to future disturbance (Dubé et al., 2013). Using this knowledge and models developed from the retrospective analysis, **prospective analysis is about predicting and evaluating how indicators or conditions** (e.g., caribou population, fish health) might respond to additional stress in the future—stress caused by two or more projects or activities in the regional environment (e.g., fragmentation, river crossings) in addition to natural disturbances. Where data are limited, Gunn and Noble (2012) explain that prospective analysis can involve "summing up" individual effects such that the total effects on indicators of concern are evaluated and summarized into trend information, focusing on regional environmental issues and whether they will grow worse or better, based on broad regional change agents such as "surface disturbance" that are, by definition, cumulative and provide a measure of ecosystem health.

Scenario-based approaches are especially valuable to regional CEA and to understanding cumulative change associated with current and future disturbance regimes. The use of scenarios in impact prediction was discussed in Chapter 6. Although scenario-based modelling is only one means by which cumulative effects can be approached, several dynamic system-modelling tools are available to support CEA, including ALCES and MARXAN. Introduced in Chapter 6 as a tool for use in impact prediction, ALCES (A Landscape Cumulative Effects Simulator) is a system dynamics simulation tool for exploring the behaviour or response of resource systems to disturbances. It is designed to forecast and explore alternative outcomes and conditions and allows users to examine the area and length of different land uses or disturbances, such as roads or forest cut blocks, within partitioned landscape units. ALCES has been applied in numerous resource and land-use planning and cumulative effects assessment exercises in Canada, including the Upper Bow Basin Cumulative Effects Study. MARXAN (Marine Spatially Explicit Annealing), a landscape and marine optimization tool, uses a site-selection algorithm to explore options for conservation and regional land-use planning (Watts et al., 2009). It divides a landscape into small parcels termed planning units and then attempts to minimize the "cost" of conservation while maximizing attainment of conservation goals. It is a useful tool for exploring alternative land-use and conservation scenarios and management options. MARXAN has not been used widely in CEA, but its utility was demonstrated in the Great Sand Hills Regional Environmental Study (GSH SAC, 2007) where it was used to support a strategic approach scenario-based analysis of effects on biodiversity caused by multi-sector land uses. See Chapter 6 for further discussion of ALCES and MARXAN and Chapter 12 for a discussion of the Great Sand Hills case.

Monitoring

Cronmiller and Noble (2018) report that long-term monitoring provides the foundation for cumulative effects science, verifying actual cumulative effects outcomes, and informing cumulative effects management actions, yet monitoring is among the most deficient aspects of cumulative effects initiatives (Dubé et al., 2013). Monitoring to support CEA must include:

1. reference condition monitoring to set the benchmark for evaluating changes;
2. effects-based monitoring to evaluate actual (versus predicted) changes in environmental conditions or indicators;
3. stressor-based monitoring to track disturbance or changing stress conditions, refine predictive models, and regulate project permitting.

This requires a monitoring framework that provides a consistent regional approach in terms of sampling strategy, endpoints, and protocols such that monitoring efforts become more coordinated, approaches become standardized, and data become regionally available (Dubé et al., 2013). Such a framework might provide explicit direction to project-specific monitoring activities under EA approvals (see Chapter 9) such that the terms of reference for project EAs in a given region might include a specific suite of indicators that all projects must monitor and report—regardless of whether those indicators are of relevance to their individual project. Wong, Noble, and Hanna (2019) suggest that these indicators should provide an indication of departure from normal conditions and signal to both project proponents and governments that other indicators may need to be examined or that certain indicators need to be monitored more intensely in certain regions.

At the same time, monitoring program design must be relevant to regulatory decisions and provide the information needed to make decisions about the cumulative effects, and required cumulative effects mitigation strategies, for individual project applications in a timely fashion. As Cronmiller and Noble (2018) explain, the pursuit of cumulative effects monitoring questions that are outside the scope of, or that do not coincide with, the needs of decision-making may result in short-lived monitoring programs or in monitoring programs that are data-rich but information-poor and that have limited influence in land-use and regulatory outcomes.

Wong, Noble, and Hanna (2019) identify several attributes of monitoring programs to support CEA, arguing that monitoring programs must be able to meet the needs of their users by providing information to assess, evaluate, and make decisions regarding the cumulative effects of development proposals. The attributes are as follows:

- *Consistency:* the presence (or absence) of parameters monitored across different programs is consistent.
- *Compatibility:* it is possible to integrate or compare monitoring data across programs or datasets—i.e., consistency in the approach to data collection/analysis.
- *Observability:* monitoring data are of sufficient scale and resolution to allow tracking of baseline change over time and/or across space.
- *Detectability:* monitoring data or parameters allow for the early detection of change in baseline conditions or potential threats to water quality.
- *Adaptability:* parameters monitored are useful for understanding or detecting change at multiple spatial scales and applicable to multiple project or disturbance types.
- *Accessibility:* monitoring data are accessible, retrievable, and available in a usable format.
- *Usability:* monitoring data meet multiple end-user needs, including those of project proponents and regulatory decision-makers.

Management

The best way to manage cumulative effects is to avoid them. However, this is not always possible—most development happens in areas that have already been subject to some level of past disturbance. There is little point to monitoring cumulative change if there is no authority or mechanism to implement management action. Cumulative effects management is about managing land use and development within limits, managing across environmental media and resource sectors, and managing risk and opportunity. This requires, at a minimum, that thresholds, benchmarks, or limits that trigger management action are set and there is an opportunity, and political will, to simply say "no" to development actions that may cause significantly adverse cumulative effects or to implement compensatory measures or enhanced mitigation practices for already-existing projects and land uses to "make room" for new development. For example, a cumulative effects management strategy may deem a new access road in a highly fragmented landscape "acceptable" only if there is a mechanism to require the restoration of multiple existing and abandoned access roads. Cumulative effects management thus requires not only collaboration across sectors and government agencies but strong leadership with the mandate and authority to implement cumulative effects management measures and influence regulatory decisions.

Governance for Cumulative Effects Management

Institutional challenges, more so than scientific or technical ones, are often the most significant constraints to CEA and management. Cronmiller and Noble (2018) report that the institutions established to support regional assessment initiatives, such as SEA, are often created in response to a crisis or specific trigger and do not always accommodate longer-term CEA and monitoring needs. Assessment and monitoring initiatives in any given region occur across a variety of different actors, each with different goals and objectives, and what is monitored and assessed under CEA programs does not always align with the needs of planners and decision-makers. A further challenge to regional approaches to CEA is that rarely is there authority to implement recommendations or to carry forward CEA findings to specific project-based assessments (Spaling et al., 2000). Many CEAs have been "one-offs," disconnected from regulatory-based development decision-making (Sheelanere, Noble, & Patrick, 2013). Under the Impact Assessment Act (2019), even with the commitment to regional assessments to address cumulative effects, there are no specific mechanisms to ensure that the results directly influence decisions about individual project actions.

CEA monitoring and management require institutional commitment and collaboration that extends beyond that which is normally present in project EA applications—it requires committed leadership and good governance. Promising practices and frameworks are emerging, such as the Elk Valley Cumulative Effects Framework, introduced earlier in this chapter, and British Columbia's province-wide cumulative effects initiative (Box 11.5). For CEAs conducted outside the scope of the legislated project EA process, Sheelanere, Noble, and Patrick (2013) and Kristensen, Noble, and Patrick (2013) suggest a number of requisites for implementing, sustaining, and ensuring influential CEA, namely:

- an agency with the authority and mandate for CEA, including the means to direct monitoring programs and influence decisions about land use and project development in the region;
- clearly defined stakeholder roles and responsibilities for undertaking the CEA, implementing the results, and monitoring and following up for continual learning and improvement;
- sharing of monitoring data, both spatial and aspatial and in common data formats, among all stakeholders;
- a means of implementing CEA initiatives, enforcing monitoring programs and compliance, and ensuring influence over development decisions taken at the individual project level;
- sufficient financial and human resources to implement and sustain, over the long term, CEA programs and requirements (e.g., monitoring programs, landscape modelling, reporting, communication and data management, and coordination).

Sheelanere, Noble, and Patrick (2013) report that assessing and managing cumulative effects beyond the scope of individual projects is essential but challenging—scientifically

Box 11.5 British Columbia's Cumulative Effects Framework

As in most provinces and territories, the traditional approach to managing natural resources in British Columbia has been on a sector-by-sector and project-by-project basis, using a variety of tools and regulatory processes. In 2010, the government of British Columbia embarked on a process to develop a province-wide cumulative effects framework that would apply across natural resource sectors and provide sector-wide, and regionally informed, support to land-use and project-based decision processes. The framework outlines a stepwise process for CEA that includes the following components: i) selecting values for assessment; ii) defining standard assessment protocols; iii) assessing the current condition of values, including identification of key drivers of change, and management tools and responses; and iv) reporting on current and potential future conditions of values and management responses. The cumulative effects framework does not replace existing policy and regulatory processes, including project EA; rather, it is intended to inform and strengthen them. Once fully implemented, the framework is intended to inform and support strategic decisions, such as land-use planning, objective setting, and other forms of management and direction; tactical decisions, such as defining priorities for research, monitoring, and development planning; and operational decisions, such as informing authorizations for natural resource activities and project-specific EAs. That said, implementation of the framework is proving to be challenging, and the framework has yet to be fully integrated into provincial review and decision processes.

Source: Council of Canadian Academies, 2019; Government of British Columbia, 2016

and institutionally. They argue the need to rethink our assumptions about cumulative effects, think outside the project-specific approach, move toward the integration of cumulative effects knowledge in land-use planning and regulatory decision-making, and invest in the capacity-building requirements and institutional frameworks needed to support long-term cumulative effects initiatives.

Key Terms

accumulated state
administrative scale
analysis scale
dynamic systems-based modelling
effects-based CEA
phenomenon scale
project- or activity-centred

prospective analysis
regional assessment
retrospective analysis
spatial scale
stressor-based CEA
valued component–centred

Review Questions and Exercises

1. Do requirements exist under your provincial, territorial, or state EA system for proponents to assess the cumulative effects of their projects?
2. What is the difference between effects-based and stressor-based approaches to cumulative effects assessment?
3. Using an example, explain how a proponent might use spatial bounding to their advantage to make the effects of their project on the landscape seem insignificant.
4. It has been said that cumulative effects assessment is simply EA done right. Do you agree?
5. Cumulative effects often result from multiple and often unrelated project developments in a single region—such as oil and gas, forestry, roads, recreation, and hydroelectric developments. Individually, each project is often approved based on a determination that it will not generate significant environmental effects. Cumulatively, however, all of these activities are contributing to cumulative effects.
    a. Assume a new project is proposed in the region, a large-scale solar energy facility that will cause additional impacts on VCs that are already stressed beyond their limit. Who should be responsible for managing the impacts of the solar energy project?
    b. Who should be responsible for managing overall cumulative environmental effects that have built up over time?
    c. Should the solar energy facility be allowed to proceed—even though it will create additional disturbance in the region?
    d. What actions might government take to "make room" for new development in an already stressed environmental system?

References

Beanlands, G., & Duinker, P. (1983). An ecological framework for environmental impact assessment. *Journal of Environmental Management* 18, 267–77.

Brydon, J., et al. (2009). Evaluation of mitigation methods to manage contaminant transfer in urban watersheds. *Canadian Journal of Water Quality Research* 44(1), 1–15.

Canter, L.W., & Ross, B. (2010). State of practice of cumulative effects assessment and management: the good, the bad and the ugly. *Impact Assessment and Project Appraisal* 28(4), 261–8.

COMEX (Environmental and Social Impact Review Committee). (2006). *Eastmain 1-A and Rupert Diversion hydropower project*. Report by the Provincial Review Committee to the Administrator of Chapter 22 of the James Bay and Northern Quebec Agreement. Government of Quebec.

Council of Canadian Academies. (2019). *Greater than the sum of its parts: toward integrated natural resource management in Canada*. Ottawa, ON: The Expert Panel on the State of Knowledge and Practice of Integrated Approaches to Natural Resource Management in Canada.

Cronmiller, J., & Noble, B. (2018). Integrating environmental monitoring with cumulative effects management and decision making. *Integrated Environmental Assessment and Management* 14(3), 407–17.

Culp, J., Cash, K., & Wrona, F. (2000). Cumulative effects assessment for the Northern River Basins Study. *Journal of Aquatic Ecosystem Stress and Recovery* 8, 87–94.

Dubé, M. (2003). Cumulative effect assessment in Canada: a regional framework for aquatic ecosystems. *Environmental Impact Assessment Review* 23(6), 723–45.

Dubé, M., et al. (2006). Development of a new approach to cumulative effects assessment: a northern river ecosystem. *Environmental Monitoring and Assessment* 113, 87–115.

Dubé, M., Duinker, P., Greig, L., Carver, M., Servos, M., McMaster, M., . . . & Munkittrick, K. (2013). A framework for assessing cumulative effects in watersheds: an introduction to Canadian case studies. *Integrated Environmental Assessment and Management* 9(3), 363–9.

Dubé, M., & Munkittrick, K. (2001). Integration of effects-based and stressor-based approaches into a holistic framework for cumulative effects assessment in aquatic ecosystems. *Human Ecology Risk Assessment* 7(2), 247–58.

Duinker, P., & Greig, L. (2006). The impotence of cumulative effects assessment in Canada: ailments and ideas for redeployment. *Environmental Management* 37(2), 153–61.

Elk Valley Cumulative Effects Working Group. (2018). *Elk Valley cumulative effects assessment and management report*. Elk Valley Cumulative Effects Management Framework Working Group. Cranbrook, BC.

Environment Canada. (2010). Integrated watershed management. http://www.ec.gc.ca/eau-water/default.asp?lang1/4en&n1/413D23813-1.

Environment Canada. (2011). National Environmental Effects Monitoring Office. http://www.ec.gc.ca/esee-eem/default.asp?lang1/4En&n1/4453D78FC-1.

Federal Review Panel for the Eastmain 1-A and Rupert Diversion Project. (2006). Environmental assessment of the Eastmain 1-A and Rupert Diversion Project. Ottawa, ON: Government of Canada.

FLNROD (Forest, Lands, Natural Resource Operations and Rural Development). (2019). Elk Valley cumulative effects management framework: project background. Victoria, BC: FLNROD.

Francis, S.R., & Hamm, J. (2011). Looking forward: using scenario modeling to support regional land use planning in northern Yukon, Canada. *Ecology and Society* 16(4), 18.

Gergel, S.E., Turner, M.G., Miller, J.R., Melack, J.M., & Stanley, E.H. (2002). Landscape indicators of human impacts to riverine systems. *Aquatic Sciences* 64(2), 118–28.

Government of British Columbia. (2016). Cumulative effects framework: interim policy. Victoria, BC: Government of British Columbia.

Government of Canada & Government of Newfoundland and Labrador. (2019). Agreement to conduct a regional assessment of offshore oil and gas exploratory drilling east of Newfoundland and Labrador, between Her Majesty the Queen in Right of Canada as represented by the federal minister of the environment and the federal minister of natural resources and Her Majesty the Queen in Right of Newfoundland and Labrador as represented by the provincial minister of natural resources and the provincial minister for intergovernmental and indigenous affairs.

Greig, L. (2019). Reflections on what's needed for effective impact assessment. Presentation at Cumulative Effects 2019. Calgary, AB: The Canadian Institute.

GSH SAC (Great Sand Hills Scientific Advisory Committee). (2007). *Great Sand Hills regional environmental study*. Regina, SK: Canadian Plains Research Centre.

Gunn, J., & Noble, B.F. (2012). *Critical review of the cumulative effects assessment undertaken by Manitoba Hydro for the Bipole III project*. Prepared for the Public Interest Law Centre, Winnipeg, MB. www.cecmanitoba.ca.

Gunn, J., & Noble, B.F. (2015). Sustainability considerations in regional environmental assessment. In A. Morrison-Saunders, J. Pope, & A. Bond (eds), *Sustainability assessment handbook* (pp. 79–102). Cheltenham, UK: Edward Elgar.

Gunn, A., Russell, D., & Greig, L. (2014). Insights into integrating cumulative effects and collaborative co-management for migratory tundra caribou herds in the Northwest Territories, Canada. *Ecology and Society* 19(4), 4.

Hackett, P., Liu, J., & Noble, B.F. (2018). Human health, development legacies, and cumulative effects: environmental assessments of hydroelectric projects in the Nelson River watershed, Canada. *Impact Assessment and Project Appraisal*. doi: org/10.1080/14615517.2018.1487504.

Kennedy, A.J. (ed.). (1995). *Cumulative effects assessment in Canada: from concept to practice.* Calgary, AB: Alberta Society of Professional Biologists.

Kristensen, S., Noble, B.F., & Patrick, R. (2013). Capacity for watershed cumulative effects assessment and management: lessons from the Lower Fraser Basin, Canada. *Environmental Management*. doi: 10.1007/s00267-013-0075-z.

Manitoba Clean Environment Commission. (2013). *Report on public hearing: Bipole III transmission project*. Winnipeg, MB: Manitoba Clean Environment Commission. http://www.cecmanitoba.ca.

McCold, L., & Saulsbury, J.W. (1996). Including past and present impacts in cumulative impact assessments. *Environmental Management* 20(5), 767–76.

Munkittrick, K., et al. (2000). *Development of methods for effects-driven cumulative effects assessment using fish populations: Moose River project*. Pensacola, FL: Society of Environmental Toxicology and Chemistry.

Nasen, L., Noble, B.F., & Johnstone, J. (2011). Environmental effects assessment of oil and gas lease sites on a grassland ecosystem. *Journal of Environmental Management* 92, 195–204.

Noble, B.F. (2008). Strategic approaches to regional cumulative effects assessment: a case study of the Great Sand Hills, Canada. *Impact Assessment and Project Appraisal* 26(2), 78–90.

Noble, B.F., Liu, G., & Hackett, P. (2017). The contribution of project environmental assessment to assessing and managing cumulative effects: individually and collectively insignificant? *Environmental Management* 59(4), 531–45.

Noble, B., Martin, J., & Olagunju, A. (2016). *A review of the application of cumulative effects assessment in the context of Section 22 environmental assessments conducted in the James Bay territory*. Report to the James Bay Advisory Committee on the Environment. Saskatoon SK: Aura Environmental Research and Consulting Ltd.

Odum, W. (1982). Environmental degradation and the tyranny of small decisions. *BioScience* 32(9), 728–9.

Parkins, J.R. (2011). Deliberative democracy, institution building, and the pragmatics of cumulative effects assessment. *Ecology and Society* 16(3), 20.

RAMP (Regional Aquatic Monitoring Program). (2010). *Regional Aquatic Monitoring Program (RAMP) scientific review*. http://www.ramp-alberta.org.

Ross, W. (1994). Assessing cumulative environmental effects: both impossible and essential. In A. J. Kennedy (ed.), *Cumulative effects assessment in Canada: from concept to practice*. Papers from the Fifteenth Symposium held by the Alberta Society of Professional Biologists, Calgary, AB.

Seitz, N.E., Westbrook, C.J., & Noble, B.F. (2011). Bringing science into river systems cumulative effects assessment practice. *Environmental Impact Assessment Review* 31(3), 172–9.

Sheelanere, P., Noble, B.F., & Patrick, R. (2013). Institutional requirements for watershed cumulative effects assessment and management: lessons from a Canadian trans-boundary watershed. *Land Use Policy* 30, 67–75.

Spaling, H., et al. (2000). Managing regional cumulative effects of oil sands development in Alberta, Canada. *Journal of Environmental Assessment Policy and Management* 2(4), 501–28.

Spaling, H., & Smit, B. (1994). Classification and evaluation of methods for cumulative effects assessment. In A.J. Kennedy (ed.), *Cumulative effects assessment in Canada: from concept to practice* (pp. 47–65). Papers from the Fifteenth Symposium held by the Alberta Society of Professional Biologists, Calgary, AB.

Squires, A., & Dubé, M.G. (2012). Development of an effects-based approach for watershed scale aquatic cumulative effects assessment. *Integrated Environmental Assessment and Management*. doi: 10.1002/ieam.1352.

Squires, A., Westbrook, C., & Dubé, M. (2010). An approach for assessing cumulative effects in a model river, the Athabasca River Basin. *Integrated Environmental Assessment and Management* 6(1), 119–34.

Therivel, R., & Ross, W. (2007). Cumulative effects assessment: does scale matter? *Environmental Impact Assessment Review* 27(5), 365–85.

Watts, M.E., Ball, I.R., Stewart, R., Klein, C.J., Wilson, K., Steinback, C., . . . & Possingham, H. 2009. MARXAN with zones: software for optimal conservation based land- and sea-use zoning. *Environmental Modelling and Software* 24(12) 1513–21.

Wong, L., Noble, B.F., & Hanna, K. (2019). Water quality monitoring to support cumulative effects assessment and decision making in the Mackenzie Valley, Northwest Territories, Canada. *Integrated Environmental Assessment and Management* 15(6), 988–99.

12 Strategic Environmental Assessment

Higher-Order Assessment

The Enbridge Line 3 project, a proposal to build a new 1600-kilometre pipeline to replace the current pipeline, built in 1968, linking Alberta oil sands with Wisconsin pipelines (Enbridge Pipelines Inc., 2014), was met with mixed reactions. For many, the project was an opportunity to speak out about national energy policy and the future of a resource sector. For others, the project was an opportunity to raise larger issues concerning Indigenous title and consultation obligations over resource development. Also caught up in the review of the Enbridge Line 3 project were concerns about other pipeline proposals being considered across Canada, including Energy East, Trans Mountain, and Northern Gateway, as well as oil sands expansion in general and Canada's role in fossil-fuel export (Noble, 2017).

Projects are the result of policies or strategies, or lack thereof, playing out on the ground. When development unfolds in the absence of a higher-order assessment to inform the nature of project choices and development decisions, issues such as those in the Enbridge case often emerge at the time projects are proposed. The problem is that project reviews are not designed to ask the bigger questions such as "should a particular resource sector be developed?" or "where is development appropriate and under what conditions?" Many of the decisions that affect the environment and local communities are made long before project developments are proposed, and environmental, social, economic, and political conditions have significant implications for the success of project developments. Actions taken, or not taken, at the level of policies or land-use plans can affect the nature and type of development initiatives that emerge and their implications for sustainable development.

Defining Strategic EA

Strategic environmental assessment (SEA), while variably defined, refers to the EA of strategic initiatives—typically policies, plans, and programs (PPPs). SEA is about identifying and assessing a range of PPP options at an early stage when there is flexibility with respect to future actions and the decisions to be taken. First introduced as a tool to assess the impacts of proposed PPPs (Wood & Djeddour, 1989), SEA is increasingly promoted as an instrument to also shape the formulation and implementation of PPPs, to provide for a better understanding of the complex institutional environments that influence decision-making, and to facilitate transitions toward sustainable outcomes (Table 12.1). Partidário (2012) explains that SEA is about understanding the context of a PPP being developed and assessed, identifying problems

Table 12.1 Definitions of Strategic Environmental Assessment—Past to Present

The systematic process of evaluating at the earliest possible stage the environmental effects of a PPP and its alternatives (Therivel & Partidário, 1996).

The proactive assessment of alternatives to proposed or existing PPPs, in the context of a broader vision, set of goals, or objectives to assess the likely outcomes of various means to select the best alternative(s) to reach desired ends (Noble, 2000).

A decision support tool, designed to integrate environmental and social issues into PPP decision making processes, bringing together different aspects of problems, different perspectives, and providing possible solutions in an accessible form to the decision maker (Sheate et al., 2003).

A process designed to systematically assess the potential environmental effects, including cumulative effects, of alternative strategic initiatives for a particular region . . . and in doing so inform the development of policies, plans or programs (CCME, 2009).

A strategic framework instrument that helps to create a development context toward sustainability, by integrating environment and sustainability issues in decision-making, assessing strategic development options and issuing guidelines to assist implementation (Partidário, 2012).

Source: Noble & Nwanekezie, 2017

---

and potentials, addressing key trends, and assessing environmental and sustainable viable options that will help to achieve strategic objectives. As such, SEA is focused on asking such questions as: what are the objectives, what are the key drivers, what are the strategic options, and what are the critical policies to be met? Project EA, in contrast, tends to focus on asking: what are the characteristics of the project, what are the local baseline conditions, what are the project's alternative design options, and what are the major impacts and mitigation measures?

Policies, Plans, and Programs

Strategic environmental assessment aims to integrate *environment* into PPP development and decision-making. In doing so, SEA facilitates an early, overall analysis of the relationships between the environment and PPPs and the potential effects of the projects that might emerge from those PPPs. The nature of SEA will vary depending on the regulatory system and specific assessment context and tier of application (Figure 12.1).

SEA for policies typically asks "why" and "what" questions, such as: why is a national energy policy needed, what is the policy intended to achieve, what is the likely future energy profile, what are the possible energy mixes, and what are the institutional opportunities and constraints associated with different policy directions? Large-scale government policies, such as national energy policies, for example, can set the context for subsequent planning and development decisions across multiple sectors, from energy development projects to transportation infrastructure (Figure 12.2). Policy SEA explores the potential opportunities, impacts, and

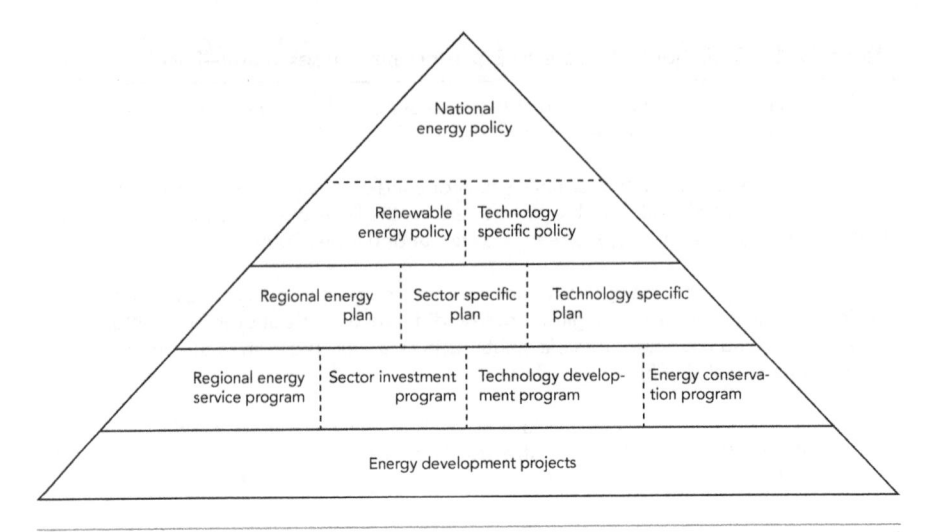

Figure 12.1 Example of assessment tiers for energy policies, plans, programs, and projects

Figure 12.2 Concentration of oil refineries, referred to locally as refinery low, east of the city of Edmonton, Alberta. Energy infrastructure projects such as refineries are subjected to review and approvals under numerous project-specific regulations to mitigate potentially adverse impacts. SEA, in contrast, asks much bigger-picture questions, such as what is the appropriate mix of fossil fuels for Canada's energy future?

implications of a proposed or existing policy, including the exploration of policy options, and can play an important role in shaping policy development.

In some instances, SEA at the policy level might tackle much broader, cross-cutting issues and seek to provide guidance for assessment and evaluation at subsequent levels or tiers of decision-making. The SEA is thus focused on providing strategic direction for how a policy or set of rules or objectives will be implemented or achieved. One example is Canada's recent SEA of climate change (Box 12.1), which sets out guidance and expectations for addressing climate change commitments in federal impact assessments. Although a policy-level SEA, the SEA has been criticized for focusing only on providing guidance to project EA and not also on exploring broader policy options for addressing climate change. That said, the SEA falls short of addressing competing climate policies and investment programs across Canada's major resource sectors and federal and provincial jurisdictions and does not assess strategic options for transitioning Canada's energy sector toward a low carbon future.

SEA for plans may focus on specific sectors or regions and explore viable options or scenarios for achieving specific development goals or conservation objectives that are set out in higher-order policies. SEA at the plan level sometimes adopts a spatially explicit approach, especially when the plan in question is a land-use plan. The results are intended to tier-forward to inform programs of development, such as the location or timing of activities, or to set specific terms for project developments (Box 12.2).

Programs are a defined series of future events or a specific set of related activities. In the context of SEA, programs often refer to "programs of development," such as a group of projects or the rules, terms, and conditions for developing certain classes of projects, infrastructure, or technology. SEA at this tier often focuses on maximizing positive outcomes and minimizing negative impacts associated with a program of development and exploring the implications of a specific program in terms of infrastructure requirements or economic or environmental impacts (Box 12.3)

Box 12.1   Canada's Climate Change Strategic Environmental Assessment

Under the new Impact Assessment Act (2019), decision-makers must consider the extent to which each assessed project "contributes to sustainability" and "hinders or contributes to" meeting Canada's climate commitments. In 2018, the Government of Canada released a draft of its Strategic Assessment of Climate Change. The report provided limited strategic assessment of climate change strategies; instead it provided guidance on how federal impact assessments will consider a project's greenhouse gas (GHG) emissions and its resilience to climate change impacts. The assessment provides guidance to proponents and others on the information requirements related to climate change throughout the EA process for those projects undergoing a federal impact assessment for the purpose of addressing public policy discussions beyond the scope of a single project assessment.

Box 12.2 SEA in Atlantic Canada's Offshore Oil and Gas Sector

The Canada–Newfoundland and Labrador Offshore Petroleum Board (C-NLOPB) is responsible for oil and gas activity offshore Newfoundland and Labrador. Offshore petroleum activities that require authorization by the C-NLOPB may also be subject to EA under federal (and often provincial) legislation. In 2002, the C-NLOPB adopted a policy decision to start conducting SEAs before exploration projects are considered and in conjunction with the call for bids process in offshore areas. The objectives of SEA under the C-NLOPB are to inform licensing in prospective offshore areas and to help streamline issues and considerations for subsequent project EAs. Several SEAs have been completed by the C-NLOPB, but the three major production facilities currently operating offshore Newfoundland and Labrador all exist in a "non-SEA" region.

Informing offshore planning and rights issuance are the major intents of SEA under the C-NLOPB. SEAs are initiated *only* in areas where no offshore oil and gas operations currently exist. As such, SEA is intended to establish a baseline condition for a potential licensing area. The results are used by the C-NLOPB for licensing decisions and, in principle, by industry to augment baseline assessments in their project EAs. Using seismic surveying as an example, if there are ecologically sensitive areas identified through an SEA, the intent is that a future EA would focus in detail on these areas as opposed to focusing on areas that may not be regionally significant. The Orphan Basin SEA findings, for example, demonstrate how SEA was designed to inform prospective activity in the study region whereby special, non-standard, or strict mitigation measures have been identified to be applied to future developments because of the need for special planning around sensitive marine habitats (see LGL Ltd, 2003).

Box 12.3 International Atomic Energy Agency's Nuclear Power Program SEA Guidance

The International Atomic Energy Agency's (IAEA) (2018) SEA guidelines for nuclear power programs establish principles and guidance for the application of SEA for nuclear power programs. The SEA process and guidance are not designed to assess *why* nuclear energy could be an option for a country. At the program level, the assumption is that such questions are addressed at an earlier stage, for example, through an SEA specifically targeting the energy policy of a country. The guidelines focus on SEA for assessing the impacts of nuclear power programs on health, society, and environmentally relevant economic considerations for seven key nuclear impact areas: main siting and technological considerations; power plant construction, operation, and decommissioning; nuclear fuel cycle; spent fuel management strategy / radioactive waste storage and disposal; physical protection and security; emergency preparedness and response; and wider physical infrastructure requirements. The guidance will be particularly valuable to Canada as the federal government explores recommendations from the Small Modular Reactor Roadmap to develop an action plan for small modular reactor use in remote communities, industry application, and grid-scale generation (see https://smrroadmap.ca).

Source: International Atomic Energy Agency, 2018

Origins and Evolution

Assessing the impacts of PPPs is not new. The US National Environmental Policy Act (NEPA) of 1969, coming into force in 1970, is often identified as the earliest provision for higher-level impact assessment, even though NEPA makes no specific reference to SEA. It was not until after some high-profile international developments—such as the World Bank's (1999) recommendation for the environmental assessment of policy; the report of the World Commission on Environment and Development, *Our common future* (1987); and the United Nations 1992 Earth Summit—that SEA gained international attention.

The term "strategic environmental assessment" was not formally introduced until 1989, first mentioned in a research report to the European Commission:

> The environmental assessments appropriate to policies, plans, and programmes are of a more strategic nature than those applicable to individual projects and are likely to differ from them in several important respects.... We have adopted the term "strategic environmental assessment" (SEA) to describe this type of assessment. (Wood & Djeddour, 1989)

The report described SEA simply as a type of EA that is appropriate to PPPs but of a more *strategic nature* than assessment processes applied to individual projects. Wood and Djeddour (1992, p. 10) argued that there "is no fundamental methodological reason why SEA should not be introduced . . . utilising a form of SEA basically similar in its basic nature to that employed for projects." The growth of SEA throughout the 1990s and 2000s was informed largely by the principles of project EA (Fundingsland-Tetlow & Hanusch, 2012): they shared the same objective—to identify and assess environmental impacts—but the object of assessment differed—PPPs versus projects. The guidance that emerged for SEA was thus based largely on project EA principles and methodology (Fundingsland-Tetlow & Hanusch, 2012; Gachechiladze & Fischer, 2012; Glasson, Therivel, & Chadwick, 2005).

Today, SEA is in place in some 60 countries (Fundingsland-Tetlow & Hanusch, 2012). However, SEA remains far less advanced than EA. SEA provisions also vary from one jurisdiction to the next. In the United States, provisions for SEA fall under NEPA in which SEA is narrowly interpreted to be **programmatic environmental assessment**, or area-wide EA. Hundreds of programmatic assessments are completed in the United States each year, which essentially involve the direct application of EA to "programs of development," such as offshore oil and gas infrastructure in the Gulf of Mexico. In the United Kingdom, SEA was initially carried out through a less formalized policy and plan environmental appraisal process. Formal requirements for SEA were first adopted by the United Kingdom in 2004 under European SEA Directive 2001/42/EC. In the Czech Republic, SEA was introduced by means of a reform to formal EA legislation, whereas in Australia, SEA was at first adopted informally as part of resource management programs and then in 1999 was introduced through formal legislated requirements under the broad umbrella of EA that includes planning strategies and programs.

SEA in Canada

In Canada, SEA was formally established at the federal level in the early 1990s by way of a cabinet directive and separate from project-based EA, making it one of the "first of the new generation of SEA systems that evolved in the 1990s" (Dalal-Clayton & Sadler, 2005, p. 61) (Table 12.2). Procedural guidance for SEA was first provided in *The environmental assessment process for policy and programme proposals*. In 1999,

Table 12.2 Timeline of SEA Development in Canada

1990	A bill is introduced to establish the Canadian Environmental Assessment Act; policies, plans, and programs are not included within the scope of the proposed act.
	Canadian Environmental Assessment Research Council (CEARC) releases guidelines for environmental assessment of policy and program proposals.
	National Round Table on Environment and Economy (NRTEE) and CEARC hold a workshop on the integration of environmental considerations into government policy.
1991	Federal government reform package introduces Canada's first initiative in the development of a system of strategic environmental assessment: *Environmental assessment in policy and program planning: a sourcebook*.
1992	Canadian Environmental Assessment Act receives legislative approval. Section 16(2) emphasizes the role and value of regional studies outside the act, but there is no reference to SEA.
1993	Federal Environmental Assessment Office (FEARO) procedural guidelines are released to federal departments on the EI process for policy and program proposals (FEARO, 1993).
1995	Amendments to the Auditor General Act require that all federal departments and agencies prepare a sustainable development strategy.
	Federal government releases *Strategic environmental assessment: a guide for policy and program officers*.
	Canadian Environmental Assessment Act comes into force.
1999	Update to the 1990 cabinet directive is released as well as guidelines for implementation.
	Minister of the environment launches a five-year review of the Canadian Environmental Assessment Act.

2000	Frameworks for regional environmental effects assessment appear in the Canadian Environmental Assessment Agency's research and development priorities for 2000.
2001	Bill C-19 is introduced to amend the Canadian Environmental Assessment Act.
2003	Bill C-9, formerly Bill C-19, receives royal assent, and the amended Canadian Environmental Assessment Act comes into force.
2004	Guidelines on the cabinet directive on SEA are updated, requiring federal departments and agencies to release a public statement when an SEA has been completed.
2007	Minister of the Environment's Regulatory Advisory Committee, Subcommittee on SEA, commissions a report on the state of SEA models, principles, and practices in Canada.
2009	Canadian Council of Ministers of the Environment releases *Regional strategic environmental assessment in Canada: principles and guidance*.
2010	Updated federal guidelines for implementing the cabinet directive on SEA are released.
2012	Federal budget implementation bill, the Jobs, Growth and Long-Term Prosperity Act, is introduced. National Round Table on the Environment and the Economy is eliminated. Canadian Environmental Assessment Act is repealed and the Canadian Environmental Assessment Act, 2012 enacted, but there is no reference to SEA in the new act.
2017	Federally commissioned Expert Panel for the Review of Environmental Assessment Processes (2017) recommends that federal EA legislation require the use of SEA to guide project EA and that the federal authority responsible for impact assessment conduct an SEA when a new or existing federal policy, plan, or program would have consequential implications for EA. No amendments to the existing cabinet directive are recommended.
2019	Canada's new Impact Assessment Act (2019) receives royal assent. Discretionary provisions are included for the minister of environment and climate change to conduct an assessment of a federal PPP or issue that is relevant to conducting a project EA under the act. SEA remains a non-legislated process under the cabinet directive.

Source: Based on Noble 2002; 2003; 2008

the Cabinet Directive on the Environmental Assessment of Policy, Plan and Program Proposals was released, requiring that SEA be conducted when i) a proposal is submitted to an individual minister or cabinet for approval; and ii) implementation of the proposal may result in important environmental effects, either positive or negative. In was not until 2004 that federal departments and agencies were required to prepare a public statement whenever an SEA had been completed. The cabinet directive was updated in 2010, linking SEA to the Federal Sustainable Development Act, establishing that each federal minister of a department or agency subject to the act is responsible for ensuring that PPPs are consistent with the Federal Sustainable Development Strategy and for reporting on department and agency performance under the strategy (Noble et al., 2019).

Noble et al. (2019) report on the findings of SEA audits conducted by the Commissioner of the Environment and Sustainable Development (CESD) between 2015 and 2018. Results of initial audits showed that ministers were not provided with information about the potential environmental effects for most PPP proposals (OAGC, 2015); however, subsequent audits indicate some signs of improvement (Table 12.3). Compliance with the cabinet directive is increasing among federal organizations, but audits based on procedural compliance and whether the directive was applied are not indicative of the value of the efforts and outcomes.

Canada's Impact Assessment Act recently introduced new provisions for SEA, independent of the cabinet directive. The act refers to "strategic assessment" as opposed SEA but provides no rationale for the new terminology. The act provides the minister of environment and climate change with the ability to establish a committee or instruct the Impact Assessment Agency to conduct an SEA. The IAA establishes that such an SEA can assess:

1. any federal policy, plan, or program—proposed or existing—that is relevant to conducting impact assessments; or
2. any issue that is relevant to conducting impact assessments of designated projects or of a class of designated projects.

This is a modest step forward for SEA in Canada, providing a legislative opportunity to conduct SEA. At the same time, however, the conduct of SEA under the Impact Assessment Act is entirely at the discretion of the minister, and the provisions are somewhat restrictive in that the SEA must be "relevant" (sec. 95(1)) to the conduct of a project EA under the scope of federal jurisdiction. This potentially limits the ability of SEA to tackle some of the most pressing strategic issues in Canada, such as shaping energy policy and providing strategic direction for institutional transition to address climate change. However, the act does not specify what is meant by "relevant," and it does not specify the criteria that the minister must use when making decisions about which SEAs will receive priority. This may introduce some flexibility for SEA to function beyond the scope of project-based assessment issues and needs. The act does not replace the cabinet directive.

There are no formal systems of SEA in Canada's provinces and territories, but some jurisdictions do provide for EA application above the project level on a case-by-case basis. Newfoundland and Labrador's *Strategic environmental review guideline for policy and program proposals*, for example, provides for a separate review process for policy and program proposals; under Saskatchewan's Environmental Assessment Act, 20-year forest management plans are subject to review—but there is no specific reference to SEA. In Quebec, the Ministère des ressources naturelles et de la faune recently established a formal SEA program for offshore oil and gas exploration in the Gulf of St Lawrence under which two SEAs have been initiated. In 2009, the Canadian Council of Ministers of the Environment released *Regional strategic environmental assessment in Canada: principles and guidance*, a voluntary SEA process designed to inform regional land-use planning and assessment. Some of the most advanced and exemplary examples of SEA in Canada have been informal applications, outside the scope of the cabinet directive (Noble et al., 2019).

Table 12.3 Summary of 2015–18 Federal SEA Audits by the Auditor General of Canada, Commissioner of the Environment and Sustainable Development[1]

Performance metrics	2015 Audit[2] (2011–14)	2016 Audit[3] (2013–15)	2017 Audit[4] (2013–16)	2018 Audit (2017)
Applying the cabinet directive[5]	PPP proposals: 1955 Directive applied: 115 (6%)	PPP proposals: 506 Directive applied: 98 (19%)	PPP proposals: 359 Directive applied: 80 (22%)	PPP proposals: 283 Directive applied: 263 (93%)
Conducting preliminary scans of PPP proposals, when the directive was applied	Ministers were not provided with information about potential environmental effects.	Environmental effects considered and the scope of assessment commensurate with the level of anticipated effects.	Environmental effects considered and the scope of assessment commensurate with the level of anticipated effects.	Environmental effects considered and the scope of assessment commensurate with the level of anticipated effects.
Timelines	From a sample of 34 preliminary scans reviewed in detail, only 13 proposals were assessed early. For most proposals, it was difficult to determine due to lack of documentation.	From a sample of 31 preliminary scans reviewed in detail, 8 were conducted early. For most proposals, it was difficult to determine due to lack of documentation.	From a sample of 43 preliminary scans reviewed in detail, 5 were conducted early. For most proposals, it was difficult to determine due to lack of documentation.	Not reported in this audit.
Public reporting	No organizations consistently reported on their SEA practices or prepared public statements.	Parks Canada was the only organization that conducted detailed SEAs and issued public statements.	All but one organization reported each year on its SEA practices.	Not reported in this audit.

(continued)

Table 12.3 (continued)

Meeting Sustainable Development Strategy (SDS) commitments	No organizations made satisfactory progress toward meeting SDS commitments.	More than 50% of the preliminary scans reviewed considered SDS goals and targets, but only Parks Canada issued public statements on how PPPs affected progress toward SDS goals.	90% of the preliminary scans reviewed considered SDS goals, but only the Public Health Agency of Canada made satisfactory progress in meeting SDS commitments to strengthen their SEA practices.	All 26 organizations made satisfactory progress in strengthening their SEA practices, implementing recommendations from previous audit reports.

1. The 2018 audit reported on proposals submitted to cabinet but excluded proposals submitted to individual ministers for approval.
2. Federal entities included in audit: Agriculture and Agri-Food Canada, Canada Revenue Agency, Canadian Heritage, Fisheries and Oceans Canada.
3. Federal entities included: Department of Justice, National Defence, Parks Canada, Public Services and Procurement Canada, Veterans Affairs Canada.
4. Federal entities included: Atlantic Canada Opportunities Agency, Canada Border Services Agency, Canada Economic Development for Quebec Regions, Public Health Agency, Public Safety, Western Economic Diversification Canada.
5. Results include a number of PPP proposals submitted to an individual minister for approval, to cabinet, and submissions to Treasury Board.

Source: Noble et al., 2019

Foundational Principles of Strategic EA

SEA extends EA upstream—but at the same time, SEA reflects characteristics different from those of EA (Box 12.4). System-based characteristics of SEA refer to the basic provisions and requirements for SEA and the position of SEA in the broader planning and decision-making environment. Procedural characteristics of SEA concern the various methodological and process elements of SEA—in other words, the practice of SEA. Result-based characteristics of SEA capture the overall influence of SEA on decision-making and subsequent actions, including the opportunity for broader system-wide learning and SEA process improvement.

Box 12.4 Characteristics of Strategic Environmental Assessment

System-based characteristics	Description
1. Provisions	• clear provisions, standards, or requirements to undertake the SEA and implement the results
2. Proactive	• application early enough to address deliberation on the root of the problem and opportunities related to PPP priorities and choices, including determining the conditions necessary to achieve desired ends
3. Integrative	• integrative of biophysical, social, economic, cultural, and political knowledge and understanding of issues • integral part of the PPP formulation process
4. Tiered	• assessment undertaken within a tiered system of policy, planning, and decision-making • ability to influence, if not direct, proposals, actions, and decision at lower tiers of planning and development, including project EA
5. Sustainability	• sustainability/sustainable development a guiding principle and integral concept

Process-based characteristics	Description
6. Flexible to PPP context	• can accommodate a range of PPP issues and is sensitive to the particular policy or planning culture or agency or organization undertaking the SEA
7. Responsibility and accountability	• clear delineation of assessment roles and responsibilities • mechanisms to ensure impartiality/independence of assessment review • opportunity for appeal of process or decision output
8. Purpose and objectives	• assessment purpose and objectives clearly stated • centred on a commitment to sustainable development principles

9. Future-oriented	• focused on identifying desirable outcomes and on what is required to achieve those outcomes • about building a desirable future, not attempting to know or predict what the future will be
10. Scoping	• opportunity to develop and apply more or less onerous streams of assessment sensitive to the context and issue • consideration of related strategic initiatives, including other PPPs • identification and narrowing of possible valued ecosystem components to focus on those of most importance based on the assessment context
11. Alternatives-based	• comparative evaluation of potentially reasonable alternatives or scenarios
12. Impact evaluation	• identification of potential impacts or outcomes resulting from each option or scenario under consideration • integration or review of sustainability criteria specified for the particular case and context
13. Cumulative effects	• assessment of potential cumulative effects and life-cycle issues associated with alternatives or scenarios
14. Monitoring program	• includes procedures to support monitoring and follow-up of process outcomes and decisions for corrective action
15. Participation and transparency	• opportunity for meaningful participation and deliberations throughout the process • transparency and accountability in assessment process
Result-based characteristics	**Description**
16. Decision-making	• identification of a "best" option or strategic direction to guide PPP implementation • authoritative decisions, position of the authority of the guidance provided
17. PPP and project influence	• defined linkage with assessment and review or approval of any anticipated lower-tier initiatives, including project EA • demonstrated PPP influence, modification, or downstream initiative • identification of indicators or objectives for related or subsequent strategic initiatives or activities
18. System-wide	• opportunity for learning and system improvement through learning • regular system or framework review, such as a mandated five-year public review process

For SEA to be effective, it must reflect the most basic, defining attributes of a *strategic* assessment (Box 12.5). The basic concept of "strategy" originated in the context of military operations and is believed to be first documented in *The art of war*—written nearly 2500 years ago. *The art of war* distinguished between the *art* of conducting war and the *tasks* of directing individual battles. Wallington, Bina, and Thissen (2007) suggest that we can similarly distinguish between the art of exploring or facilitating strategic change toward sustainability, which they define as the strategy of SEA, and the task of assessing or influencing individual actions. Strategy is thus transformative—the purpose is not to inform a single decision (i.e., a project) but to inform the entire sequence of connected decisions or actions toward broader sustainability goals (Cherp, Watt, & Vinichenko, 2007).

SEA's most commonly noted strategic attributes can be summarized in four points (Noble et al., 2019):

- Strategic is not solely about the object of assessment (i.e., PPPs); rather, it's about ways of thinking about initiatives (Partidário, 2012), the enabling conditions for PPPs to proceed in a more sustainable way, and steering or directing their design and implementation.
- SEA is about building a more sustainable and resilient future (Slootweg & Jones, 2011), meaning that the exploration of alternatives or strategic options is core (Gonzalez & Therivel, 2014).
- Key to effective SEA is how it relates to the planning and decision processes it is intended to inform (Pope, Bond, Morrison-Saunders, & Retief, 2013) and its ability to provide and consider guidance both from higher

---

Box 12.5 What Is "Strategic"?

Strategic: Activities related to broader institutional, societal, economic, and environmental objectives, long-term goal formulation, and the evaluation of strategic choices, futures, and opportunities.

Reflexive: Activities related to monitoring, assessments, and evaluation of ongoing strategies, operations, and change (social, economic, cultural, and biophysical).

Tactical: Interest-driven directing activities that relate to the dominant structures of a system, including rules and regulations, institutions, organizations, and networks, infrastructure, and routines.

Operational: The activities, tests, and actions that have a short-term horizon and support implementation.

Source: Based on Loorbach, 2010

to lower and lower to higher tiers of decision-making (Sinclair, Doelle, & Duinker, 2017).
- Effective SEA is integrated within the institutional, legislative, and administrative contexts in which PPPs are developed and implemented, thus informing and improving decision-making culture (Hilding-Rydevik & Bjarnadóttir, 2007).

Approaches to SEA

There is no single best model of SEA. SEA is flexible, leaving ample space for interpretation (Fischer, 2002). Noble and Nwanekezie (2017) suggest a spectrum of SEA, from less to more strategic (Figure 12.3). At one end of this spectrum, SEA reflects the tradition of assessing the environmental impacts of a proposed or existing PPP. At the other end of the spectrum, SEA is an opportunity to shape institutional processes and drive change. Noble and Nwanekezie (2017) argue that the object of assessment at either end of the spectrum may be the same, PPPs. The difference is in the purpose(s) of the application and the extent to which SEA adopts the strategic principles discussed above.

Compliance-based SEA is about whether, and to what extent, a proposed PPP complies with, or supports, specified objectives, policies, or commitments and, if necessary, identifies and explores options to ensure compliance. The compliance-based approach ensures that certain environmental factors have been considered in the PPP's development, or in its approval, and that the PPP supports or does not contradict other legislation or policy (Noble & Nwanekezie, 2017). SEA under this model serves a risk management role for government departments and agencies prior to PPP implementation. Partidário (2009) describes this approach as "a mechanism of control of compliance with the existing legislation and policy requirements" and cautions that although the object of assessment may be a policy, the ability to ultimately influence strategic directions can be limited (Box 12.6).

Project EA–like SEA tends to approach the generation of an impact assessment report as the end itself. SEA is removed from the formulation of a PPP and is focused on assessing its potential environmental impacts and, at best, comparing the PPP's impacts to those of viable alternatives (Noble & Nwanekezie, 2017). This is the

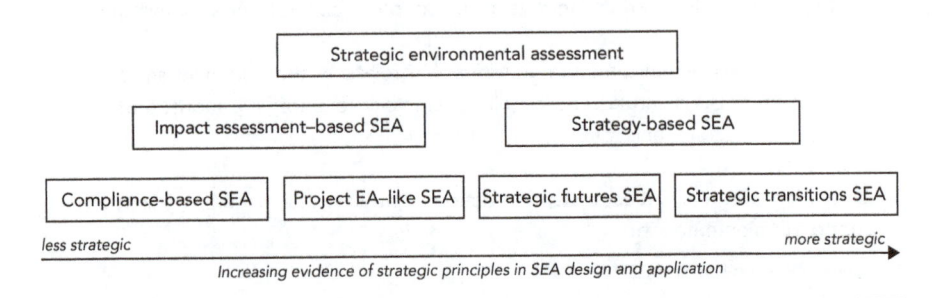

Figure 12.3 Spectrum of approaches to SEA from less to more strategic

Source: Based on Noble & Nwanekezie, 2017

Box 12.6 Compliance-Based SEA

The Greening Regional Development Programmes Network (GRDPN, 2006) *Handbook on SEA for cohesion policy 2007–2013*, an initiative financed in part by the European Union, promotes a compliance-based approach to SEA. The EU Cohesion Policy provides about one-third of the total budget of the European Community for initiatives that increase economic and social cohesion throughout the EU. As per Directive 2001/42/EC, plans and programs prepared for Cohesion Policy funding are subject to SEA. The GRDPN handbook provides guidance on how to conduct SEA for Cohesion Policy plans and programs. Its main message is that SEA is a tool for "greening" plans and programs and for improving their overall logic, consistency, and chances for success within the overall Cohesion Policy objectives. The handbook does indicate that environmental issues associated with the proposed plan or program are to be considered in SEA, but specific attention is focused on an evaluation of "the way in which the environmental protection objectives, established at international, EU or Member State level and relevant to the plan or programme, and any environmental considerations have been taken into account during the preparation of the programming document" (GRDPN, 2006, p. 32).

---

primary approach to SEA under many directive-based systems, such as under the EU directive on SEA and the Canadian cabinet directive, both of which establish a typical project EA-like approach to scoping, impact identification, and impact mitigation. Though important for ensuring that environmental factors are considered in PPPs, and potential impacts mitigated, new PPPs or strategic directions that fundamentally differ from what is initially proposed rarely emerge.

Strategic futures SEA is about shaping or formulating strategic initiatives or PPPs; it's often the approach to SEAs of land-use policies or plans in resource regions. The focus is on identifying and assessing the potential implications of alternative future scenarios or land uses, evaluating the relative risks and opportunities, and establishing a preferred strategic direction or PPP approach. SEA is often intended to influence land-use choices by shaping and providing direction to next-level decisions (Noble & Nwanekezie, 2017). SEA may even become the substitute for regional planning processes and the SEA report the land-use plan (see Environmental Assessment in Action: Strategic Futures SEA in Saskatchewan's Great Sand Hills). In the Canadian context, the strategic futures approach to SEA is gaining much traction external to the cabinet directive to explore regional land-use and development scenarios in regions such as Ontario's mineral-rich Ring of Fire, Alberta's south Athabasca oil sands, and resource development in the Northwest Territories and Yukon.

Strategic transitions SEA is focused on the institutional environment in which PPPs are formulated and implemented and the conditions that either constrain or enable their success. Partidário (2012) explains that SEA can be used to better understand the governance context of strategic initiatives, including PPPs, and influence institutional transitions toward more sustainable outcomes. SEA is a driver of fundamental change in decision-making structures and processes (Kirchhoff, McCarthy, Crandall, & Whitelaw, 2011). Noble and Nwanekezie (2017), for example, explain that SEA can be applied to ensure that a proposed climate change policy or strategy is in compliance with other policy and regulatory priorities or to assess the potential impacts of the climate policy on projects, but SEA can also play a much more strategic role—influencing the climate change policy development process, identifying opportunities for institutional innovations, and facilitating changes in governance or decision-making cultures to ensure the successful implementation of the climate change policy or strategy. This approach to SEA is rare in practice. It requires deep commitment on behalf of governments and integration of SEA as a routine part of institutional evaluation and transformation.

Environmental Assessment in Action

Strategic Futures SEA in Saskatchewan's Great Sand Hills

The Great Sand Hills is located in the southwest region of the province of Saskatchewan, about 375 kilometres west of the city of Regina. The Great Sand Hills comprises approximately 1942 square kilometres of native prairie overlaying a more or less continuous surface deposit of unconsolidated sands, with five sand-dune complexes that total 1500 square kilometres. The northern region of the Great Sand Hills is a designated representative area ecological reserve (RAER). Open grasslands and patches of trees and shrubs characterize the Great Sand Hills, and the region is home to several game species and endangered, threatened, and sensitive species.

Natural gas has been exploited in the region since the early 1950s and intensively since the 1980s, with approximately 1500 gas wells now in the area. Livestock grazing has exerted a much longer-term and widespread influence on the landscape, the result of which is a network of permanent trails and vegetative trampling and erosion. In 2004, following decades of concern over the effects of natural gas development and ranching activities on the biological diversity and ecological integrity of the region and several less than successful land-use planning initiatives, the Government of Saskatchewan appointed an independent scientific advisory committee to undertake a regional environmental study of human-induced disturbances in the Great Sand Hills. The terms of reference for the regional environmental study directed that the study be based on an SEA framework. The overall objective of the assessment was to ensure that the long-term ecological integrity of the area is maintained while economic benefits are realized.

The spatial scale of the assessment was multi-tiered, considering biophysical, socio-economic, and cultural boundaries, as well as the reach of existing PPPs that have the potential to affect any proposed land-use scenario for the region. The biophysical scale of assessment was based primarily on the Great Sand Hills' dunes and grasslands. The socio-economic boundary was based on a larger area

12 | Strategic Environmental Assessment

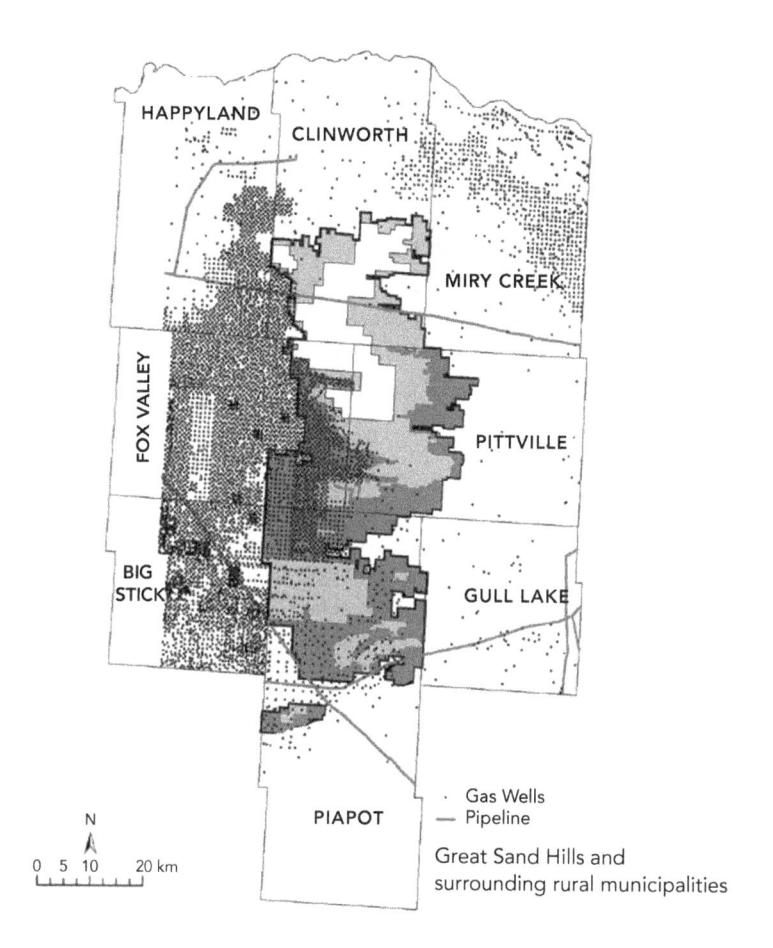

[Map showing Great Sand Hills and surrounding rural municipalities: Happyland, Clinworth, Miry Creek, Fox Valley, Pittville, Big Stick, Gull Lake, Piapot. Legend: Gas Wells, Pipeline. Scale: 0 5 10 20 km.]

of eight regional municipalities that surround the Great Sand Hills. The cultural boundary, capturing Indigenous interests, extended into the neighbouring province of Alberta. The temporal scale of the assessment was from the 1950s, the beginnings of gas development in the region, projected forward to 2020, at which time gas reserves would be fully tapped.

The assessment consisted of three phases: a baseline assessment, trends identification, and evaluation of alternative scenarios of human-induced surface disturbance. The baseline assessment characterized the biophysical, socio-economic, and cultural environment of the region; identified broad-scale cumulative change; and collected data for identification of stressors, trends projection, and scenarios. The underlying objectives of the baseline assessment were to identify those human activities that have the greatest potential for disturbance and for affecting ecological integrity and VC sustainability. VCs were identified through an open scoping process involving local stakeholders and First Nations and previous land-use planning initiatives. Data collection was based on field studies, secondary sources, focus groups,

continued

participatory mapping, and interviews with more than 250 community members and other interests. The biophysical component of the baseline assessment focused much attention on biodiversity, specifically delineating concentrated biodiversity using a site selection algorithm, MARXAN, based on collected species and habitat data. A total of 37 core biodiversity areas were identified, those with the most to lose if not managed, representing various levels of biological irreplaceability.

In the second phase of the assessment, trends in surface disturbance were identified across the landscape using retrospective analysis of aerial photography and land-use and vegetation/species databases. Rates of change were established for each of the three main sources of stress in the region—roads and trails, gas wells, and cattle watering holes. Significant statistical correlations were found to exist between gas wells and the distribution of roads and trails, and roads and trails were found to be a reasonable surrogate for human-induced surface disturbance in the region. In 1979, for example, there were 76 gas-well surface leases in the region. By 2005, 1391 new wells had been established. Associated with the increase in gas-well development was a growth in roads and trails, which had increased from 2497 kilometres in 1979 to approximately 3175 kilometres by 2005, with new roads and trails 153 times more likely to be built in association with a new gas-well pad than elsewhere in the region. Annual rates of change in development and associated patterns of surface disturbance were used to build statistical models to quantify trends in surface disturbances. Parallel to the biophysical assessment was a regional economic assessment, a telephone-based survey to assess local perceptions of current impacts of ranching and gas activity, and a series of participatory GIS workshops with stakeholders to identify goals for the region, including preferred land-use patterns and designations.

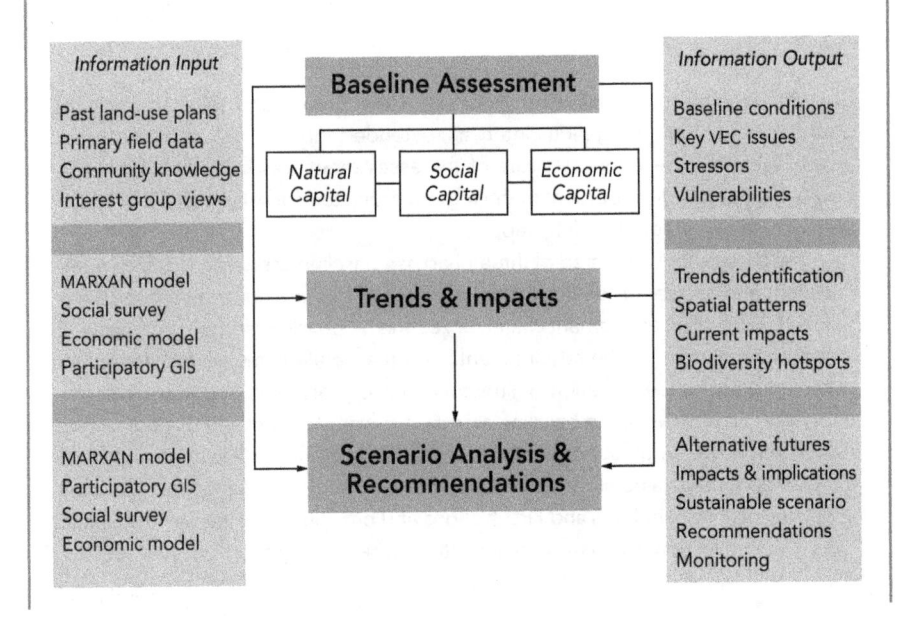

Information Input	Baseline Assessment	Information Output
Past land-use plans Primary field data Community knowledge Interest group views	Natural Capital — Social Capital — Economic Capital	Baseline conditions Key VEC issues Stressors Vulnerabilities
MARXAN model Social survey Economic model Participatory GIS	Trends & Impacts	Trends identification Spatial patterns Current impacts Biodiversity hotspots
MARXAN model Participatory GIS Social survey Economic model	Scenario Analysis & Recommendations	Alternative futures Impacts & implications Sustainable scenario Recommendations Monitoring

In the third phase of the assessment, alternative land-use scenarios were developed based on possible futures and rates of change in human disturbance, primarily roads and trails, gas wells, and cattle watering holes. A GIS was then used to project future growth rates, based on the statistical models developed, and spatial patterns of disturbance across the landscape. Species, range, and biodiversity responses to disturbances under each scenario were modelled using statistical and spatial relationships determined in the baseline and trends analysis phases, and the patterns of disturbance relative to the locations of biodiversity hotspots were mapped. The first two scenarios were based on past trends in development and represented two variations of the future; the main difference between the scenarios was a more ambitious natural gas development agenda under the second such that all proven, probable, and possible reserves would be developed by the end of the projection period. The third scenario was predicated on a conservation-based approach and designed to conserve biodiversity through the protection of biodiversity hotspots and by further reducing surface disturbance outside the core biodiversity areas. Under all three scenarios, disturbances due to cattle watering holes were projected using a random pattern and minimum spacing criteria. An example of the scenario projections for natural gas well development is depicted in the figure below.

The third scenario, a conservation-based scenario that delineated sites of enhanced biodiversity protection and best-practice management for developments

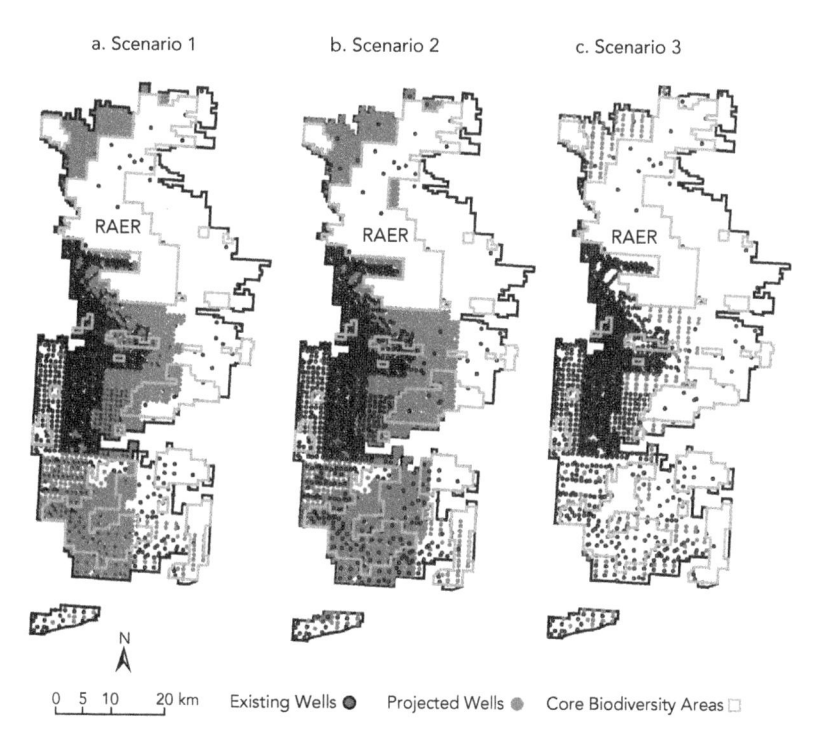

a. Scenario 1   b. Scenario 2   c. Scenario 3

0 5 10 20 km   Existing Wells ●   Projected Wells ●   Core Biodiversity Areas ▢

continued

outside those core biodiversity areas, was identified as the recommended development scenario. More than 60 recommendations emerged from the assessment regarding future development planning, environmental monitoring, and biodiversity targets and thresholds. The Great Sand Hills assessment concluded nearly 15 years of regional planning to provide a strategic direction for effects management and future land use. What was different about the regional Great Sand Hills assessment, in comparison to past planning and assessment initiatives in the region, was its grounding in an SEA framework. This enabled the assessment to reach beyond the constraints of individual project-based issues, many of which are not subject to any form of impact assessment, in order to address the nature and underlying sources of cumulative change and to identify desirable futures and outcomes.

That said, several broader challenges emerged from the Great Sand Hills experience and important lessons for future SEA applications elsewhere. First, the Great Sand Hills assessment, like many regional studies or assessments in Canada, was a "one-off" initiative with no real mechanism to sustain it as an integral part of regional planning and downstream project assessment. Second, the intent in the Great Sand Hills assessment was that it would inform and guide future development activities, land-use zoning, and decision-making. However, as in most other jurisdictions in Canada, there is no formal tiered system of policy, plan, and program assessment to effectively carry the results forward from the strategic down to the project scale. Third, although the Ministry of the Environment commissioned the Great Sand Hills assessment, it lacked the institutional capacity and the authority to fully implement the Great Sand Hills plan and recommendations. Many recommendations were beyond the scope of the ministry, particularly issues that relate to land-use governance, gas royalties, and socio-economics. Delineating roles and responsibilities for implementing the results of SEA, and creating an enabling institutional environment, is an important prerequisite for its success.

Sources: Noble, 2008; figures and maps adapted from GSH SAC, 2007

SEA Benefits

SEA is based on the premise that project EA, which reacts to a proposed development, is not sufficient by itself to ensure sustainable development. SEA attempts to integrate *environment* into higher-order decision-making processes. This early integration of environmental considerations into PPPs ensures a more proactive process whereby strategic options and futures are identified and assessed early before irreversible project decisions are taken (Box 12.7). SEA is thus a way of ensuring that downstream project planning and development occurs within the context of the *desirable* outcomes that society wants to achieve, effectively managing the *sources* rather than the *symptoms* of environmental change.

Box 12.7 SEA Benefits and Opportunities

- Streamlining project EA by early identification of potential impacts and cumulative effects
- Allowing more effective analysis of cumulative effects on broader spatial scales
- Facilitating more effective consideration of ancillary or secondary effects and activities
- Addressing the causes of impacts rather than simply treating the symptoms
- Offering a more proactive and systematic approach to decision-making
- Facilitating the examination of alternatives and the effects of alternatives early in the decision process before irreversible decisions are taken
- Facilitating consideration of long-range and delayed impacts
- Providing a suitable framework for assessing overall, sector-wide, and area-wide effects before decisions to carry out specific project developments are made
- Providing focus for project EA on how proposed actions fit within the broader context of the region
- Verifying that the purpose, goals, and direction of a proposed plan or initiative are environmentally sound and consistent with broader policy, plan, and program objectives for the region
- Saving time and resources by setting the context for subsequent regional and project EAs, making them more focused, effective, and efficient

SEA Design

There is no single, agreed-upon framework for SEA, but based on recent international experiences, several common steps can be identified. As with EA, specific design requirements are necessary for each SEA application—including specific methods and techniques. Most of the methods and techniques used in SEA are readily available from project EA, land-use planning, and policy appraisal (Noble, Gunn, & Martin, 2012). The following sections present the basic components of an SEA framework (Figure 12.4).

CONTEXT
1. Establish the SEA purpose
2. Scope the strategic context and objectives
3. Establish baseline conditions

ASSESSMENT
4. Identify strategic options
5. Assess impacts, opportunities, and risks
6. Identify a strategic direction

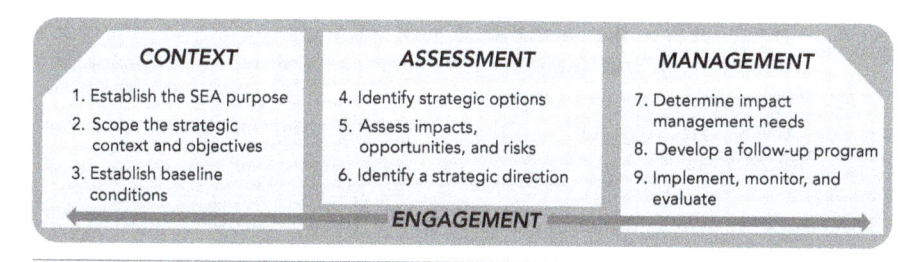

MANAGEMENT
7. Determine impact management needs
8. Develop a follow-up program
9. Implement, monitor, and evaluate

ENGAGEMENT

Figure 12.4 Generic SEA process components

The framework, based on Noble (2017) and CCME (2009), is intended to provide the structure to SEA that allows decision-makers to create and evaluate strategic initiatives and their impacts, without in any way constraining the choice of methods or tools or the tier of application.

Establish the SEA Purpose

Context-setting is essential to any impact assessment process. The first stage of SEA is to establish the purpose of the assessment and the **governance framework**. Several key questions must be asked, including:

- What is the purpose of the SEA—its intended role(s) or objective(s)—and why is it important?
- Who is responsible for the SEA, and what stakeholders need to be involved and at what stages of the process?
- What is it that the agency or organization hopes to achieve, inform, or influence by undertaking the SEA?
- Who are the intended knowledge-users?
- What are the needs and opportunities for linking up with other PPPs and with project EA?
- What is the timeline for the strategic decision?

SEA is often triggered by specific legislative or directive-based requirements, but it can also be proactively pursued to tackle complex environmental and societal challenges. For example, SEA may be applied to explore the opportunities and risks of transitioning northern, off-grid, communities to environmentally sustainable energy options or to inform a nationwide framework for recovery planning for species at risk. The agencies and stakeholders involved must see value in the SEA process to assist in its contributing actions toward a specific purpose or set of objectives. SEA thus can be valuable under a variety of situations and opportunities, including:

- conducting analyses to inform strategies to address specific objectives or priorities;
- the development and (re)assessment of priorities or performance objectives to address environmental issues or challenges;
- institutional, policy, or knowledge-gap analysis to identify key issues affecting environmental outcomes and achievement of objectives and to recommend institutional strengthening actions;
- the pursuit of upstream analytical work, including baseline assessment and visioning, to inform institutional or policy reform or strategy formation; and
- exploring sector-wide opportunities by examining environmental priorities, exploring future options, and facilitating sustained stakeholder interaction.

A clear governance framework is important to the success of SEA and to its ability to inform decision-making. The governance framework refers to the network of interrelated government and/or non-government organizations and institutional arrangements that will inform, and be informed by, the SEA. The role of a governance framework is important for setting SEA priorities but also for ensuring the support of those departments or agencies whose cooperation, and future PPPs, is important to realizing the strategic objectives set out in the SEA.

Scope the Strategic Context and Objectives

Scoping is about characterizing the condition(s) or context for the assessment. Scoping ensures that the SEA is focused on a manageable set of issues that are of critical importance to decision-making, to the stakeholders engaged, and to the development of PPPs. Scoping involves describing the institutional context and identifying the critical decision factors.

Describe the institutional context

It is necessary to identify the policies, mandates, and targets or objectives that need to be met or that may influence the SEA and the adoption of SEA outcomes. Institutional context refers to the various policies, objectives, or legal obligations that are set out in broad agency mandates or commitments that are relevant to the purpose or focus of the SEA (Partidário, 2012; World Bank et al., 2011). These policies, objectives, or obligations may include such matters as national climate commitments or energy policies and may affect the scope of strategic options that can be reasonably considered and help to frame specific objectives to guide the assessment.

Identify critical decision factors

Depending on the level of assessment (policy, plan, or program), emphasis may focus on broad policy issues or on the state of the region and valued components. These issues or factors are the **critical decision factors** (CDFs) for the SEA (Partidário, 2012)—what matters for decision-making. CDFs may be identified based on policies, agency mandates, or sustainability goals and informed by stakeholder engagement (Gallardo, Duarte, & Dibo, 2016). CDFs are often broad-based, such as ecological functioning or inter- and intra-generational equity, and must be translated into more specific objectives (or assessment criteria) that can provide direction for assessing the opportunities and risks associated with different strategic options (Figure 12.5). For example, in an SEA focused on electricity supply futures, CDFs such as inter- or intra-generational equity may be translated into a more specific objective, such as ensuring reliable and affordable

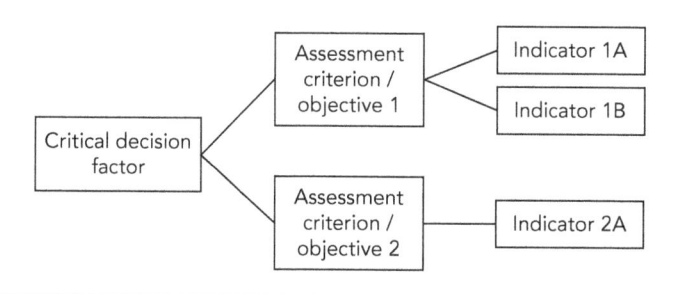

Figure 12.5 Critical decision factors, objectives, and indicators

access to electricity to meet expected demands over the next 50 years. Objectives often reflect agency, ecological, or societal *outcomes* to be pursued or achieved. Objectives may be expressed as quantifiable targets, or they may be broad qualitative and aspirational statements. Objectives need to be made operational by indicators or metrics to assess conditions, opportunities, and risks and to help draw conclusions about the acceptability of different options in achieving objectives (see Partidário, 2012).

Establish the Baseline Condition(s)

SEA requires a baseline or reference condition against which options can be evaluated (World Bank et al., 2011). Establishing the baseline condition(s) usually consists of the following:

- *Characterizing conditions and trends:* Indicators or metrics are used to characterize current conditions and trends. Attention is focused on what is known, knowledge gaps and uncertainties, and the potential relationships between driving forces and changes or trends in CDFs. It can sometimes be difficult at the strategic level to quantify condition change; depending on the objective of the SEA and data available, only a general sense of direction is needed in baseline conditions—i.e., getting better or getting worse; increasing or decreasing (CCME, 2009)—but some sense of directional change is important.
- *Assessing distance from objectives:* When data are limited for trends analysis, attention is focused on assessing conditions in relation to targets, intended objectives, or desirable state(s). The purpose is to understand the current situation relative to achieving specified objectives and with respect to key policies, objectives, or obligations—not simply to create a data inventory (Partidário, 2012).
- *Benchmarking capacity:* The ability or capacity of existing policies, strategies, or institutional arrangements to realize desired objectives, or to direct change toward their achievement, needs to be understood. The focus may be on current institutional policies, strategies, mandates, or management frameworks—depending on the scope and purpose of the SEA. Consideration is also given to the capacity to transform or to facilitate change, if needed, toward the achievement of objectives.

Identify Strategic Options

What is the range of possibilities or options available? SEA is fundamentally about identifying and evaluating the opportunities and risks associated with alternative pathways; it is about the consideration of the consequences and most appropriate responses under different circumstances (Duinker & Greig, 2006). This typically involves some consideration of what *could* happen, what is *most likely* to happen based on the current situation and recent trends, and what is *preferred* to happen in terms of meeting specific development goals.

There are different ways to approach the development of strategic options, depending on the purpose or objective(s) of the SEA. Generally, one of two basic approaches is adopted (Box 12.8):

1. Future scenarios are developed that represent different visions of a plausible future. Scenarios are especially useful when multiple end points or futures are possible and when the intent is to understand how to prepare for or manage a range of future possibilities—or whether specific management objectives can even be achieved. The future states may represent different mandates, different policy environments or outcomes, different assumptions about environmental or climate conditions, or different natural resource sector structures or compositions, to name a few.
2. Strategic options are identified that represent different paths to achieve agreed-upon goals, commitments, or mandates or to reach a desired condition or outcome. Strategic options are policy or planning actions, directions, or strategies that help transition from the current state to a more desired one (Partidário, 2012). A range of reasonably plausible, but structurally different, strategic options should be considered, in addition to the current situation, if applicable, and carried forward. The strategic options approach is appropriate when there is a shared vision regarding the issue(s) at hand. Attention is focused on how to bridge the gap to get there (Partidário, 2012).

Box 12.8 Future Scenarios and Strategic Options in SEA

Future scenarios present outcomes—what things could look like. Sometimes future scenarios can be developed to help determine where we want to end up, based on the objectives defined. Consider a recent proposal for SEA of Brazil's sugarcane industry (Gallardo et al., 2016). Sugarcane is among the most important sources of energy in Brazil and used for a variety of other purposes—but mainly sugar production. Most sugarcane mills operating in Brazil have the option, and flexibility, to shift between the production of sugar and ethanol based on emerging market opportunities. Given projected demands for biofuels globally, a significant increase in the production of ethanol from sugarcane in Brazil is anticipated. As such, Brazil's Decennial Energy Plan presents a single future—business as usual—developed from a series of mathematical models and based on macroeconomic projections to show what the future will most likely look like given anticipated market conditions. But is this the most environmentally and socio-economically sustainable future? The SEA framework identifies a competing sustainability scenario against which the implications can be explored and decisions made about what is desirable and what actions might be needed to reverse trends in indicators, or at least mitigate them. An example from the SEA framework is provided on page 264.

Strategic options present pathways—things we could do or opportunities we can explore. Strategic options are formulated to help determine how best to achieve a desired future or at least to explore the implications of different ways to move forward. This was the approach adopted by the Scottish Environment Protection Agency in an SEA for flood risk management strategies. Flood risk management

continued

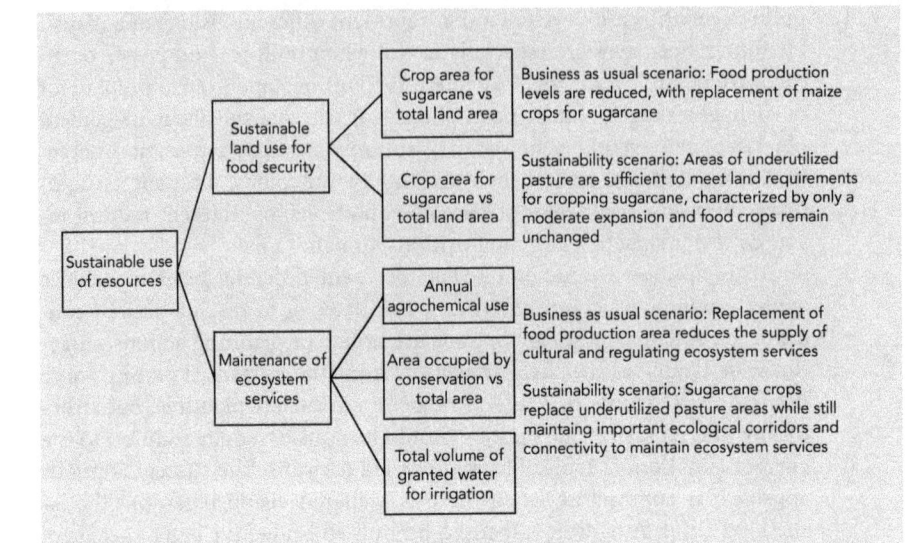

```
Sustainable use          Sustainable        Crop area for       Business as usual scenario: Food production
of resources             land use for       sugarcane vs        levels are reduced, with replacement of maize
                         food security      total land area     crops for sugarcane

                                            Crop area for       Sustainability scenario: Areas of underutilized
                                            sugarcane vs        pasture are sufficient to meet land requirements
                                            total land area     for cropping sugarcane, characterized by only a
                                                                moderate expansion and food crops remain
                                                                unchanged

                                            Annual
                                            agrochemical use
                                                                Business as usual scenario: Replacement of
                         Maintenance of     Area occupied by    food production area reduces the supply of
                         ecosystem          conservation vs     cultural and regulating ecosystem services
                         services           total area
                                                                Sustainability scenario: Sugarcane crops
                                                                replace underutilized pasture areas while still
                                            Total volume of     maintaining important ecological corridors and
                                            granted water       connectivity to maintain ecosystem services
                                            for irrigation
```

strategies set the direction and priorities for flood risk management and aim to reduce flood risk. The SEA of Scotland's Flood Risk Management Strategies does not address the effects of individual planning actions; rather, it focuses on groups of potential actions, or strategies, to help direct decisions about future investments in specific flood risk management initiatives. The SEA focuses on several topics, from population and human health to climate to material assets, each of which is defined by an objective and a set of qualitative indicators based on ecosystem services. A range of non-structural runoff reduction and river and floodplain restoration flood risk management strategies are then presented and the risks and opportunities for ecosystem services qualitatively assessed to better understand the desirability of different types of actions and to inform decisions about different paths to sustainably managing flood risks. An example from the SEA framework is provided below.

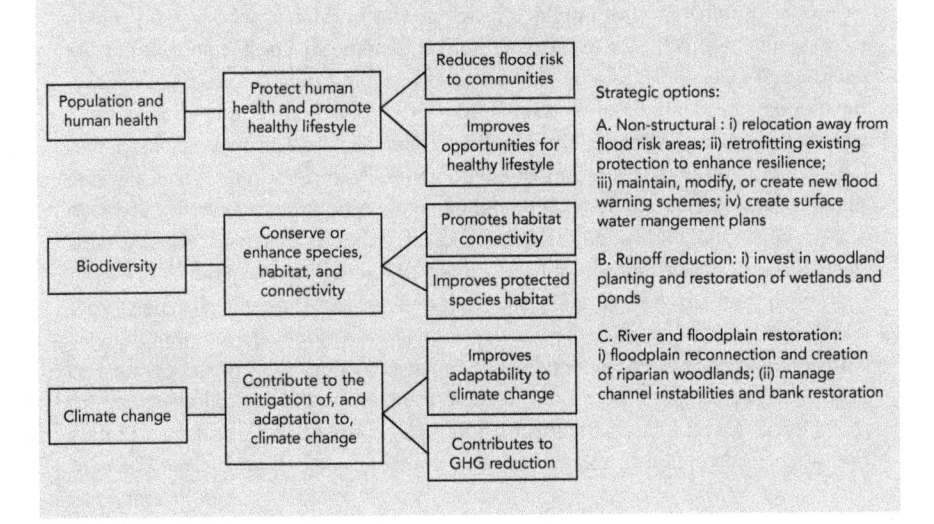

```
                         Protect human         Reduces flood risk
Population and           health and promote    to communities        Strategic options:
human health             healthy lifestyle
                                               Improves              A. Non-structural: i) relocation away from
                                               opportunities for     flood risk areas; ii) retrofitting existing
                                               healthy lifestyle     protection to enhance resilience;
                                                                     iii) maintain, modify, or create new flood
                         Conserve or           Promotes habitat      warning schemes; iv) create surface
                         enhance species,      connectivity          water mangement plans
Biodiversity             habitat, and
                         connectivity          Improves protected    B. Runoff reduction: i) invest in woodland
                                               species habitat       planting and restoration of wetlands and
                                                                     ponds

                                               Improves              C. River and floodplain restoration:
                         Contribute to the     adaptability to       i) floodplain reconnection and creation
                         mitigation of, and    climate change        of riparian woodlands; (ii) manage
Climate change           adaptation to,                              channel instabilities and bank restoration
                         climate change
                                               Contributes to
                                               GHG reduction
```

A combination of the future scenario- and strategic options-based approach is also possible whereby a range of future scenarios is developed, followed by a range of strategies for achieving each scenario and the risks and opportunities for CDFs assessed, or a range of strategies is developed, the possible futures they may lead to is explored, and the implications for CDFs are assessed.

Assess Impacts, Opportunities, and Risks

What are the impacts, opportunities, and risks under each future or strategic option? What are the potential condition changes or responses in indicators? Are thresholds or limits likely to be exceeded? Will the future goal or situation bring us closer to achieving specified targets or sustainability objectives? What are the risks and distribution of impacts associated with the status quo versus alternative options? By focusing the assessment on opportunities and risks, SEA can help to find better pathways and help policy and decision-makers to work toward better outcomes.

Depending on the types of questions asked, available resources, and the information available, assessment may be quantitative and based on modelling or simulation approaches. This may be the case when the SEA is focused on resource sectors (e.g., energy production futures or options) and where systems models can be developed. In other cases, assessment may be qualitative and based on reasoned argumentation about the opportunities and risks presented by each option or strategic direction. There is no universal method, or set of methods, that is best, and methods may involve spatial-analytical tools, such as MARXAN or ALCES where appropriate to the objectives and data; decision analytical tools such as multi-criteria analysis when systematic comparisons of options against criteria is desirable; Bayesian belief networks when graphically tracing and depicting inferences and outcomes is desirable; or simple goals-achievement matrices and reasoned argumentation when assessments are meant to be qualitative and rapid.

Identify a Strategic Direction

SEA should result in some form of strategic direction for agencies or decision-makers or for policy or strategy formulation. It might be that a preferred future scenario or strategic option is identified and recommended for adoption—this could be the case when the SEA is focused on, for example, developing agency priorities or mandates. If so, recommendations often centre on how to maximize opportunities and minimize risks under that specific option or future scenario. In many cases, however, the SEA is exploratory of futures and options, and no specific strategic direction or future is extracted as *the* solution. In such cases, the recommendations that emerge concern how to better manage, or formulate more resilient policies, under a range of possibilities such that objectives can be achieved. The recommendations or direction emerging depend, in large part, on the reasons for pursuing SEA in the first place.

There is a range of different types of recommendations or directions that *could* emerge, for example:

- the adoption of a specific option as a strategy or foundation for a PPP;
- the recognized need to formulate certain PPPs to better achieve objectives;

- the need for more resilient strategies or PPPs that are adaptive to a range of future trajectories and outcomes;
- capacity-building recommendations for overcoming institutional or agency constraints on achieving desired outcomes;
- transformations in priorities, policies, or mandates to better align with desired futures and objectives achievement;
- rethinking of mandates and objectives to better align with macro-level goals or commitments; or
- the need to fill critical knowledge gaps to better understand the uncertainties associated with the adoption of different strategic options.

Determine the Impact Management Needs

Even the "preferred" strategic direction may have potentially adverse effects that cannot be avoided and must be managed. They may be impacts on certain environmental components, resource sectors, societal groups, or other PPPs. Consideration must also be given to institutional conditions or governance. For example, there may be institutional or policy constraints to the strategic direction identified or to achieving certain desired objectives. At the strategic level, it is typically institutional challenges, more than technical ones, that need to be managed. It may be that a desired option simply cannot be implemented given institutional or policy constraints (e.g., regulatory constraints, competing policy objectives, agency mandates), and efforts must focus on whether such constraints can be managed. It may be necessary to revisit objectives or even the chosen strategy.

The purpose is to consider the policy and the institutional, legal, regulatory, and capacity opportunities and gaps to address the opportunities and risks associated with the strategic course of action (World Bank et al., 2011). Understanding institutional opportunities and constraints can help to set in motion actions that lead to fundamental transitions in policies, mandates, and management strategies—or, at a minimum, help to identify institutional constraints and the capacity-building requirements to realize objectives or to achieve desired environmental sustainability outcomes. Several questions may be posed, including:

- What are the institutional capacities and constraints that need to be addressed?
- Are changes needed in agency or institutional strategies (i.e., existing PPPs) or mandates to realize sustainable development goals and objectives?
- Are the potential risks conflicting with other policies, mandates, and commitments?

Develop a Follow-Up Program

No assessment is complete without a follow-up program. Follow-up is essential to ensuring that SEA, and the resulting PPP, is delivering its anticipated or desired outcomes and that any mitigation measures prescribed have been implemented and are working. The rationales for follow-up at the strategic tier are generally the same as those at the project level: derived from the notions of uncertainty and risk associated with decision-making and on the need for feedback and learning. However, the complexities and uncertainties of determining post-implementation environmental implications are exacerbated at the strategic level in that:

- PPPs are often formulated in abstract terms, resulting in vague directions for acting;
- whereas significant deviations from the original plan are abnormal at the project level, they are typical of strategic-level processes; and
- there are fewer direct linkages between decisions at the strategic level and actual PPP impacts (Cherp, Watt, & Vinichenko, 2007).

It is important to be constantly scanning for significant exogenous influences, or "game-changers," that may affect the strategic course of SEA—including the reasonableness of the options considered. Such issues or circumstances are what Cherp, Watt, and Vinichenko (2007) refer to as emergent or unexpected issues or events—they cannot be controlled, but they can be instrumental to the nature and shape of the SEA process. Such exogenous influences, or emergent factors, may include, for example, new agency, national, or international policy or institutional commitments or obligations; significant changes in market conditions; technological innovations that may provide new options or solutions to environmental challenges; or new discoveries that may require revisiting certain assumptions and objectives or reassessing the opportunities and risks.

Implement the Strategy, Monitor, and Evaluate

Regardless of the process efficiency and effectiveness, SEA is of little benefit if results are not implemented. However, implementation of a new strategy or PPP is easier said than done. There is no single best implementation style; rather, implementation depends on the environment within which the strategy or PPP is being implemented (Noble & Harriman, 2008). The more complex the strategy and the greater the uncertainty or potential conflict involved, the more preferred is an adaptive approach. Strategies and PPPs must be sufficiently adaptive to system changes, bifurcation, and external and emergent stressors and responsive to new knowledge gained through monitoring and follow-up processes (Cherp, Watt, & Vinichenko, 2007).

Engagement

Meaningful engagement is foundational to EA and equally important in SEA, but such engagement must be strategic and smart. Strategic means that engagement occurs early, shaping the purpose of SEA and establishing the governance framework, and at critical points throughout the SEA process, but it need not be comprehensive. Smart means transparent communication about trade-offs, risks, and uncertainties associated with the strategic options or futures under consideration.

The stakeholders that need to be engaged, and in different capacities, can vary significantly and yet still be appropriate, depending on the objectives of the SEA. Based on recent experiences with SEA, for example, participation is sometimes restricted to government agencies, and public input is requested only at critical decision points, if at all. This is often the case when the SEA is intended to give strategic direction to government and may involve, among other things, agency mandates or sensitive information. In other cases, there is widespread public consultation with all stakeholders from an early stage through to SEA completion. This is often the case when the SEA is designed, in part, to gain public input on an issue that is of strategic or societal importance or when SEAs are applied in regional planning contexts (Hirji & Davis, 2009). In between are SEA applications for which Indigenous governments or communities, or the private sector, engaged as key interests who are potentially affected by the strategic options or futures being considered, can provide information that will allow a better understanding of the distribution of opportunities and risks or whose opposition would be detrimental to the success of any strategic initiative.

There is no specific formula for engagement that is appropriate for all SEA applications, and the nature of stakeholders involved, and their level of engagement, will depend on the strategic issue(s) being addressed and the intended purpose of the SEA. Key is to ensure a dialogue platform with stakeholders at the relevant administrative levels, both public and private. Partidário (2012) explains that this will, importantly, enable discussion on strategic priorities and objectives and a shared vision, identification of CDFs and situation assessment, and validation of the assessment of opportunities and risks under different futures or strategic options for the achievement of sustainable development goals.

Enduring Challenges

The utility of SEA rests largely on its ability to improve and streamline next-level decisions and to fill important policy gaps. A major challenge to SEA in Canada, however, is its often–ad hoc nature (Doelle, 2009) and disconnect from a larger system of integrated policy, planning, and project decision-making (Noble et al., 2019). SEA applications have been described as "one-off" activities, often with limited influence over policy choices or project EA design or approvals (Council of Canadian Academies, 2019). There is also reluctance among political decision-makers to fully implement SEA (Noble et al., 2019)—especially when the issues at hand are politically sensitive (Lobos & Partidário, 2014). The problem is that most strategic issues, from climate change to energy policy to regional land-use designations, are politically sensitive. Good SEA means shaping the formation of PPPs and priorities; informing strategic-level decisions; and fostering transitions in decision-making practices toward more sustainable futures (Noble et al., 2019). Doing so is difficult in the absence of the political appetite to implement SEA and to align governance structures to ensure that SEA can play a meaningful role in decision-making.

Key Terms

compliance-based SEA
critical decision factors
governance framework
programmatic environmental assessment
project EA–like SEA
strategic environmental assessment (SEA)
strategic futures SEA
strategic transitions SEA

Review Questions and Exercises

1. What are the potential benefits and challenges of applying environmental assessment to policy-level issues?
2. Identify the different models of SEA approaches, and provide an example of the type of problem or situation to which each might apply.
3. What are the provisions, if any, for policy, plan, or program assessment in your jurisdiction? What do you see as the main challenges to SEA?
4. Visit the website for the Impact Assessment Agency of Canada at https://www.canada.ca/en/impact-assessment-agency.html where you can search for the Cabinet Directive on the Environmental Assessment of Policy, Plan and Program Proposals. Visit the European Commission website at https://ec.europa.eu/environment/eia/sea-legalcontext.htm where you can find the EU SEA directive. Compare and contrast the directives in terms of their objectives, scope, requirements for reporting, and public involvement.
5. An important part of SEA is the identification of critical factors that are important for decisions and strategic options. Following one of the examples in Box 12.8, identify a problem (e.g., energy transition, climate policy, land-use zoning) relevant to your region, and scope the critical decision factors, objectives, indicators, and possible strategic options or scenarios.
6. Obtain a completed SEA from your local library or government registry, or access one online. Using Box 12.4 as a guide, explore the SEA for evidence of "strategic" characteristics. Are the characteristics listed in Box 12.4 evident in the SEA? Are there certain characteristics that seem to be missing? Compare your findings to those of others.

References

CCME (Canadian Council of Ministers of the Environment). (2009). *Regional strategic environmental assessment in Canada: principles and guidance*. Winnipeg, MB: CCME.

Cherp, A., Watt, A., & Vinichenko, V. (2007). SEA and strategy formation theories: from three Ps to five Ps. *Environmental Impact Assessment Review* 27, 624–44.

Council of Canadian Academies. (2019). Greater than the sum of its parts: toward integrated natural resource management in Canada. Ottawa, ON: The Expert Panel on the State of Knowledge and Practice of Integrated Approaches to Natural Resource Management in Canada.

Dalal-Clayton, B., & Sadler, B. (2005). *Strategic environmental assessment: a sourcebook and reference guide to international experience*. London, UK: Earthscan.

Doelle, M. (2009). Role of strategic environmental assessments in energy governance: a case study of tidal energy in Nova Scotia's Bay of Fundy. *Journal of Energy and Natural Resources Law* 27(2), 112–44.

Duinker, P., & Greig, L. (2006). The impotence of cumulative effects assessment in Canada: ailments and ideas for redeployment. *Environmental Management* 37(2), 153–61.

Enbridge Pipelines Inc. (2014). Line 3 replacement program project description. National Energy Board Filing no. A61876. https://apps.neb-one.gc.ca/REGDOCS/Item/View/2487792.

FEARO (Federal Environmental Assessment Review Office). (1993). *The environmental assessment process for policy and programme proposals*. Hull, QC: FEARO.

Fischer, T.B. (2002). Strategic environmental assessment performance criteria—the same requirements for every assessment? *Journal of Environmental Assessment Policy and Management* 4(1), 83–99.

Fundingsland-Tetlow, M.F., & Hanusch, M. (2012). Strategic environmental assessment: the state of the art. *Impact Assessment and Project Appraisal* 30(1), 15–24.

Gachechiladze, M., & Fischer, T. (2012). Benefits and barriers to SEA follow-up: theory and practice. *Environmental Impact Assessment Review* 34, 22–30.

Gallardo, A., Duarte, G., & Dibo, A.P. (2016). Strategic environmental assessment for planning sugarcane expansion: a framework proposal. *Ambiente and Sociedade* 19(2). http://dx.doi.org/10.1590/1809-4422ASOC127007V1922016.

Glasson, J., Therivel R., & Chadwick, A. (2005). *Introduction to environmental impact assessment*. London, UK: Routledge.

Gonzalez, A., & Therivel, R. (2014). Alternatives in strategic environmental assessment of plans and programs. *Fastips* 7. March. Fargo, ND: IAIA.

GRDPN (Greening Regional Development Programmes Network). (2006). *Handbook on SEA for cohesion policy 2007–2013*. EU, INTERREG IIC, GRDP.

GSH SAC (Great Sand Hills Scientific Advisory Committee). (2007). *Great Sand Hills regional environmental study*. Regina, SK: Canadian Plains Research Centre.

Hilding-Rydevik, T., & Bjarnadóttir, H. (2007). Context awareness and sensitivity in SEA implementation. *Environmental Impact Assessment Review* 27, 666–84.

Hirji, R., & Davis, R. (2009). *Strategic environmental assessment: improving water resources governance and decision making*. Washington, DC: The World Bank.

International Atomic Energy Agency. (2018). *Strategic environmental assessment for nuclear power programs: guidelines*. Vienna, Austria: IAEA.

Kirchhoff, D., McCarthy, D., Crandall, D., & Whitelaw, G. (2011). Strategic environmental assessment and regional infrastructure planning: the case of York Region, Ontario, Canada. *Impact Assessment and Project Appraisal* 29(1), 11–12.

LGL Ltd. (2003). *Orphan Basin strategic environmental assessment*. St John's, NL: Canada–Newfoundland and Labrador Offshore Petroleum Board.

Lobos, V., & Partidário, M. (2014). Theory versus practice in strategic environmental assessment (SEA). *Environmental Impact Assessment Review* 48, 34–46.

Loorbach, H.D. (2010). Transition management for sustainable development: a prescriptive, complexity-based governance framework. *Governance* 23, 161–83.

Noble, B.F. (2000). Strategic environmental assessment: what is it and what makes it strategic? *Journal of Environmental Assessment Policy and Management* 2(2), 203–24.

Noble, B.F. (2002). The Canadian experience with SEA and sustainability. *Environmental Impact Assessment Review* 22(1), 3–17.

Noble, B.F. (2003). Auditing strategic environmental assessment practice in Canada. *Journal of Environmental Assessment Policy and Management* 5(2), 127–47.

Noble, B.F. (2008). Strategic approaches to regional cumulative effects assessment: a case study of the Great Sand Hills, Canada. *Impact Assessment and Project Appraisal* 26(2), 78–90.

Noble, B.F. (2017). *Getting the big picture: how regional assessment can pave the way for more inclusive and effective environmental assessments*. Ottawa, ON: Macdonald-Laurier Institute.

Noble, B., Gibson, R., White, L., Blakley, J., Nwanekezie, K., & Croal, P. (2019). Effectiveness of strategic environmental assessment in Canada under directive-based and informal practice. *Impact Assessment and Project Appraisal* 37(3–4), 344–55.

Noble, B.F., Gunn, J., & Martin, J. (2012). Survey of current methods and guidance for strategic environmental assessment. *Impact Assessment and Project Appraisal* 30(3), 139–47.

Noble, B.F., & Harriman, J. (2008). *Regional strategic environmental assessment: methodological guidance and good practice*. Report prepared for the Canadian Council of Ministers of the Environment, Winnipeg, MB.

Noble, B.F., & Nwanekezie, K. (2017). Conceptualizing strategic environmental assessment. *Environmental Impact Assessment Review* 62, 165–73.

OAGC (Office of the Auditor General of Canada). (2015). *Report 2—Departmental progress in implementing sustainable development strategies: 2015 Fall reports of the Commissioner of the Environment and Sustainable Development*. Ottawa, ON: Office of the Auditor General of Canada. http://www.oag-bvg.gc.ca/internet/English/parl_lp_e_925.html.

Partidário, M.R. (2009). Does SEA change outcomes? International Transport Research Symposium, Discussion Paper 2009-31. Paris, France: OECD/ITF.

Partidário, M.R. (2012). *Strategic environmental assessment: better practice guide*. Lisbon, Portugal: Portuguese Environment Agency & Redes Energeticas Nacionais, SA.

Pope, J., Bond, A., Morrison-Saunders, A., & Retief, F. (2013). Advancing the theory and practice of impact assessment: setting the research agenda. *Environmental Impact Assessment Review* 41, 1–9.

Sheate, W.R., et al. (2003). Integrating the environment into strategic decision-making: conceptualizing policy SEA. *European Environments* 13, 1–18.

Sinclair, A.J., Doelle, M., & Duinker, P. (2017). Looking up, down, and sideways: reconceiving cumulative effects assessment as a mindset. *Environmental Impact Assessment Review* 62, 183–94.

Slootweg, R., & Jones, M. (2011). Resilience thinking improves SEA: a discussion paper. *Impact Assessment and Project Appraisal* 29(4), 263–76.

Therivel, R., & Partidário, M.R. (1996). *The practice of strategic environmental assessment*. London, UK: Earthscan.

Wallington, T., Bina, O., & Thissen, W. (2007). Theorising strategic environmental assessment: fresh perspectives and future challenges. *Environmental Impact Assessment Review* 27, 569–84.

Wood, C., & Djeddour, M. (1989). *Environmental assessment of policies, plans and programs*. Interim report to the Commission of European Communities. Manchester, UK: EIA Centre.

Wood, C., & Djeddour, M. (1992). Strategic environmental assessment: EA of policies, plans and programmes. *Impact Assessment Bulletin* 10(1), 3–23.

World Bank. (1999). *Environmental assessment*. Operational Policy and Bank Procedures, no. 4.01. Washington, DC: World Bank.

World Bank, University of Gothenburg, Swedish University of Agricultural Sciences, & Netherlands Commission for Environmental Assessment. (2011). *Strategic environmental assessment in policy and sector reform: conceptual model and operational guidance*. Washington, DC: World Bank.

World Commission on Environment and Development. (1987). *Our common future*. Oxford, UK: Oxford University Press.

13 Professional Practice and Ethics

Professional Practice

As discussed in Chapter 1, several parties are involved in the EA process, including the project proponent, regulators and decision-makers, public interest groups, and in some cases review panels, consultants, and advisors to each of these parties. There are many roles for the EA professional. An EA professional may be an industry employee responsible for managing the company's EAs and day-to-day environmental regulatory processes; a government employee responsible for implementing EA regulations, determining the need for an EA, and following up on industry commitments; an active member of an environmental non-government organization responsible for EA awareness and playing a watchdog role during the EA process; an Indigenous lands manager responsible for EAs and permitting of development projects on lands under Indigenous jurisdiction; an independent consultant to proponents hired to conduct fieldwork in support of the EIS and assisting with and the preparation of the EIS; or an expert advisor on EA to industry, governments, and public interest groups, including review panels, on the quality of the EA, the nature and potential risks associated with a project's impacts, and the quality of the project's overall EA process.

The EA Practitioner

Perhaps the most diverse role in the EA process is that of the consulting practitioner. Morrison-Saunders and Bailey (2009) explain that practitioners, often employed as consultants to project proponents, advise on relevant EA policies and procedures; assist proponents in fulfilling the regulatory and administrative obligations of EA; undertake the technical work necessary to identify, assess, and mitigate impacts; prepare the EIS; and in some cases engage in consultation processes on behalf of the proponent. Practitioners typically work in interdisciplinary teams. They must be technically competent and understand the EA system and regulatory requirements. Perhaps most important, they must be able to tackle complex problems, manage projects, process and integrate large amounts of information, and communicate effectively. EA regulators, in contrast, implement EA policy and procedures in accordance with the legislative framework and applicable regulatory requirements. They sign off on EA scoping requirements, check to ensure that the EIS complies with the terms of reference established for the assessment, and verify that public comments have been adequately addressed. Morrison-Saunders and Bailey (2009) describe regulators as the gatekeepers for proponents seeking approval of new proposals.

Skills and Qualifications

EA practitioners come from a diverse range of social science, natural science, and technical backgrounds, including geography, law, business, engineering, public health, biology, political studies, sociology, anthropology, chemistry, geology, and hydrology, to name a few. There is no internationally recognized certification for EA professionals, though some jurisdictions do have certification requirements for EA practitioners (e.g., South Africa), whereas most recognize certain professional designations (e.g., professional biologist, geographer, engineer, or geoscientist) but have no EA certification requirements per se. There are, however, minimum international guidelines or standards for EA professionals. In 2010, the International Association for Impact Assessment (IAIA) adopted *Guideline standards for IA professionals* (IAIA, 2010). The guidelines establish the minimum standards for the profession for both impact assessment practitioners (Table 13.1) and administrators (Table 13.2). The standards for impact assessment administrators are similar to those for practitioners but place much more emphasis on knowledge of relevant environmental and related institutions, legislation, policies, and administrative procedures.

Two important skill sets and characteristics for EA practitioners not addressed in the IAIA standards are communication skills and the ability to work in a highly interdisciplinary environment. First, communication is critical for EA practitioners to effectively carry out their responsibilities (Kreske, 1996), including communication between EA team members and the ability to facilitate meaningful communications between proponents and regulators, proponents and communities, and regulators and communities. EA practitioners must also be able to gather, process, integrate, and effectively communicate complex and often highly technical information to non-disciplinary experts in a form that is meaningful to those who require that information to make decisions about the project and the acceptability of its impacts. Second, the EA process is by nature interdisciplinary—it involves the combining of knowledge from multiple disciplines and assembling dynamic and interdisciplinary teams of specialists and generalists to undertake the work required (Morrison-Saunders & Bailey, 2009). The interdisciplinary nature of the teams that carry out EAs, and subsequently review and evaluate the EIS, is of critical importance to a quality EA process. Erickson (1994) explains that EA depends upon the cooperation and coordination of individuals within a team of interdisciplinary practitioners, meaning that good EA often comes down to "getting things done through people" and not specialities or technical aspects alone.

The individuals involved in EA must have access to the complete set of expertise necessary to carry out baseline studies, make predictions of impacts and assess them, propose mitigation and management measures, design follow-up procedures, involve the publics, and evaluate the quality and integrity of the findings. If important skills are lacking, critical impacts or possible mitigation measures can be missed, adverse impacts could occur, and conflict may arise among the affected publics. No individual possesses all the skills necessary to carry out all aspects of an EA—some practitioners bring technical, scientific, and legal skills, and others bring project management and systems skills. However, each individual must appreciate the interdisciplinary nature of the impact assessment process and be willing to work outside disciplinary boundaries. This includes a willingness to adapt disciplinary-based

Table 13.1 Guideline Standards for Impact Assessment Practitioners

Category	Practitioner	Senior Practitioner	Lead Practitioner
Education and training	Relevant degree from an accredited university or a member in good standing of a relevant professionally accredited organization	Relevant degree from an accredited university or a member in good standing of a relevant professionally accredited organization	Relevant degree from an accredited university or a member in good standing of a relevant professionally accredited organization
Experience	Minimum 2 years' experience in undertaking and reporting on impact assessment studies	Minimum 5 years of progressively senior experience and responsibility in designing, undertaking, and reporting on at least component impact assessment studies, including public participation	Minimum 10 years of progressively senior experience and responsibility in designing, undertaking, and reporting on at least component impact assessment studies, including public participation
Understanding of impact assessment methods	A good understanding of impact assessment methods, including cumulative and strategic assessment	A thorough working knowledge of impact assessment methods, including cumulative and strategic assessment	A thorough working knowledge of impact assessment methods, including cumulative and strategic assessment
Impact assessment study management	Demonstrated, under direction, an ability to effectively plan and carry out specialist impact assessment studies	Demonstrated an ability to effectively lead at least component impact assessment studies and, under direction, some multi-disciplinary studies and to look beyond compliance to develop and promote best practice	Demonstrated an ability to effectively lead and integrate comprehensive, multi-disciplinary impact assessment studies at all scales and to look beyond compliance to develop and promote best practice
Sustainable development	A good understanding of the structure, functioning, and interrelatedness of ecological, socio-economic, health, and political systems that support sustainable development	A good understanding of the structure, functioning, and interrelatedness of ecological, socio-economic, health, and political systems that support sustainable development	A good understanding of the structure, functioning, and interrelatedness of ecological, socio-economic, health, and political systems that support sustainable development and a demonstrated ability to apply this understanding to project planning and impact assessment

Impact assessment administrative systems	Familiar with the relevant impact assessment administrative systems and guidelines	Working knowledge of the relevant impact assessment administrative systems and guidelines and a demonstrated ability to interpret and fulfill their requirements	A broad working knowledge of impact assessment administrative systems and guidelines and a demonstrated ability to interpret and fulfill their requirements
Professional development	Actively engaged in continuing professional development through readings, publications/presentations, and/or training	An active commitment to best practice and continuing professional development through readings, publications/presentations, training, and/or mentoring	An active commitment to best practice and continuing professional development through readings, publications/presentations, training, and/or mentoring
Mentoring		An active commitment to mentoring less experienced practitioners	An active commitment to mentoring less experienced practitioners

Source: Adapted from IAIA guidelines, available at www.iaia.org

Table 13.2 Guideline Standards for Impact Assessment Administrators

Category	Administrator	Senior Administrator	Lead Administrator
Education and training	Relevant degree from an accredited university or a member in good standing of a relevant professionally accredited organization	Relevant degree from an accredited university or a member in good standing of a relevant professionally accredited organization	Relevant degree from an accredited university or a member in good standing of a relevant professionally accredited organization
Experience	Minimum 2 years of impact assessment experience, with an emphasis on the administration of public sector impact assessment processes	Minimum 5 years of progressively senior experience and responsibility, with an emphasis on the administration of public sector impact assessment processes, including some experience with conducting integrated impact assessment studies and related public participation	Minimum 10 years of progressively senior experience and responsibility, with an emphasis on the administration of public sector impact assessment processes, including some experience with conducting integrated impact assessment studies and related public participation
Understanding of impact assessment methods	A good understanding of impact assessment methods, including cumulative and strategic assessment	A thorough working knowledge of impact assessment methods, including cumulative and strategic assessment	A thorough working knowledge of impact assessment methods, including cumulative and strategic assessment
Impact assessment administrative systems	Familiar with the relevant impact assessment administrative systems and guidelines	Good working knowledge of the relevant impact assessment, environmental, and related institutions, legislation, policies, and administrative procedures	Detailed working knowledge of the relevant impact assessment, environmental, and related institutions, legislation, policies, and administrative procedures

Review of impact assessment documents	Capable of drafting, under direction, integrated impact assessment requirements for projects, of evaluating the adequacy of assessment documents, of drafting project approval conditions, and of following up on the implementation of conditions	Demonstrated ability to establish integrated impact assessment requirements for projects in at least a few sectors, to evaluate the adequacy of assessment documents, to draft project approval conditions, and to follow up on the implementation of conditions	Demonstrated ability to establish integrated impact assessment requirements for a full range of project types and scales, to evaluate the adequacy of assessment documents, to draft project approval conditions, and to follow up on the implementation of conditions
Sustainable development	A good understanding of the structure, functioning, and interrelatedness of ecological, socio-economic, health, and political systems that support sustainable development	A good understanding of the structure, functioning, and interrelatedness of ecological, socio-economic, health, and political systems that support sustainable development	A good understanding of the structure, functioning, and interrelatedness of ecological, socio-economic, health, and political systems that support sustainable development and a demonstrated ability to apply this understanding to impact assessment reviews and decision-making
Professional development	Actively engaged in continuing professional development through readings, publications/presentations, and/or training	An active commitment to best-practice and continuing professional development through readings, publications/presentations, training, and/or mentoring	An active commitment to best-practice and continuing professional development through readings, publications/presentations, training, and/or mentoring
Mentoring		An active commitment to mentoring less experienced administrators	An active commitment to mentoring less experienced administrators

Source: Based on IAIA guidelines, available at www.iaia.org

methods, find new ways of approaching problems, and being open to new methods and different forms of knowledge to assist in understanding the nature and the severity of potential impacts. EA team leaders, in particular, must have strong teamwork skills, communication skills, and interdisciplinary skills.

Ethical Conduct

The values, norms, experiences, and interrelations of practitioners and other actors engaged in the EA process play a significant role in the quality and integrity of both a project's assessment and the decisions taken. EA is a form of applied research; thus, practitioners must apply ethical research standards (Lawrence, 2005). Ethics refers to moral duty or obligation, which typically gives rise to values or governing principles that are used to judge the appropriateness of behaviour or conduct. More specifically, ethics is a branch of philosophy that is focused on the structuring and defending of what constitutes "right" and "wrong." Lawrence (2005) suggests that **normative ethics**, which seeks to arrive at moral conduct standards, and applied or **practical ethics**, which studies specific practical problems and involves a commitment to action, are especially relevant to EA.

Professional ethics concerns the conduct of professionals in practice, typically set out in codes of conduct (Box 13.1). These codes of conduct assist members of a professional organization in understanding appropriate conduct (the difference between "right" and "wrong") and in applying a certain standard to their professional behaviour and actions. In many instances, failure to comply with a code of conduct can result in expulsion from the professional organization or withdrawal of professional licences to practise.

Ethical Responsibilities

It is the role of the practitioner to ensure that, to the extent possible, complete, unbiased, and accurate information is available to all parties involved in the EA process. It is not professional to produce an EA report solely to meet a legal requirement when such a report must be submitted. Fuggle (2012) explains that the social contract between impact assessment professionals and civil society and decision-makers is such that impact assessments will be conducted with integrity and will be free from misrepresentation or deliberate bias, and impact assessments will respect citizen rights to participate in decisions that affect them. Free from bias does not mean that impact assessment is value-free. Values and good judgment are core to the EA process, from determining the need for an assessment and scoping the valued ecosystem components to interpreting the significance of potential environmental effects and determining whether a project is in the best interest of the environment and society. Ethical responsibilities and how the practitioner reacts to ethical dilemmas is the focus of this chapter's Environmental Assessment in Action feature.

Based on the mission and values statement of the International Association for Impact Assessment (see www.iaia.org), included among the ethical and professional responsibilities of those engaged in environmental assessment is to:

- compile or review impact assessments with integrity and honesty and free from misrepresentation or deliberate bias;

Box 13.1 International Association for Impact Assessment Code of Conduct

As a self-ascribed professional member of the International Association for Impact Assessment, the information and services that I provide must be of the highest quality and reliability. I consequently commit myself:

- To conduct my professional activities with integrity, honesty, and free from any misrepresentation or deliberate bias.
- To conduct my professional activities only in subject areas in which I have competence through education, training, or experience. I will engage, or participate with, other professionals in subject areas where I am less competent.
- To take care that my professional activities promote sustainable and equitable actions as well as a holistic approach to impact assessment.
- To check that all policies, plans, activities, or projects with which I am involved are consistent with all applicable laws, regulations, policies, and guidelines.
- To refuse to provide professional services whenever the professional is required to bias the analysis or omit or distort facts in order to arrive at a predetermined finding or result.
- To disclose to employers and clients and in all written reports any personal or financial interest that could reasonably raise concerns as to a possible conflict of interest.
- To strive to continually improve my professional knowledge and skills and to stay current with new developments in impact assessment and my associated fields of competence.
- To acknowledge the sources I have used in my analysis and the preparation of reports.
- To accept that my name will be removed from the list of self-ascribed professional members of IAIA should I be found to be in breach of this code by a disciplinary task-group constituted by the IAIA Board of Directors.

Source: See https://www.iaia.org/ethical-responsibilities.php

---

- not condone the use of violence, harassment, coercion, intimidation, or undue force in connection with any aspect of impact assessment or implementation;
- conduct impact assessments in the awareness that different groups in society experience benefits and harm in different ways;
- take gender and other social differences into account and be especially mindful of the concerns of Indigenous peoples;
- strive to promote considerations of equity as a fundamental element of impact assessment;
- give due regard to the rights and interests of future generations; and
- not advance one's own private interests to the detriment of the public, clients, or employing institutions.

Environmental Assessment in Action

What Would You Do?

The stakes are often high in EA. The fate of projects and, depending on the severity of the potential impacts, the sustainability of environments and communities can hang in the balance. Impact assessment professionals are thus faced with ethical dilemmas on a regular basis. Typical scenarios are presented below for discussion and consideration. There is no "answer" provided—these are ethical challenges and meant to generate discussion with your peers or fellow practitioners.

Scenario 1

You are a consultant hired by a project proponent to manage an environmental assessment process for a large pulp and paper mill. As the EIS manager, you are responsible for reviewing the technical reports generated by the sub-consultants, presenting the relevant information in the EIS, and concluding whether the predicted impacts will be significant or of concern to environmental quality or public health. Part of the assessment process was based on a predictive model of how chemical X, discharged from pulp mill operations into the river system, might bio-concentrate in sturgeon—a species of considerable importance to a local Indigenous community. The model used to predict the impacts is an off-the-shelf tool, which itself is known among the scientific community to be imperfect based on its internal assumptions and simplifications of ecological systems. But it is the best available modelling tool. Further, there was limited baseline data available to populate the model, thus introducing considerable uncertainty into the predictive outcome. As a trained expert, you are aware of the limitations and uncertainty, as well as the lack of data (and limited quality of data) used to populate the model and predict the concentration and significance of chemical X. The model output, however, shows that the predicted levels are within the regulatory limits and not of concern.

As a practitioner, hired by the project proponent, what would you do?

- Would you accept the model's results as is and report the impact as insignificant?
- Would you report the model results but disclose the uncertainties and poor data involved in making the prediction, thus making the prediction unreliable from a regulator's perspective?
- What are the possible ethical or career implications of your action?

Scenario 2

The following case was developed based on G. Fred Lee and Anne Jones-Lee's "Practical environmental ethics: is there an obligation to tell the whole truth?"

Landfill sites can pose a potential threat to local and regional groundwater. To manage the risk of contamination, there are two types of standards for landfill design: first, performance standards that must be achieved by the design, such as the prevention of contamination to groundwater over the life cycle of the project; second, minimum design standards, such as the thickness and permeability of the liner at the bottom of the landfill pit, used to separate soil and groundwater from waste.

Regulations governing disposal of wastes emphasize the need to protect groundwater from use-impairment for as long as the wastes represent a threat. Regulations governing landfill operations identify a "minimum design standard" for the liner thickness to ensure a maximum impermeability. The regulations *do not* state that the minimum design standard will meet the performance standard and ensure the long-term protection of groundwater. In previous projects, the minimum design standard model had been breached by leachate in only a few months, posing a risk of groundwater contamination.

i) A practitioner hired by the project proponent to assess the landfill project claims in her technical report for the EIS that the proposed landfill "will, without question, be protective of groundwater quality over the long term." Note that the practitioner did not claim that the groundwater will be protected from impact forever.

- Is this unethical practice?

ii) A practitioner hired by the project proponent to assess the landfill project skirts the ethical problem by limiting his evaluation to whether or not the minimum design standard, as set out in regulation, has been met. The practitioner does not say whether the regulations are sufficient to protect health and environmental quality.

- Is this unethical practice?

Scenario 3

Discuss how you would react to each of the following situations:

- The terms of reference you are provided with to undertake the assessment unreasonably constrains the study, likely resulting in results that are not representative of the baseline conditions or of the range of potential impacts of the project.
- As a consultant for a project proponent, you are pressured to limit the scope of the assessment or to present only part of your findings.
- You are asked to change your conclusions about the significance of a predicted impact to make it appear less significant than you believe is suggested by your data.
- Communities refuse to meet with you and participate in the study, yet you are required as per the study terms of reference to consult with local communities.
- You are asked not to report on the uncertainties associated with your models or technical studies used to inform impact predictions.
- In reviewing the EIS on behalf of a proponent, you find inaccuracies or errors in the reported information.
- As an independent expert, you are asked to submit a favourable (or unfavourable) review of a proponent's EIS.

Key Terms

normative ethics
practical ethics

professional ethics

Review Questions and Exercises

1. In addition to the dilemmas presented in Environmental Assessment in Action, what other ethical issues might a professional encounter in the practice of EA?
2. Are there certain ethical issues that may emerge in the professional practice of EA that are specific to working with Indigenous communities? Hint: Consider such matters as traditional knowledge and ownership of information.
3. Using the Internet, search the "codes of conduct" for various professional organizations or associations, such as professional engineers, professional geoscientists, or professional biologists. Are there similarities in the codes of conduct between these professions and the code of conduct presented in Box 13.1 for impact assessment professionals?

References

Erickson, P. (1994). *A practical guide to environmental impact assessment*. San Diego, CA: Academic Press.

Fuggle, R. (2012). Ethics. *Fastips* 2. April. www.iaia.org.

IAIA (International Association for Impact Assessment). (2010). *Guideline standards for IA professionals*. Fargo, ND: IAIA.

Kreske, D.L. (1996). *Environmental impact statements: a practical guide for agencies, citizens, and consultants*. New York, NY: John Wiley & Sons.

Lawrence, D.P. (2005). *Environmental impact assessment: practical solutions to recurrent problems*. Hoboken, NJ: John Wiley & Sons.

Morrison-Saunders, A., & Bailey, M. (2009). Appraising the role of relationships between regulators and consultants for effective EIA. *Environmental Impact Assessment Review* 29, 284–94.

14 Environmental Assessment Prospects

The world of EA has evolved considerably since the first edition of this book was published in 2006. At the time, nobody would have thought that "impact assessment" or "environmental assessment" would become a household term. Some of the reason it has can be attributed to major reforms to EA at the federal level, namely, the introduction of the Canadian Environmental Assessment Act, 2012 and the Impact Assessment Act of 2019—both of which were surrounded by conflict between environmental organizations, industry, and provincial, federal, and Indigenous governments. Canada's energy sector was at the centre of both reforms (Lauerman, 2018), with public information and views about EA fuelled by "alternative facts" (Fischer, 2018) concerning the implications that EA might pose for the nation's largely fossil fuel–based energy economy (Olszynski, 2019).

At the same time, public expectations about what a project-based review and impact management instrument can and should accomplish have changed (Sinclair, Doelle, & Gibson, 2018; Fischer, 2017). Gibson, Doelle, and Sinclair (2016) suggest that the context in which EA "is now promoted and eroded" is increasingly complex and very different from the context into which EA was first introduced in the 1970s. It's also very different from what it was just a decade ago. Impact assessment has been forced to tackle increasingly complex issues and to do so with greater engagement of those interests affected most by the decisions taken—typically, local communities and, in most cases, Indigenous communities.

The demands now placed on EA reflect the urgency of issues facing society, from climate change and biodiversity loss to human rights infringements and impacts to Indigenous lands and traditional livelihoods. It also reflects mounting expectations for governments to respond and perceptions about the efficacy of impact assessment as an appropriate tool for the job (Noble et al., 2019). However, an emerging, and very much practical, challenge to EA across Canada is that many of the most important issues raised when resource development projects are tabled are beyond the scope and scale of the assessment of any one project—and also beyond the means and authority of any one project proponent (and sometimes government) to resolve (Udofia, Noble, & Poelzer, 2017).

The need to tackle complex and global environmental challenges is urgent, but project EA is not well suited for the job (Hegmann & Yarranton, 2011)—it was never designed to address issues of such strategic scope and scale. In a 2019 letter to Impact Assessment and Project Appraisal, I suggest that impact assessment in Canada is at a critical juncture and that we are faced with two possible paths. One path is to continue with the status quo, approaching EA on a project-by-project basis and expecting it to do more and more in response to even bigger challenges.

This will likely result in a project EA and regulatory system that is over-burdened with issues that it simply isn't designed to address, leading to further dissatisfaction with EA, conflict, stalled project decisions, and only superficial treatment of the issues that matter most.

A more optimistic path is to re-imagine project EA as embedded in a nested and integrated system of strategic and regional assessment, supported by the vast science and traditional knowledge base that exists across Canada. Under this scenario, strategic and regional assessments are the norm rather than the exception, paving the way for projects that ensure more sustainable outcomes and rethinking those that do not. Development initiatives (i.e., projects) are triggered and shaped based not *solely* on economic opportunity but also on strategic policy and planning objectives intended to proactively address complex environmental and societal challenges. Thinking and practice thus shifts from using EA as the default tool for assessing and resolving the impacts of everything toward more strategic processes designed to explore policy options and inform regional development trajectories that generate and enhance environmental and social values.

Of course, such shifts are contentious—they require that government policies and plans be up for debate through transparent impact assessment and that project proponents and those who set the terms and conditions of EA and project approvals are accountable to strategic and regional policies, plans, and priorities—and to science. Results of two consecutive reforms to federal EA in Canada suggest that new project-based legislation is unlikely to be the source of innovation needed to truly advance impact assessment. Rather, the greatest opportunities for advancement and transformation of impact assessment are likely rooted in governance solutions.

The *first* opportunity is the growing capacity and desire of many Indigenous governments to leverage strategic partnerships and drive their own assessment processes, coupled with strengthening government commitments to the United Nations Declaration on the Rights of Indigenous Peoples, which articulates the rights of Indigenous peoples "to determine and develop priorities and strategies for the development or use of their lands or territories and other resources" (Article 32). Consider two cases introduced earlier in this book: the Ktunaxa Nation Council's partnership with Teck Coal Ltd to develop, from the bottom up, a cumulative effects framework for the Elk Valley, which would later be adopted as part of a province-wide cumulative effects decision-support framework (Chapter 11), and the Squamish Nation–led Woodfibre EA, which exerted significant influence over project design and impact management (Chapter 10). In a more recent example, in response to concerns about the cumulative pressures of development and climate change, the Mikisew Cree First Nation successfully lobbied the World Heritage Committee to pressure Canada, as a UNESCO member state, to conduct a strategic assessment of the Wood Buffalo region, Alberta, and Peace-Athabasca Delta—home to Wood Buffalo National Park (a UNESCO World Heritage Site) and traditional territory of the Mikisew Cree (IEC, 2018). All three of these examples illustrate innovation and the growing capacity of Indigenous peoples to leverage impact assessment to influence decision-making and address complex environmental and social challenges.

The *second* opportunity is investing in, and making better use of, the science capacity that exists across Canada's post-secondary institutions and university-based research centres. There is a mass of underutilized expertise and data to support impact assessment—data that can save both time and resources for proponents and governments, that can add value to impacted communities, and that have been acquired independent of proposed developments. The science and research community, in collaboration with communities, industry partners, non-profits, and governments, have the capacity and expertise to support impact assessment in ways that are not always possible in the hallways of government. This includes, for example, long-term monitoring and assessment of regional change, model development to support impact prediction, and the maintenance of open access and quality-controlled data for those involved in assessment processes. An increased role for the scientific and research communities is not only a means to strengthen impact assessment but a means to increase independence in the generation and interpretation of evidence used to support regulatory decisions and to re-establish public trust in impact assessment. Of course, this requires some serious rethinking about what and whose evidence is weighed in impact assessment, and decision-makers must be prepared to make decisions based on that evidence over political aspirations.

Project EA has proven itself as an effective instrument—when implemented early, and in collaboration with those communities most affected, and when focused on what project EA was designed to focus on: identifying, assessing, mitigating, and managing the impacts of individual project proposals. It is the task of more strategic and regional processes to ensure that sufficient evidence is available to support decisions, that the "right" projects are being considered and that the types of development proposed are appropriate for the region and align with higher-level policy commitments and plans. Many of the controversies that surround project EA are not because of the failings of project EA per se but because there is no other venue to address the issues of greatest importance or because EA is left to function in a silo and in the absence of any regional or strategic guidance.

References

Fischer, T.B. (2017). Editorial. *Impact Assessment and Project Appraisal* 35(2), 117.
Fischer, T.B. (2018). Editorial: IA, alternative facts and fake news—is the post-factual turn starting to turn? *Impact Assessment and Project Appraisal* 36(3), 207.
Gibson, R.B., Doelle, M., & Sinclair, A.J. (2016). *Next generation environmental assessment for Canada: basic principles and components of generic design.* The Next Generation Environmental Assessment Project. Waterloo, ON: University of Waterloo.
Hegmann, G., & Yarranton, G. (2011). Alchemy to reason: effective use of cumulative effects assessment in resource management. *Environmental Impact Assessment Review* 31(5), 484–90.
IEC (Independent Environmental Consultants). (2018). *Volume 1: Milestone 3—final SEA report. Strategic environmental assessment of Wood Buffalo National Park World Heritage Site.* Edmonton, AB: IEC.
Lauerman, V. (2018). Analysis: Bill C-69—an economy and energy security killer. *Daily Oil Bulletin* 9(24). https://www.dailyoilbulletin.com/article/2018/9/24/analysis-bill-c-69-an-economy-and-energy-security-.

Noble, B.F., Gibson, R., White, L., Blakley, J., Nwanekezie, K., & Croal, P. (2019). Effectiveness of strategic environmental assessment in Canada under directive-based and informal practice. *Impact Assessment and Project Appraisal* 37(3–4), 344–55.

Olszynski, M. (2019). Proposed Bill C-69 amendments undermine science. *Policy Options*, 27 May. https://policyoptions.irpp.org.

Sinclair, J., Doelle, M., & Gibson, R.B. (2018). Implementing next generation assessment: a case example of a global challenge. *Environmental Impact Assessment Review* 72, 166–76.

Udofia, A., Noble, B.F., & Poelzer, G. (2017). Meaningful and efficient? Exploring the challenges to Aboriginal participation in environmental assessment. *Environmental Impact Assessment Review* 65, 164–74.

Glossary

accumulated state With reference to baseline studies, the current condition or state of a valued component or system, reflecting the total change or disturbance that has occurred from past to present.

accuracy The closeness of a measurement or prediction to a specific value.

active publics Those who affect decisions, such as industry associations, environmental organizations, quasi-statutory bodies, and other organized interest groups.

activity information Information about the different actions, components, or activities associated with a proposed undertaking that may lead to environmental effects.

adaptive management (AM) A multi-step, deliberative process that involves exploring alternative management actions and making explicit forecasts about their outcomes, carefully designing monitoring programs to provide reliable feedback and understanding of the reasons underlying actual outcomes, and then adjusting objectives or management actions based on this new understanding.

additive effects Effects that result from individual but separate actions that accumulate over time or across space.

administrative scale Spatial boundaries or scale defined by jurisdictional or management responsibilities for a project, environmental effect, or valued component.

ALCES A Landscape Cumulative Effects Simulator—a system dynamics simulation tool for exploring the behaviour or response of resource systems to disturbances.

alternative means Different ways of carrying out a proposed project—typically alternative locations, timing of activities, or engineering designs.

alternatives to Different ways of addressing a problem at hand or meeting the proposed project objectives; renewable energy, for example, would be considered an "alternative to" a proposed coal-fired generating station.

ambient environmental quality monitoring A pre-project assessment of the surrounding environment inclusive of biophysical and socio-economic factors; collected information is used as a baseline in comparing a project's environment during development, operation, and post-operation against unaffected control sites in order to monitor the impact of a project.

analogue approaches An approach to impact prediction based on the use of existing data or information, including the lessons learned from previous and similar projects and experiences.

analysis scale The scale applied to examine valued components or indicators and impacts across space.

antagonistic effects Individual adverse effects that have potential to partially cancel each other out when combined.

auditing An objective examination or comparison of observations with predetermined criteria.

balance model Model designed to identify inputs and outputs for specified environmental components; these models are commonly used to predict change in environmental phenomena.

baseline study Analysis of the biophysical and socio-economic state of the environment at a given time, including key trends and drivers, that can be used for the assessment, prediction, and interpretation of environmental change.

benchmark A standard or point of reference against which an environmental change or effect can be measured or judged.

Berger Inquiry A federal royal commission in the early 1970s, led by Justice Thomas Berger, into the potential effects of a proposal to develop an energy pipeline corridor from the Mackenzie River Delta in the Beaufort Sea through the Northwest Territories to tie into gas pipelines in northern Alberta.

bridging knowledge systems Maintaining the integrity of Indigenous and Western science knowledge systems while creating settings to ensure the two-way exchange of understanding and learning.

buffer zone An area of undisturbed environment, usually separating a project's actions or disturbance from background conditions such as a riparian buffer zone.

case-by-case screening Determining the need for assessment based on the specific project characteristics and local context, usually in the absence of any formal project lists.

cautionary threshold The level of change or set condition at which monitoring efforts should be increased to more closely monitor valued component or indicator conditions and the effectiveness of best management practices verified to prevent any further adverse change.

climate-resilient projects Projects designed to adapt to changing climate conditions and climate risks.

climate risk The impacts of climate change on a project, defined in terms of climate hazard, exposure, and vulnerability

community health and well-being A holistic concept comprised of a series of social, economic, health, environmental, and other indicators that provides an overall indication of the state or functioning of a local community.

compensation The measures taken by the proponent to make up for adverse environmental impacts of a project that exist after mitigation measures have been implemented.

compliance-based SEA A strategic assessment focused on whether, and to what extent, a proposed policy, plan, or program complies with, or supports, specified objectives, policies, or commitments and, if necessary, identifies and explores options to ensure compliance.

compliance monitoring Monitoring to ensure that all project regulations, agreements, laws, and specific guidelines have been adhered to.

condition-based indicator An indicator (e.g., phosphorus concentrations, benthic invertebrate abundance) that provides direct, measurable information about the condition or state of a valued component.

confirmatory analysis Used to test for uncertainty in impact-predictive techniques and to ensure similar predictive outcomes from different types of techniques.

context area Established during scoping and baseline assessment to delineate the background conditions, such as ecosystem health or functioning, for components likely affected by a project.

continuous effects Effects that are ongoing or persistent over space or time.

control-impact design A monitoring design for comparing project-exposed sites to reference or control sites.

control site A reference point where the environment is not affected by the project, used to monitor the nature and extent of project-induced change in areas that are affected by the project.

cost-benefit analysis An assessment method that expresses project impacts in monetary terms, measuring the relative costs of a project against its potential or total benefits.

critical decision factors In the context of strategic environmental assessment, the issues or factors that matter most to decision-making, such as those factors defined in policies, agency agendas, or sustainability goals, and serve to focus or guide the assessment.

critical threshold Maximum acceptable change, socially or ecologically, beyond which impacts may be long-term or irreparable.

cumulative effect A change in the environment caused by the combined or interacting effects of multiple actions, including natural disturbances, that accumulate across space and time.

cumulative effects monitoring Monitoring the accumulated state of environmental conditions or indicators of cumulative stress associated with developments or other disturbances in a region.

cumulative impact See **cumulative effect**.

Delphi technique An iterative survey-type questionnaire that solicits the advice of a group of experts, provides feedback to all participants on the statistical summaries of the responses, and gives each expert an opportunity to revise her or his judgments.

designated activities Projects or activities that are contained on a project list or specified in regulations, or as determined by a responsible minister, that may require environmental assessment.

determinants of health and well-being Underlying factors, such as physical health, education, mental health, health services, coping skills, and social support networks, that collectively provide an indication of an individual's health.

direct effects First-order impacts from project actions or activities.

direction of change Whether a valued component or indicator condition is trending "up" (i.e., positive) or "down" (i.e., negative), indicating the nature of change but not necessarily the magnitude of change.

disturbance-based indicators Indicators of human disturbance, such as human access or linear feature density, typically used as a proxy for stress to or effects on environmental components.

draft EIS audit Review of the project environmental impact statement according to its terms of reference.

duty to consult A formal, legal obligation in Canada for governments to consult with Indigenous people in cases where Indigenous rights, claims, or titles are known and may be affected by a development or decision, even in cases where those rights, claims, or titles have not yet been proven in court.

dynamic systems-based modelling Modelling focused on describing and predicting the interactions over time between multiple components of a system, including how the mechanics of components of the system evolve or respond over time under different conditions.

early warning indicators Indicators that can be measured to detect the possibility of adverse stress on valued components before they are adversely affected.

ecosystem services The benefits (goods, services, values) provided by/derived from healthy functioning ecosystems, such as flood control, carbon sequestration, habitation provision, and recreation.

effectiveness monitoring Monitoring of mitigation actions implemented to manage anticipated impacts and whether those actions are working to hold impacts to acceptable levels.

effects-based CEA Measures multiple environmental responses to stressors; the results are then compared to some control point to determine the actual measure of cumulative change.

effects-based monitoring Monitoring the condition or performance of the receiving environment, based on the premise that measuring change in environmental indicators, or early warning indicators of potentially affected valued components, is the most direct and relevant means of assessing change.

environmental assessment (EA) A systematic process designed to identify, predict, and propose management measures concerning the impacts of a proposed undertaking on the biophysical or human environment.

Environmental Assessment Review Process (EARP) The first Canadian federal environmental assessment process, formally introduced in 1973 by federal guidelines order.

environmental baseline The past, present, and likely future state of the environment without the proposed project or activity.

environmental change Measurable change in an environmental parameter over time.

environmental effect The difference in the condition of an environmental parameter with as opposed to without a proposed development activity.

environmental impacts Environmental effects that have an estimated societal value placed on them.

environmental impact statement (EIS) The formal documentation produced from the environmental assessment process that provides a non-technical summary of major findings, statement of assessment purpose and need, and a detailed description of the proposed action, impacts, alternatives, and mitigation measures.

environmental management plans (EMPs) Plans prepared by a proponent that detail the specific impact mitigation strategies for a project and the ways in which they are to be implemented.

environmental preview report An early report that is used to determine whether or not an environmental assessment is needed and the level of assessment required; it outlines the project, potential environmental effects, alternatives, and management measures.

environmental protection plans Mandatory management plans that result from project-based EAs; these management plans are tailored to the project as a result of the identification of key impacts and issues and management measures through the environmental assessment process.

eutrophication Excessive nutrients in a water body, frequently caused by runoff, that causes dense algae growth and often oxygen depletion of the water body.

experimental monitoring Research into environmental systems and their impacts for the purpose of gathering information and knowledge and testing hypotheses.

ex-post evaluation Acting and making decisions based on the result of structure, analysis, and appraisal of information concerning project impacts.

fly-in fly-out Projects located in remote areas with high amounts of air traffic to and from the site; typically associated with remote mining projects where temporary project work camps are constructed to house workers, who commute by charter air service on a basis, for example, of two weeks at the work site and two weeks off during the lifespan of the project.

follow-up The monitoring, auditing, and ex-post evaluation activities that occur after a project's approval, focused on whether a project has had or is continuing to have environmental effects.

functional scale Scale relationship based on how different environmental components function across space.

fuzzy sets A class of objects with a continuum of grades of membership or a mathematical model of vague qualitative or quantitative data.

governance framework The network of interrelated government and/or non-government organizations and institutional arrangements that will inform, and be informed by, an impact assessment.

gradient-to-background monitoring Monitoring system that measures the effects caused by the impact source at an increasing distance from the impact origin to the point of background assimilation; an "artificial" control point is established.

health impact assessment (HIA) A systematic process designed to identify, predict, and propose management measures concerning the impacts of a proposed undertaking on human health and well-being.

human access The combined land surface anthropogenic disturbance caused by industrial activities.

hybrid screening A screening approach that combines the characteristics of case-by-case and list-based screening.

Impact Assessment Act Coming into force in 2019, replacing the Canadian Environmental Assessment Act, 2012, sets out the requirements for impact assessment in Canada at the federal level and the role(s) of the Impact Assessment Agency.

impact avoidance A form of impact management whereby impacts are avoided at the outset by way of alternative project designs, timing, or location rather than managed or mitigated after they occur.

impact benefit agreement (IBA) Legal agreement between a proponent and a community or group that will potentially be affected by a project; generally applied to ensure that the resources for maximizing the benefits associated with the development are fully capitalized on.

impact magnitude matrices Impact matrices that provide some indication of the relative importance or significance of the affected components.

impact matrices A tool for communicating assessment information, comprised of a two-dimensional checklist of project activities on one axis and potentially affected environmental components on the other.

impact meaning In the context of significance determination, the context within which impact characteristics are viewed and interpreted (e.g., regulatory, social, ecological, sustainability).

impact measurement In the context of significance determination, the characteristics of the impact (e.g., magnitude, spatial extent, duration).

impact mitigation Minimizing adverse environmental change associated with a project by implementing environmentally sound construction, operating, scheduling, and management principles and practices within project design.

impact significance The degree of importance of an impact based on the characteristics of the impact, the receiving environment, and societal values.

implementation monitoring An evaluation of whether or not the recommendations presented in a project's environmental impact statement were actually put into practice.

inactive publics Publics not normally involved in environmental planning, decisions, or project issues—yet they may be affected.

incremental effects Marginal changes in the condition of an environmental component caused by human actions over time.

Indigenous and local knowledge A cumulative body of knowledge, practice, and belief, evolving by adaptive processes and handed down through generations by cultural transmission, about the relationship of living beings (including humans) with one another and with their environment.

Indigenous-led impact assessment Environmental assessment processes governed by, designed by, and implemented by an Indigenous group.

induced development Development that is triggered or enabled by other developments, such as mineral exploration or economic activity resulting from a new road being built in a previously remote inaccessible area.

induced effects Effects that result from induced development or from the effects of effects.

industrial concessions Agreements between individuals/companies and provincial, territorial, and federal governments that allow for the exploration and/or exploitation of renewable and non-renewable natural resources.

initial environmental examination A preliminary study prepared to establish whether an environmental assessment is needed and what level of assessment should be implemented.

intervenors An individual or formal group or interest who formally intervenes or participates as a third party in a legal or public hearing proceeding.

Inuvialuit Final Agreement Coming into effect in 1984 as the first land-claim agreement settled in the Northwest Territories, signed between the Inuvialuit and the Government of Canada and defining the Inuvialuit Settlement Region.

irreversible impact An environmental impact that cannot be reversed because of economic, ecological, or technological limitations.

James Bay and Northern Quebec Agreement Signed in 1975, the first comprehensive land-claim agreement in modern times between the Governments of Quebec and Canada and Cree and Inuit peoples of northern Quebec.

ladder of participation A concept proposed by Sherry Arnstein in 1969, suggesting that citizen engagement in planning processes can be classified on a gradient or "ladder" from manipulation (low participation) to citizen control (high participation).

Leopold matrix An environmental assessment matrix for identifying first-order project–environment interactions, consisting of a grid of possible project actions along a horizontal axis and environmental considerations along a vertical axis.

life-cycle assessment The cradle-to-grave assessment of projects from their inception and start-up to post-operation.

list-based screening Determining the need for assessment based on a checklist of projects or activities or defined undertakings, usually defined in a project list or regulations.

local study area Established during scoping and baseline assessment to delineate where a project's immediate or local effects are likely to occur.

Mackenzie Valley Resource Management Act An act implemented by the federal government to give decision-making authority to northerners concerning environment and resource development activities within the Mackenzie Valley region of the Northwest Territories; proclaimed in 1998, the act governs environmental assessment in the region.

magnitude The amount of change in a measurable parameter relative to baseline conditions (e.g., percentage of habitat change, amount of change in concentration of a contaminant, percentage change in land use) or other target.

management targets Thresholds or standards that reflect the desired target condition for an environmental component toward which management actions are directed.

MARXAN Marine Spatially Explicit Annealing, a landscape and marine optimization tool that uses a site-selection algorithm to explore options for conservation and regional biodiversity protection planning.

maximum allowable effects level An approach to impact prediction based on specifying certain desired limits or thresholds that a certain impact is not to exceed.

meaningful engagement Those potentially affected by development, or who have a vested interest in development, are enlisted into the planning, assessment, and decision process to contribute to it, thus providing opportunities for the exchange of information, opinions, interests, and values.

models Conceptual diagrams or mathematical equations used to simplify real-world environmental systems.

monitoring A systematic process of data collection or observations used to identify the cause and nature of environmental change.

monitoring for knowledge The monitoring used after impacts occur; data are collected and used for future impact prediction and project management.

monitoring of agreements Monitoring and auditing of agreements between project proponents and affected groups to ensure compliance.

multi-criteria evaluation A structured analytical approach that involves the assessment of competing alternatives or options against multiple criteria.

National Environmental Policy Act (NEPA) The US legislation of 1969 that required certain development project proponents to demonstrate that their projects would not cause adverse environmental effects; the beginning of formal environmental assessment.

network or system diagrams Models based on box-and-arrow diagrams that consist of environmental components linked by arrows indicative of the nature of energy flow or interaction between them.

non-point-source stress Environmental stress from diffuse sources that cannot be traced back to a particular project origin, such as runoff from urban areas or pollution introduced to streams from groundwater.

normative ethics The notion that "right" and "wrong" are found within an individual's behaviour; the focus is on arriving at moral conduct standards.

Nunavut Land Claims Agreement Canada's largest land-claims settlement and land claims–based environmental assessment process; signed in 1993, giving the Inuit self-governing authority and leading to the establishment of a new territory, Nunavut, in 1999.

off-site impacts Impacts that occur at a distance or removed from the recognized project area.

Oldman River dam A dam constructed at the confluence of the Oldman, Castle, and Crowsnest rivers in Alberta to support large-scale irrigation development, triggering legal challenges by environmental groups that would pave the way for federal EA legislation.

on-site impacts Impacts that occur directly in the immediate project area.

participant funding programs Programs established by governments (though they can also be established by project proponents) that provide financial support to the public (usually interest groups, Indigenous groups) to participate in impact assessment processes.

performance audit An assessment of a proponent's capability to respond to environmental incidents and of its management performance.

Peterson matrix A multiplicative environmental assessment matrix consisting of project impacts and causal factors, resultant impacts on the human environment, and the relative importance of those human components used to derive an overall project impact score.

phenomenon scale The scale used to determine the spatial extent within which certain environmental components and valued components operate and function.

point-source stress Environmental stress that can be traced back to a particular project origin, such as discharge or emissions from a development activity.

practical ethics The notion that "right" and "wrong" can be found within scenarios; the focus is on studying specific practical problems to derive a commitment to action.

precautionary principle The principle that when information is incomplete but there is threat of an adverse effect, the lack of full certainty should not be used as a reason to preclude or postpone actions to prevent harm.

precision The closeness of measurements or predictions to each other.

predictive technique audit A type of environmental audit in which a project's predicted effects are compared to the actual effects.

probability analysis An analysis that uses quantified probability to classify the likelihood of an impact occurring and under what environmental conditions.

professional ethics The proper conduct of professionals in practice, typically set out in codes of conduct.

programmatic environmental assessment Under the US NEPA, the application of environmental assessment to multiple projects or to programs of development.

project EA-like SEA An approach to strategic assessment focused on assessing the potential impacts of a proposed or existing policy, plan, or program and comparing impacts to those of viable alternatives.

project impact audit A type of auditing that focuses on determining whether the actual project impacts were predicted in the environmental impact statement.

project lists A list of activities or projects (e.g., designated projects), usually defined in regulations and based on specified project size or thresholds, for which an assessment may be required.

project- or activity-centred Assessment processes focused on the effects of a project or activity on each of the valued components of concern, whereby multiple projects or activities may be affecting the same valued components but the assessment is conducted one project at a time.

prospective analysis Predicting and evaluating how indicators or conditions (e.g., caribou population, fish health) might respond to additional stress in the future.

public participation Involvement of individuals and groups that are positively or negatively affected by a proposed intervention subject to a decision-making process or are interested in it.

range of natural variability The notion that ecosystems are dynamic and fluctuate within a certain range of normalcy beyond which, when disturbed, adverse effects occur.

reasoned argumentation Sifting through information, data, perspectives, and expressed values using structured methods (e.g., decision support aids, matrices, network diagrams) to focus on matters of most importance to decision-making and to build reasoned arguments that support a claim or position.

receptor information Information pertaining to the processes resulting from project-induced effects, such as habitat fragmentation, that is important to consider when characterizing the environmental setting.

regional assessment An approach to assessment concerned with region-based environmental planning or development and assessing the impacts of area-specific plans and program initiatives.

regional study area Established during scoping and baseline assessment to delineate the potential zone of influence of project effects within the project's regional environment, typically also capturing the effects of other projects and activities.

regulatory permit monitoring Site-specific monitoring that includes regular documentation of requirements necessary for permit renewal or maintenance.

remediation The process of post-industrial or post-development site cleanup, which typically involves the removal of contaminants or pollution from soil and water.

remote sensing Detecting and monitoring the physical characteristics of an area by measuring its reflected and emitted radiation at a distance from the targeted areas.

residual effects Effects that remain after all management and mitigation measures have been implemented.

restoration An impact management action focused on restoring environmental quality, rehabilitating certain environmental features, repairing ecological functions, or restoring environmental components to varying degrees.

retrospective analysis See **retrospective assessments**.

retrospective assessments Another term used for baseline assessments, where emphasis is placed on identifying and analyzing past trends and conditions.

reversible impact A change to the environment caused by a project that can be reversed with proper impact management or restoration action.

review panel An independent panel appointed to undertake an environmental assessment, or a type of environmental assessment (i.e., assessment by review panel) that is applied to projects with uncertain or potentially significant effects or if warranted by public and stakeholder concern.

risk The probability of an adverse event occurring, with an analysis of the severity of the consequences associated with that event.

risk assessment The process of accumulating information, identifying possible risks, risk outcomes, and the significance of those outcomes and assessing the likelihood and timing of their occurrence.

scenario A coherent, internally consistent and plausible description of a possible future state.

scenario uncertainty A level of uncertainty where how an impacted system might change is fairly understood but the likelihood and extent of change are not known.

scoping An early component of the environmental assessment process to identify important issues and parameters that should be the focus of the assessment.

screening The selection process used to determine which projects need to undergo an environmental assessment and to what extent.

secondary effects Effects resulting from a direct impact.

sensitivity analysis Examination of the sensitivity of an impact prediction to minor differences in input data, environmental parameters, and assumptions.

social impact assessment (SIA) A systematic process designed to identify, predict, and propose management measures concerning the impacts of a proposed undertaking on the human environment, including social systems, structures, relations. and well-being.

social licence The ongoing acceptance of a company or industry's business practices and operations by communities or the general public.

spatial model A type of model used to depict and understand the spatial relationship between phenomena.

spatial scale The geographic scale used to define the spatial extent of a project environmental assessment or its effects on valued components.

statistical model A type of model used to test relationships between variables and to extrapolate data.

statistical significance In statistical hypothesis testing, helps to quantify whether a result is likely to have occurred because of chance or because of some factor or variable of interest.

statistical uncertainty A level of uncertainty where calculated error and probabilities are known and the decision risk can be calculated.

stranded assets Project infrastructure or investments that have suffered from unanticipated or premature devaluations or conversions to liabilities.

strategic environmental assessment (SEA) The environmental assessment of initiatives, policies, plans, and programs and their alternatives.

strategic futures SEA A strategic assessment focused on identifying and assessing the potential implications of alternative future scenarios or land uses, evaluating the relative risks and opportunities, and establishing a preferred strategic direction or approach.

strategic transitions SEA A strategic assessment focused on the institutional environment in which policies, plans, and programs are formulated and implemented and the conditions that either constrain or enable their success.

stress-based indicator Also referred to as a disturbance-based indicator, such as human access or linear feature density, typically used as a proxy for effects on or level of risk to environmental components.

stress-based monitoring Monitoring project actions or stress, such as the concentration and volume of water discharged from an operation.

stressor-based CEA Assessment that predicts cumulative effects associated with a particular agent of change.

sustainability assessment (SA) A systematic process designed to evaluate the contributions or detractions of a proposed undertaking toward sustainability objectives or outcomes.

synergistic effects When the total effects are greater than the sum of the separate, individual effects.

systemic uncertainty When uncertainties cannot be estimated by any current method or technique—we simply don't know.

target threshold Typically, a politically or socially defined limit, a margin of safety, and a mandatory trigger for management action.

terms of reference A document normally prepared by a government agency that sets out for the proponent the required contents of an environmental impact statement or monitoring program.

threshold The point or level at which there is an abrupt change in the condition of an environmental component or where small changes result in large responses.

threshold-based prediction Basing impact predictions on prior experiences using approaches such as maximum allowable effects levels whereby an impact is capped and not to exceed a certain threshold or level of change.

threshold-based screening A screening process whereby proposed developments are placed in categories and thresholds are set for each type of development, such as project size, level of emissions generated, or area affected.

traditional use studies Studies that combine community knowledge with ethnographic, archival, and archaeological information to highlight places, land uses, and landscapes and values of cultural, heritage, spiritual, use, or community importance.

uncertainty disclosure Open and transparent reporting of uncertainties, unknowns, and assumptions.

uncertainty matrix A matrix used to communicate uncertainty, comprised of information about the location of uncertainty, the level of uncertainty, and the nature of uncertainty.

United Nations Declaration on the Rights of Indigenous Peoples Adopted by the United Nations in 2007, an international instrument to enshrine (Article 43) the rights that "constitute the minimum standards for the survival, dignity and well-being of the indigenous peoples of the world."

valued component–centred An approach to assessment that is focused on the valued component of concern and considers the impacts of all sources of disturbance or stress—whether caused by individual projects subject to assessment, projects exempt from assessment, other types of land uses and human activities, and natural drivers.

valued components Components or attributes of the human and physical environment that are considered important or highly valued and therefore require evaluation within environmental assessment.

VC indicators Provide a measure of qualitative or quantitative magnitude for an environmental impact and might include, for example, specific parameters of air quality, water quality, or employment rates; allow decision-makers to gauge environmental change efficiently.

zone of influence Established during scoping and baseline assessment to delineate the spatial extent of a project's potential impacts on the local and regional environment.

Index

Note: Page numbers in *italics* indicate figures.

abandoned sites, 130
Aboriginal Affairs and Northern Development Canada: guidance for significance determination, 156, 160
acceptability: decision tree for, 166–7; levels of, 166; range of, 166; significance and, 160–1; social (public) values and, 160–1
accessibility: of EA documentation, 204–5; monitoring and, 231
accumulated state, 228, 287
accuracy, 108, 287
actions, 114; additive effects and, 115
activity information, 82, 287
actors: in EA process, 10–11
adaptability: monitoring and, 231
adaptive capacity, 102
adaptive management (AM), 6, 136–9, 287; cycle of, *137*
administrators, 273; *Guideline standards for IA professionals*, 276–7
affected interests, 10–11
agreements: land-claims, 23; monitoring of, 175, 291; *see also* impact benefit agreements (IBAs)
Ajax mine project, 153
Alberta: Athabasca region, 143–4, 215; EA in, 19, 28–9; Environmental Protection and Enhancement Act, 19; oil and gas wells in, 131; oil sands, 19, 117, 219, 228, 238, 253; Regional Municipality of Wood Buffalo, 186–7; terms of reference in, 67
Alberta Liabilities Disclosure Project, 131
Alberta Society of Professional Biologists, 219
ALCES (A Landscape Cumulative Effects Simulator), 112, 229–30, 265, 287; in Elk Valley CEMF, 223
"alternative facts," 283
alternative means, 38, 40, 287; James Bay Lithium Mine project and, 41–2
alternatives: Louis Riel Trail Highway Twinning Project and, 44–7; methods to support assessment of, 42–7; "no action," 39; pre-project planning and, 38–40; "to," 38, 39–40, 287
analogue approaches: to impact prediction, 108–9, 287
analytical science, 9
Antunes, P., R. Santos, and L. Jordão, 150
Arctic: climate change and, 102–4; *see also* northern Canada; *specific regions*
AREVA Resources Canada, 143
Arnstein, S., 48

Art of war, The, 251
Arts, J., P. Caldwell, and A. Morrison-Saunders, 172
Athabasca Basin: IBAs in, 143–4; oil sands in, 117, 186–7, 228, 253; uranium mining in, 143; water discharge and withdrawals in, 215
Athabasca Working Group (AWG), 143
Atomic Energy of Canada, 29
auditing and audits, 173, 287; decision-point, 174; draft EIS, 174, 288; performance, 174, 292; predictive technique, 174, 292; project impact, 174, 292; SEA, 247–8; types of, 174
Auditor General of Canada: SEA audits, 247–8
Aura Environmental, 135
AuRico, 160
Australia: EA in, 3; Environmental Protection Act, 3; level of assessment required in, 66; SEA in, 244

Baker, D. and E. Rapaport, 150–1
Ball, Ian, 112
Ball, M., B.F. Noble, and M. Dubé, 73, 205
Barnes, J., D. Marquis, and G. Yamazaki, 157, 163
baseline assessment, 7, 72–88; boundaries and, 80–5; condition and change analysis, 85–8; indicators and, 77–80; knowledge to support, 88–90; significance and, 148; valued components and, 72–80
baseline study, 72, 287
Bayda Commission Cluff Lake Board of Inquiry, 19
Beanlands, Gordon and Peter Duinker, 211
Beaufort Sea hydrocarbon review, 29
benchmarks, 80, 287; CEA and, 214, 231; follow-up and monitoring and, 182; SEA and, 262; significance and, 153–4
Berger, Thomas, 27–8
Berger Inquiry, 27–8, 287; *see also* Mackenzie Valley pipeline inquiry
Berkes, F., J. Colding, and C. Folke, 196
bias, 278
Bill C-69, xi, 26, 31, 99; *see also* Impact Assessment Act
Bipole III Transmission Line project, 87–8, 215–16
Blueberry River First Nations, 12, 156
Bocking, S., 28
Booth, A. and N. Skelton, 156
boundaries, 80–5; future, 84; jurisdictional, 84, 85; natural, 227; past, 84; spatial, 81–3; temporal, 83–4; transboundary effects, 84–5

Brazil, 263, 264
bridging knowledge systems, 196–7, 287
Briggs, S. and M.D. Hudson, 150
British Columbia: application information requirements in, 67; CEA in, 232, 233; Declaration on the Rights of Indigenous Peoples Act, 193; EA in, xi, 18–19, 100; Environmental Appeals Board, 195; Environmental Assessment Act, 19, 100; Environmental Assessment Office, 77, 79, 100–1; Greenhouse Gas Reduction Targets Act, 99; mining in, 158–60; Reviewable Projects Regulation, 19
buffer zones, 130, 287
Building common ground: a new vision for impact assessment in Canada, 26, 31
businesses or corporations, private, 11

Cabinet Directive on the Environmental Assessment of Policy, Plan and Program Proposals, 245, 246, 253
California Environmental Quality Act, 59
Cameco Corporation, 118, 143
Canada: EA in, 17–34; EA as continuous learning process, 33–4; federal EA in, 25–6; harmonization and, 17–18; human access and industrial concession in, 1, 2; map of EA systems in, *18*; the north and, 23–5; origins and development of EA in, 3, 25–33; provinces and territories in, 17–23; public participation in, 29, 33, 48; reforms to EA in, 284; SEA in, 244–8; screening in, 59, 66; UNDRIP and, 193; *see also specific acts, agencies, and jurisdictions*
Canada–Newfoundland and Labrador Offshore Petroleum Board (C-NLOPB), 242
Canada–Nova Scotia Offshore Petroleum Board, 22
Canadian Council of Ministers of the Environment (CCME), 151, 239, 260; coordination accord, 17–18; *Regional strategic environmental assessment in Canada: principles and guidance*, 246
Canadian Energy Regulator Act, 31
Canadian Environmental Assessment Act (1992), 25, 28–9
Canadian Environmental Assessment Act (2003), 26, 29–30, 218
Canadian Environmental Assessment Act (2012), xi, 26, 30, 283; critiques of, 30, 31–2; regional assessment in, 219; *see also* Impact Assessment Act
Canadian Environmental Assessment Agency, 26, 29
Canadian Impact Assessment Registry, 204
Canter, L.W. and B. Ross, 214

CanTung, 118
capacity: adaptive, 102; carrying, 105, 214; SEA and, 262
careers: in EA, 272
carrying capacity, 105, 214
Carson, Rachel, 2
Cash, D. et al., 196
Cashmore, C., R. Gwilliam, R. Morgan, D. Cobb, and A. Bond, 8
Cashmore, M., 9
causal factors, 114
C.D. Howe Institute, 131
CEA. *See* cumulative effects assessment (CEA)
Centre for Indigenous Environmental Resources, 199
Cherp, A., A. Watt, and V. Vinichenko, 267
change, rates of, 114; stressor-based CEA and, 225
Cheviot coal mine, 106–7
Churchill River Basin study, 219
climate change: costs related to, 104; demands on EA and, 283; impact prediction and, 99–101; impacts on projects and, 101–4; LNG activity in BC and, 99–100; public interest and, 32–3; SEA of, 241
climate exposure, 101
climate hazard, 101
climate-resilient projects, 101, 288
climate risk, 101, 102, 288
climate sensitivity, 102
climate vulnerability, 101
coastal infrastructure projects: climate change and, 102
code of conduct (IAIA), 279
Cold Lake Oil Sands Project, 78, 182
collaboration: CEA and, 217, 231
collaborative approach: significance determination and, 163–4
Colombia, 3
Commissioner of the Environment and Sustainable Development (CESD), 204; SEA audits by, 246, 247–8
communication skills, 273
community health and well-being, 96, 288; Lower Churchill Hydroelectric Generating Project and, 97–8
compatibility: monitoring and, 231
compensation, 131–3, 288
composite approach: significance determination and, 164–7
compounding, 107
condition changes, 85–8, 114
confirmatory analysis, 121, 288
consistency: monitoring and, 231

Constitution Act (1982), 17; *Section 35* rights, 32, 33, 155, 192
consultation. *See* Indigenous engagement; public participation
Consumers' Association of Canada, 216
context area, 83, *83*, 288
context-setting, 222
control-impact design, 183, 288
control sites, 183-4, 288
Convention on Environmental Impact Assessment in a Transboundary Context (Espoo Convention), 84
corporate change, 8
cost-benefit analysis, 163, 288
Cotton, R. and J.S. Zimmer, 17
Council of Canadian Academies, 12, 52, 196-7; Expert Panel on Integrated Natural Resource Management in Canada, 30
court decisions: duty to consult and, 193; "environmental assessment" in, *31*
Cree: Eastmain 1A project and, 218
critical decision factors (CDFs), 261-2, 288
Cronmiller, J. and B. Noble, 230, 231, 232
cross-boundary movement, 107
cultural context: Indigenous peoples and, 155-6
cumulative effects, 12, 219, 288; defining, 211; significance and, 148; on social systems, 212
cumulative effects assessment (CEA), 211-33; in Canada, 214; challenges to, 212-17; collaboration and, 217; components of, 226-31, *226*; defining, 212; effects-based, 224-5, 289; governance for, 232-3; issues scoping, 226; management of, 231; monitoring and, 230-1; overview of, 211-17; possible futures and outcomes, 214-16; prospective analysis, 229-30; retrospective analysis, 228-9; status quo thinking versus requirements, 217; stress-or-based, 224, 225, 294
Cumulative Environmental Management Association, 219
Czech Republic: SEA in, 244

data: open, 185; research capacity and, 284-5; sharing of, 185
data collection: continuity in, 184-5; scoping and, 71; targeted approach to, 181
De Beers Canada: proposed Victor diamond mine, 201-2
decision-makers, 10, 14
decision-making: SEA and, 268; significance and, 148
de Kerckhove, D.T., C.K. Minns, and B. Shuter, 30
Delphi technique, 109, 288
design and engineering change, 8

designated activities, 59, 288
detectability: monitoring and, 231
determinants of health and well-being, 96, 288
developing countries: EA in, 3
development: big-picture issues of, 12-13; induced, 12, 290; policy, planning, and decision-making around, 12; programs of, 241, 244; "victims" of, 52, 53
diagrams, network or system, 85, 86, 291
Diavik diamond mine, 117, 138, 175
Diduck, A., P. Fitzpatrick, and J. Robson, 136
direction of change, 116-17, 288
diversity: significance and, 167
Dogrib Treaty 11 Council (now Tłı̨chǫ Government), 175
draft EIS audit, 174, 288
Duinker, P. and L. Greig, 213, 214, 225
Duncan, R., 121
duration, 113, 118-19, 149
duty to consult, 193-6, 288-9
dynamic systems-based modelling, 229, 289

EA. *See* environmental assessment (EA)
Eagle Mountain-Wood Fibre EA, 100
Eastmain 1A project, 218
Eckel, L., K. Fisher, and G. Russell, 50
ecoducts, 128-9
economy: natural resources sector and, 1
ecosystem services, 96, 289
effects: additive, 114-15, 287; adverse, 32, 33; antagonistic, 115-16, 287; continuous, 119, 288; cumulative, 12, 211, 212, 219, 288; direct, 113, 288; environmental, 93, 94, 289; incremental, 114, 290; induced, 113, 114, 290; residual, 147, 293; secondary, 113, 293; synergistic, 115, 294; as term, 94, 114; transboundary, 84-5; *see also* impacts
Ehrlich, A., 200
Ehrlich, A. and W. Ross, 149, 151, 154, 166
EIS. *See* environmental impact statement (EIS)
Elk River Alliance, 222
Elk Valley, 220
Elk Valley Cumulative Effects Management Framework, 73-4, 220-4, 232, 284
emergent factors, 267
Enbridge: Line 3 project, 238; Northern Gateway project, 30
"end-of-pit-lake," 138
energy projects: climate change and, 101; holistic concepts and, 96
engagement: follow-up and monitoring and, 185-6; meaningful, 48-50, 202-6, 267, 291; in project description, 54; SEA and, 267-8; *see also* Indigenous engagement; public participation

environmental assessment (EA), 289; actors in, 10–11; approvals despite adverse impacts, 6; big-picture issues and, 12–14; in Canada, xi, 17–34; CEA, 211–33; challenges to, 212–17; class (Ontario), 20–1; components of process, 7–8; defining, 3–6; determining need for, 58–68; expectations of, 9–10; federal systems, xi, 17, 25–6; follow-up and monitoring, 172–88, *172*; formal, 5; "good" and "bad," 14; impact management, 127–44; impact prediction and characterization, 93–123; Indigenous consultation and engagement, 192–207; informal, 5; key points about, 14; level of assessment required, 65–7; opportunities for, 284–5; overview of, 1–14; pre-project planning, 37–47; process, 6; professional practice and ethics, 272–81; project description, 54–5; prospects, 283–5; provincial systems, 17–23; public engagement, 47–54; purpose and objectives of, 6–10; reform and, 8, 29–33, 284; scoping and baseline assessment, 71–90; SEA, 238–68; significance determination, 147–68; spectrum of philosophies and values, 9–10; time required for, 6; types and labels of, 3, 4; *see also* project EA
Environmental assessment process for policy and programme proposals, The, 244
Environmental Assessment Review Process (EARP), 25, 26–7, 289
Environmental Assessment Review Process Guidelines Order, 25
environmental awareness, 2
environmental baseline, 72, 289
environmental change, 93, 289
environmental design, 9
environmental governance, 9
environmental impact assessment (EIA), 2; *see also* environmental assessment (EA)
environmental impact statement (EIS), 3, 7, 32, 289
environmental management plans (EMPs), 133, 289; minimum standards, 135; typical components of, 134
environmental mitigation plans, 133
environmental preview report (EPR), 66, 289
environmental protection plans, 133, 289
Environment and Climate Change Canada, 183
Environment Canada: Environmental Effects Monitoring programs, 181
Erickson, P., 273
Espoo Convention, 84
ethical conduct, 278–81; scenarios and challenges, 280–1
ethics: normative, 278, 291; practical, 278, 292; professional, 278, 292
Europe: EA in, 3

European SEA Directive 2001/42/EC, 244, 253
European Union: Cohesion Policy, 253; Directive 85/337/EEC, 3
eutrophication, 113–14, 116, 289
evaluation: ex-post, 173, 289; multi-criteria, 42, 47, 163, 291; post-project, 6; SEA and, 267
exclusionary standard, 152
exogenous influences, 267
experts: impact prediction and, 109; roles for, 11
extrapolation, 110
Exxon Mobil: Sable offshore gas project, 201

Fair Mining Collaborative, 204
Federal Environmental Assessment Review Office (FEARO), 25, 26, 29
federal government: EA legislation and, xi, 17, 25–6; Economic Action Plan 2012, 30; follow-up requirements and, 173; Major Projects Management Office, 203; SEA and, 244–6; *see also specific acts*
Federal Sustainable Development Act, 245
Federal Sustainable Development Strategy, 245
Findlay, S., 204
First Coal Corporation, 156
First Nations: EA system in BC and, 19; *see also* Indigenous peoples
Fisheries Act, 30, 73, 181, 183
flexibility: CEA boundaries and, 228
fly-in fly-out, 117–18, 143, 289
follow-up, 172–3, 289; community engagement and, 185–6; components of, 173, *174*; control sites, 183–4; data collection and, 181, 184–5; effective, 179–86; hypothesis-based or threshold-based approaches, 182; objectives of, 179–81; open data and data sharing, 185; requirements for, 173; SEA and, 266–7; *see also* monitoring
forest resource projects, 1; climate change and, 102
FortisBC, 207
Fort McKay Berry Focus Group, 187
Fort McKay First Nation, 187
Fort Nelson First Nation, 195
Fortune Creek project, 100
Fortune Minerals: NICO poly-metallic mine project, 206
Fox Lake First Nation, 197
fragmentation, 107
frequency, 113, 118–19, 149
Friends of the Oldman River, 28–9
Fuggle, R., 278
funding: for Indigenous engagement, 203–4; for participants, 50–2, 292
future boundaries, 84
future scenarios: SEA and, 263–4, 265
fuzzy sets, 163, 289

Galaxy Lithium, 40
Galbraith, L., B. Bradshaw, and M. Rutherford, 139–40
Galore Creek Project, 142, 143–4
Geographic Information Systems (GIS), 46, 83, 110, 112, 162
Giant Mine, 130
Gibson, R.B., M. Doelle, and A.J. Sinclair, 283
Gibson, G. et al., 206
Gibson, G., D. Hoogeveen, A. MacDonald, and The Firelight Group, 205–6
Gibson, R.B., 33–4, 139
Gibson, R. et al., 156–7
Goldenville mine, 130
good judgment, 278
governance: for CEA, 232–3; framework, 260–1, 290
Grasslands National Park, 176–7
Gray, E., 100
Great Sand Hills Regional Environmental Study, 113, 230; strategic futures SEA, 254–8
greenhouse gas (GHG) emissions, 99, 100–1, 241
Greening Regional Development Programmes Network (GRDPN): *Handbook on SEA for cohesion policy 2007–2013*, 253
Greenland: IBAs in, 141–2; Mineral Resources Act, 142
Greenstone Gold Mines GP Inc.: Hardrock project, 180–1
Greig, L., 212
Greig, L. and P. Duinker, 88–9, 108
Guideline standards for IA professionals, 273; for administrators, 276–7; for practitioners, 274–5
Gunn, J. and B.F. Noble, 216, 229

Hackett, P. J. Liu, and B.F. Noble, 106, 212
Haida Nation v. British Columbia, 193, 194
Haisla Nation, 132
Hajkowicz, S.A., G.T. McDonald, and P.N. Smith, 163
Hanna, K. and B.F. Noble, 204
Hanna, K. and J. Parkins, 19
Hardrock project, 180–1
harmonization, 17–18
Haug, P.T. et al., 147, 149, 151
health and well-being, determinants of, 96, 288
health impact assessment (HIA), 3, 290
holistic concepts: impact prediction and, 96
honest broker, 11
Hostovsky, C. and S. Graci, 20
housing developments, 110
Hulse, D.W., A. Branscomb, and S.G. Payne, 111
human access, 1, *2*, 290
Hunsberger, C., R. Gibson, and S. Wismer, 185

hydroelectricity projects: EA and, 19, 58, 59, 83–4; impacts of, 101–2, 113, 114; in Manitoba, 20, 86, 152, 197–8, 215–16; in Newfoundland and Labrador, 97–8; in Nova Scotia, 22; in Quebec, 218
Hydro-Québec, 21, 218
hypotheses: approaches to impact significance and, 182; impact prediction and, 108

IAIA. *See* International Association for Impact Assessment (IAIA)
ice storm (1998), 101
impact analysis: significance and, 148
impact assessment (IA), 2, 7, 26; challenges to, 283–5; Indigenous-led, 205–6, 290; use of term in federal legislation, 32; *see also* environmental assessment (EA)
Impact Assessment Act, xi, 26, 30–3, 283, 290; "alternatives to" in, 39; basic steps in, 32; CEA and, 214, 219; climate change and sustainability in, 241; designated projects/activities under, 48, 51, 59, 66, 67, 246; EA registry, 204; environmental effects under, 33, 161; follow-up and, 51, 173, 204; Indigenous engagement and, 199; Indigenous rights and, 192; levels of assessment in, 66; participant funding programs and, 51; Physical Activities Regulations, 59; public interest and, 160, 161; public participation and, 48; regional assessment in, 232; screening requirements and, 59, 66–7; SEA in, 246; significance and, 147
Impact Assessment Agency, 26, 31, 66, 219; SEA and, 246; template for impact statements, 67
impact avoidance, 127–9, 290
impact benefit agreements (IBAs), 139–42, 290; lessons learned, 143–4; timing of, 140–1, *140*
impact characterization, 113–21; frequency and duration, 113, 118–19; impact measurement and, 149–50; likelihood, 113, 120–1; magnitude, 113, 116–17; nature, 114–16; order, 113–14; reversibility, 113, 119–20; significance and, 149–50, 151; spatial extent, 113, 117
impact magnitude matrices, 163, 290
impact management, 7, 127–44; adaptive management, 136–9; checklist for prescriptions, 133–5; mitigation hierarchy, 127–33; positive impacts and, 139–42; SEA and, 266
impact matrices, 74, 290; Leopold matrix, 74–7; sample of, *75*
impact meaning, 149, 150–61, 290; benchmarks or limits and, 153–4; cultural context and recognized and expressed rights, 155–6; legal context and, 151–2; political context and, 160; public interest and, 160–1; social (public) values and acceptability, 160–1; sustainability and, 156–60; vulnerability and, 154–5

impact measurement, 149–50, 290
impact minimization, 129–30
impact mitigation, 127, 173, 290; avoidance, 127–9; compensation, 131–3; hierarchy of, 127–33, 134; minimization, 129–30; public interest and, 32, 33; restoration, 130; significance and, 148; *see also* impact management; mitigation hierarchy
impact prediction, 93–123; analogue approaches, 108–9; approaches to, 107–13; categories of, 94–107; change and project effects, 93–4; cumulative impacts, 104–7; elements of, 107–8; of the environment on the project, 101–4; expert judgment, 109; modelling and extrapolation, 109–10; of the project on the environment, 94–101; scenarios, 111–12; uncertainty and, 121–3; threshold-based, 110–11, 294
impacts: air, 95; acceptability of, 166–7; assessing in SEA, 265; avoiding, 127–9, 290; biophysical, 94, 95; cumulative, 104–7; environmental, 94, 289; human, 94, 95; infrastructure and service, 95; irreversible, 119–20, 291; off-site, 117, 292; positive, 139–42; reversible, 119, 293; on-site, 117, 292; socio-cultural, 95; socio-economic, 95; as term, 94, 114; terrestrial, 95; water, 95; *see also* cumulative effects; effects; impact characterization
impact significance, 147–8, 290; approaches to, 162–7; collaborative approach to, 163–4; "compared-to" approach, 153; components of, 149–62; composite approach to, 164–7; determination of, 7, 147–68; impact characterization and, 149–50, 151; impact meaning and, 149, 150–61; impact measurement and, 149–50, 151; interpretations in EA process, 148; principles and concepts of, 148, 167–8; reasoned argumentation approach to, 164; relative, 153; sustainability and, 156–60; technical approach to, 162–3; vulnerability and irreplaceability and, *155*
Imperial Oil: Kearl oil sands mine project, 138–9
inclusiveness: significance and, 167
indicators: CEA and, 225, 230; condition and change analysis, 85–8; condition-based, 77, 78, 95, 288; disturbance-based, 79, 288; early warning, 181–2, 183, 289; selecting meaningful, 79–80; stress-based, 78, 79, 95, 294; valued component (VC), 72–3, 77–80, 294
Indigenous and local knowledge (ILK), 196–8, 290
Indigenous engagement, 192–207; benefits of, 192; challenges, 199–202; EA and, 195–6; intent of, 196; late timing of, 199–201; limited financial and human resource capacity, 201–2; meaningful, 202–6; misaligned expectations about scope of EA, 201; participation fatigue, 202; SEA and, 268; tools for, 53

Indigenous-led impact assessment, 205–6, 290
Indigenous peoples, 10–11; duty to consult and, 193–6; EA system in BC and, 19; engagement and, 37; IBAs and, 139; Mackenzie Valley pipeline inquiry and, 27; in the north, 21, 23; opportunities for EA and, 284; pre-project planning and, 38; public interest and, 33; Quebec development and, 21; reconciliation and, 196; Trans Mountain Expansion project and, 186; UNDRIP and, 192, 193
Indigenous rights, 155, 156, 192; Impact Assessment Act and, 33
indirect (source of change), 107
induced development, 12, 290
industrial concessions, 1, 2, 290
information: activity, 82, 287; principles of, 89–90; in project description, 54; provision, 9, 10; receptor, 82, 292
initial environmental examination (IEE), 66, 290
Institute of Environmental Assessment (UK), 4
institutional change, 8
institutional context: SEA and, 261, 266
interdisciplinary environment: EA as, 273–8
Intergovernmental Panel on Climate Change, 111
International Association for Impact Assessment (IAIA), 4, 99; code of conduct, 279; *Guideline standards for IA professionals*, 273; Principles of Environmental Impact Assessment Best Practice, 160
International Atomic Energy Agency: SEA guidelines of, 242–3
intervenors, 204, 290
Inuvialuit Final Agreement, 23, 290–1
Inuvialuit Settlement Region, 24, 202; Environmental Impact Review Board, 24; Environmental Impact Screening Committee, 24
irreplaceability: significance and, 154–5
Isaac, T. and A. Knox, 192
issue advocate, 11

Jack Pine mine project, 117
James Bay and Northern Quebec Agreement, 21, 23, 218, 291
James Bay Lithium Mine project, 40–2
João, E., 81
Jobs, Growth and Long-term Prosperity Act, 26, 30
Joint Federal–Provincial Panel on Uranium Mining Developments in Northern Saskatchewan, 143
Joseph, C., 153
jurisdiction: boundaries and, 84, 85; harmonization and, 17–18; *see also* federal government; provinces and territories

Kearl oil sands mine project, 138-9
Keeyask Generation Project, 86, 152, 197-8
Keeyask Hydropower Limited Partnership, 197
Kemess North copper-gold mine, 157, 158-60, 199
Key Lake Uranium Mine, 118
Kilgour, B. et al., 183
Kingsclear First Nation, 51
Kirchoff, D., H. Gardner, and L. Tsuji, 30
Kiruna mine (Sweden), 128-9
Kitchenuhmaykoosib Inninuwug First Nation, 201
Kitikmeot Inuit Association, 175
knowledge: bridging systems, 196-7, 287; Indigenous and local, 196-8, 290; monitoring for, 179, 291; scientific, 88-90, 196, 197
Kristensen, S., B.F. Noble, and R. Patrick, 232
Ktunaxa Nation, 220, 221; Teck Coal and, 284
Kwasniak, A.J., 139

ladder of participation, 48, 291
Land, L., 195, 199
land-claims agreements, 23
landfills, 280-1
landowners, private, 11
land-use planning, 12-13, 211, 230, 233; in Alberta, 19; in Northwest Territories, 24; in Nunavut, 65
Larsen, P. et al., 104
Lawe, L., J. Wells, and Mikisew Cree First Nations Industry Relations Corporation, 185-6
Lawrence, D.P., 149, 151-2, 163, 164, 165
Lax Kw'alaams First Nation, 199
Lee, G. Fred and Anne Jones-Lee, 280
legal designations or standards: significance and, 151-2
legislation, xi, 12; origins of EA, 2-3; *see also specific legislation*
Leopold, L.B., et al., 74
Leopold matrix, 74-7, 291; sample of, 76
Leung, W., B. Noble, J. Jaeger, and J. Gunn, 121
life-cycle assessment, 172, 291
life-cycle regulators, 66
likelihood: impact significance and, 113, 120-1, 149
limits: CEA and, 217, 231; impact meaning and, 153-4
liquefied natural gas (LNG) activities, 12, 79, 99-100; *see also* oil and gas projects
lithium, 40
LNG Canada Export Terminal, 100; wetland compensation strategy, 132-3
local study area, 83, *83*, 117, 291
Lorneville study, 21
Louis Riel Trail Highway Twinning Project, 44-7
Lower Athabasca Regional Plan, 187
Lower Churchill Hydroelectric Generating Project, 97-8

Luke, L. and B. Noble, 100
Łutsël K'e Dene First Nation, 175, 200

McCold, L. and J.W. Saulsbury, 228
Mackenzie Gas Project, 72-3
Mackenzie Valley: EA in, 24; IBAs in, 139-40; transboundary effects and, 84-5
Mackenzie Valley Environmental Impact Review Board (MVEIRB), 24, 199, 200, 202
Mackenzie Valley pipeline inquiry, 23, 27-8
Mackenzie Valley Resource Management Act (MVRMA), 23, 24, 96, 291
MacMillan Bloedel, 194
MAELs. *See* maximum allowable effects levels (MAELs)
magnitude, 291; impact significance and, 113, 116-17, 149
Maliseet First Nations, 51, 52
management: CEA and, 231, 232; in Elk Valley CEMF, 222; monitoring for, 175-9; prescriptions, 133-5; targets, 80, 291; *see also* adaptive management; impact management
mandatory standard, 152
Manitoba: EA in, 20; hydroelectricity development in, 215-16
Manitoba Clean Environment Commission, 20, 87-8, 197, 216
Manitoba Environment Act, 20
Manitoba Hydro, 215, 216; Bipole III project, 87-8; Keeyask hydroelectric project, 86, 152, 197-8
MARXAN (Marine Spatially Explicit Annealing), 112-13, 229, 230, 265, 291
matrices: impact, 74, *75*, 290; interaction, 42; Leopold, 74-7, *76*, 291; Peterson, 42-4, 292; risk classification matrix, 120
maximum allowable effects levels (MAELs), 110-11, 160, 291
maximum zones of detectable influence, 227-8
Mayer, S. et al., 61
Meadowbank gold mine, 118
meaning. *See* impact meaning
meaningful engagement, 48-50, 202-6, 267, 291; characteristics of, 49
meaningful Indigenous engagement, 202-6
Metal and Diamond Mining Effluent Regulations, 183
Mikisew Cree First Nation, 284
Mikisew Cree First Nation v. Canada, 193
Mi'kmaq Chiefs of New Brunswick, 52
mining operations: alternatives and, 39-40; diamond, 117, 138, 175, 201-2; fly-in fly-out, 117-18; IBAs and, 139; impact significance and, 147, 157; reclamation and, 130; stress-based monitoring and, 182-3; uranium, 143, 184-5

minister of environment, 24, 30, 219, 246
Miramar Giant Mine Ltd, 130
Mitchell, B., 52
mitigation. *See* impact mitigation
mitigation hierarchy, 127–33, 134; avoidance, 127–9; illustration, *128*; minimization, 129–30; compensation, 131–3; restoration, 130; *see also* impact mitigation
models, 108, 109–10, 291; ALCES, 112, 223, 229–30, 265, 287; balance, 109, 287; MARXAN, 112–13, 229, 230, 265, 291; scenarios and, 211; simulation, 109, 110, 162, 265; spatial, 42, 110, 293; statistical, 109, 110, 293; technical, 162–3
monitoring, 173, 179–86, 291; of agreements, 175, 291; ambient environmental quality, 178, 287; community engagement and, 185–6; compliance, 174–5, 288; control-impact, *184*; control sites and, 183–4; cumulative effects, 178–9, 288; CEA and, 230–1, 232; data collection and, 181, 184–5; effectiveness, 178, 289; effects-based, 182–3, 289; experimental, 179, 289; gradient-to-background, 184, *184*, 290; hypothesis-based, 182; implementation, 175, 178, 290; for knowledge, 179, 291; for management, 175–9; methods and techniques, 188; objectives of, 179–81; open data and data sharing, 185; rationale for post-decision, 174–9; regulatory permit, 175, 293; SEA and, 267; significance and, 148; stress-based, 182–3, 294; threshold-based, 182; for understanding, 179; Wood Buffalo Environmental Association and, 186–7; *see also* follow-up
Montague Mines, 130
Morris, P. and R. Therivel, 107
Morrison-Saunders, A., 5
Morrison-Saunders, A. and M. Bailey, 272
Morrison-Saunders, A. et al., 3
multi-scaled approach: spatial boundaries and, 228
Murray, F. et al., 185
Musqueam First Nation, 207

Nalcor Energy, 97
'Namgis First Nation, 203
Nasen, L., B.F. Noble, and J. Johnstone, 215
National Aeronautics and Space Administration (NASA): guidance for information, 89
National Energy Board, 100, 186, 201, 204
National Environmental Policy Act (NEPA), 2–3, 243, 291; significance and, 149
national interest/significance, 160; in New Zealand, 165
natural resources: adverse consequences of use, 1; benefits from, 1; jurisdiction in Canada, 17; use and access in Canada, 1, *2*

natural variability, range of, 80, 153–4, 292
nature: impact significance and, 113, 114–16, 149
need, for project, 37–40
negotiated agreements, 139; *see also* impact benefit agreements (IBAs)
Netherlands Environmental Assessment Agency, 122
network diagrams, 85, *86*, 291
New Brunswick: EA in, 17, 21; Clean Environment Act, 17, 21; Environmental Impact Assessment Regulations, 21
Newfoundland and Labrador: EA in, 17, 22–3; EIS guidelines in, 67; Environmental Assessment Act, 22; Environmental Protection Act, 17, 22–3; regional assessment of offshore oil and gas exploratory drilling, 219, 220; *Strategic environmental review guideline for policy and program proposals*, 246
Newmont Mining Corporation, 142
New Zealand, 165; Resource Management Act, 165
Nexen, 195
Nielsen, Delaney and Associates, and Publivate, 88
Noble, B. et al., 246
Noble, B.F., 51, 156, 195, 204, 239, 260
Noble, B.F. and J. Birk, 185
Noble, B.F. and J. Gunn, 152
Noble, B.F., G. Liu, and P. Hackett, 214
Noble, B., J. Martin, and A. Olagunju, 178
Noble, B.F. and K. Nwanekezie, 252, 254
Noble, B.F. and K. Storey, 160
Noble, B.F. and A. Udofia, 27, 51, 201, 202, 205
"no net loss" policy, 132, 133
non-government environmental organizations, 9
non-point-source stress, 104, 291
Northcliff Resources Ltd, 51–2; Sisson tungsten-molybdenum project, 205
northern Canada: EA in, 23–5
Northern frontier, northern homeland, 27
Northern Gateway project, 30
Northern Rivers Ecosystem Initiative, 219
Northgate Minerals Corporation, 158, 160
North Slave Metis Alliance, 175
Northwest Territories: EA in, 24; resource development in, 253
Northwest Territories Devolution Act, 24
NovaGold Resources, 142
Nova Scotia: EA in, 22; Environmental Assessment Act, 22; Environmental Assessment Regulations, 150; Environmental Protection Act, 22; project lists in, 60; Water Act, 22
nuclear power programs: SEA guidelines for, 242–3
Nuclear Safety and Control Act, 66
Nunavut: EA in, 23, 65–6

Nunavut Impact Review Board (NIRB), 23, 65
Nunavut Land Claims Agreement, 23, 291–2; significance and, 156, 160
Nunavut Planning and Assessment Act (NuPPAA), 65
Nunavut Planning Commission (NPC), 65

objectives: EA and, 6–10; of follow-up, 179–81; of public participation, 50; SEA and, 261–2
observability: monitoring and, 231
Odum, W., 211
offsetting, 131–3
oil and gas projects: in Alberta, 131; environmental management plans and, 134; offshore Newfoundland and Labrador, 219, 242; orphaned wells, 131; in Saskatchewan, 254–8
oil sands, 19, 117, 219, 238, 253; CEA and, 228; Wood Buffalo Environmental Association and, 186–7
Olagunju, A. and J. Gunn, 73
Oldman River Dam project, 28–9, 292
Olszynski, M., 137
"one-window approach," 18
Ontario: EA in, 17, 20–1; Environmental Assessment Act, 20; Renewable Energy Approvals, 21
Ontario Power Generation, 150
open data, 185
operational: as term 251
operational frameworks: significance and, 166
opposition, philosophical, 14
Orca project sand and gravel mine, 203
order: impact significance and, 113–14, 149
Orenstein, M., 6
Orphan Basin SEA, 242
Our common future, 243

Pacific Northwest project, 100
Page, J., 89
Paris Agreement, 99
Parkins, J.R., 213, 214
participant funding programs, 50–2, 292; Indigenous engagement and, 203–4
participation: as philosophy or value, 9; *see also* public participation
participation fatigue, 202
Partidário, M.R., 238, 252, 253, 268
past boundaries, 84
Peigan Indian Reserve, 28
permafrost thaw, 102–4, *103*
Petersen, C. et al., 122
Peterson matrix, 42–4, 292
petroleum and natural gas (PNG) projects: in Saskatchewan, 61–4; *see also* oil and gas projects
The Philippines, 3
Pielke, R.A., 11

pipelines, 12, 238
Placer gold mine, *152*
Planitex, 201
Plate, E., M. Foy, and R. Krehbiel, 203
Poder, T. and L. Luki, 89
point-source stress, 104, 292
Polaris Minerals, 203
policies, plans, and programs (PPPs): example of assessment tiers for energy, *240*; SEA and, 238–9, 239–41, 252, 253, 258, 265–6, 267
policy, 12; and context in project description, 54
political context: significance and, 160
Possignham, Hugh, 112
practice, professional, 272–8
practitioners, 11, 272; ethical conduct and, 278–81; *Guideline standards for IA professionals*, 274–5; skills and qualifications for, 273–8
precautionary principle, 68, 292
precision, 292; environmental management plans and, 134; impact prediction and, 108
pre-project planning, 7, 37–47; consideration of alternatives, 38–40; "need for" and "purpose of" project, 37–8; roles and responsibilities, 37, 38
Prince Edward Island: EA in, 22
probability analysis, 121, 292
probable standard, 152
professional practice, 272–8; ethics and, 278–81; variety of roles in, 272
programmatic environmental assessment, 244, 292
programs: of development, 241, 244; SEA for, 241
project- or activity-centred approach, 212, *213*, 292
project description, 7, 54–5
project EA, *13*, 239, 283–4, 285; CEA and challenges to, 212–17; individually insignificant actions and, 215–16
project lists, 59, 60, 292
projects: "best" possible, 38; climate-resilient, 101, 288; need for, 37–8; purpose of, 37–8, 54
proponents, 9, 10; duty to consult and, 194, 195; Indigenous engagement and, 203; mitigation and, 105–6; open data and data sharing and, 185; pre-project planning and, 37, 38
prospective analysis, 229–30, 292
prospective assessment: in Elk Valley CEMF, 222
prospects, 283–5
Prosperity mine (Taseko Mines), 18
protest: Indigenous engagement and, 199
provinces and territories: EA legislation and, xi, 3; EA systems in, 17–25; follow-up requirements and, 173; resource rights and, 17; SEA and, 246; *see also specific province or territory*
public engagement. *See* public participation

public hearings, 66
public interest: federal legislation and, 32–3; impact meaning and, 160–1
public participation, 6, 47–54, 292; benefits of early engagement, 49–50; citizen power and, 48; consulting and, 48; EARP and, 27; federal EA legislation and, 29, 33; identifying publics, 52–3; ladder of, 48; levels of, 48; meaningful engagement, 48–50, 202–6, 267, 291; "non-participation," 48; objectives of, 50; participant funding programs, 50–2; providing information and, 48; requirements for, 47–8; roles and responsibilities, 37; SEA and, 267–8; techniques for, 54; tools for, 53–4
public registries, 204–5
public(s), 11; active, 52, 287; identifying, 52–3; inactive, 52, 290; influence versus stake in outcome, 52–3; pre-project planning and, 38
Pulp and Paper Effluent Regulations, 183
pulp and paper activities: impacts and, 119; monitoring and, 181, 183
pure scientist, 11
purpose, of project, 37–8, 54

qualifications: EA practitioners and, 273–8
Quebec: EA in, 21; Eastmain 1A project, 218; Environmental Quality Act, 21, 218; Forest Act, 21; SEA in, 246

Rabbit Lake uranium mine project, 184–5
range of acceptability, 166
range of natural variability (RNV), 80, 153–4, 292
reasoned argumentation, 164, 292
receptor information, 82, 292
recommendations, 8; SEA and, 265–6
reconciliation, 196
refinery low, 240
reflexive: as term 251
reforms: to EA in Canada, 8, 29–33, 284
Regional Air Quality Coordinating Committee (RAQCC), 186; see also Wood Buffalo Environmental Association
Regional Aquatics Monitoring Program, 219
regional assessment, 219–25, 284, 292; governance and, 232; oil and gas drilling offshore of Newfoundland and Labrador, 219, 220
Regional Municipality of Wood Buffalo, 186
Regional strategic environmental assessment in Canada: principles and guidance, 246
regional study area, 83, *83*, 117, 292–3
registries, public, 204–5
regulators, 10, 272; life-cycle, 66; pre-project planning and engagement and, 37
regulatory designations or standards: impact significance and, 151–2
regulatory context: in project description, 54

remediation, 130, 293; of Alberta's oil and gas wells, 131
remote sensing, 71, 293
research community: science capacity and, 284–5
restoration, 130, 293
retrospective analysis: CEA and, 228–9
retrospective assessments, 87–8, 293; in Elk Valley CEMF, 222; *see also* baseline assessment
reversibility: impact significance and, 113, 119–20, 149
review panel, 66, 293
rights: impact meaning and, 155–6; Indigenous, 33, 155, 156, 192; *Section 35*, 32, 33, 155, 192; treaty, 155, 192, 201
Ring of Fire, 253
risk, 120, 293; climate, 288; effectiveness and implementation monitoring and, 178; SEA and, 265
risk assessment, 120, 293
risk-based methods, 211
risk classification matrix, *120*
roles: for EA practitioners, 272; pre-project planning and, 37, 38
Ross, W., 217
Rowan, M., 151, 154–5
Rumsfeld, Donald, 136

Sable offshore gas project, 201
Salmo Consulting Inc., 80
Saskatchewan: case-by-case screening in, 60; EA in, 17, 19–20; Environmental Assessment Act, 17, 19, 44, 60, 133, 246; environmental management plans in, 133–4; oil and gas development in, 61–4, 215; SEA in, 246, 254–8; uranium mining in, 143, 184–5
scale, 80; administrative, 227, 287; analysis, 227, 287; functional, 82, 289; phenomenon, 227, 292; spatial, 81–3, 226–7, 293; temporal, 83–4; *see also* boundaries
scenario-based modelling, 211; CEA and, 229–30
scenarios, 111–12, 293; future, 263–4, 265; illustration, *111*; uncertainty, 122, 293
science arbiter, 11
science capacity: EA and, 284–5
Science manual for Canadian judges: guidance for information, 89
scientific change, 8
scientific knowledge, 196, 197; baseline assessment and, 88–90; principles for, 89–90
scope: spatial boundaries and, 227
scoping, 7, 293; CEA and, 226–8; considering follow-up and monitoring during, 181; functions of, 71; SEA and, 261–2; significance and, 148; spatial boundaries and, 226–8; temporal boundaries and, 228
scoring: significance and, 163
Scotland: flood risk management in, 236–64

screening, 7, 58–64, 293; approaches to, 58–64; case-by-case, 60–1, 287; definition and purpose of, 58; hybrid, 64, 290; level of assessment required, 65–7; list-based (prescriptive), 59–60, 60–1, 291; in Nunavut, 65–6; precautionary principle and, 68; "screening out" projects, 60–1; significance and, 148; threshold-based, 64, 294

SEA. *See* strategic environmental assessment (SEA)

Section 35 rights, 155, 192; public interest test and, 32, 33; *see also* Constitution Act

Seitz, N.E., C.J. Westbrook, and B.F. Noble, 84, 214, 227, 228

sensitivity analysis, 121, 293

setbacks, 130

Sheate, W.R. et al., 239

Sheelanere, P., B.F. Noble, and R. Patrick, 232–3

shifting baseline syndrome, 87–8

significance. *See* impact significance; statistical significance

significance determination. *See* impact significance

Silent spring (Carson), 2

Sinclair, J. and A. Diduck, 205

Sippe, R., 166

Sisson tungsten-molybdenum project, 51–2

skills: EA practitioners and, 273–8

Skwelwil'em Wildlife Management Area, 207

slumping, *103*

Slvumut project, 206

Small Modular Reactor Roadmap, 243

Snell, T. and R. Cowell, 68

social (public) values: significance and, 160–1

social impact assessment (SIA), 3, 293

social licence, 192, 293

space crowding, 107

spatial bounding/boundaries, 81–3; CEA and, 226–8

spatial extent: impact significance and, 113, 117, 149

special interest groups, 11

Species at Risk Act, 73

Spectra Energy (now Enbridge Inc.), 12; Westcoast Connector Gas Transmission Project, 156, 202

Squamish Nation: EA of Woodfibre Liquefied Natural Gas Facility, 206–7, 284

Squires, A. and M.G. Dubé, 182

stakeholder empowerment, 8

standards: impact significance and, 151–2

State of the Baptiste Lake Watershed report, 112

statistical significance, 162, 293

Stelfox, Brad, 112

St Mary's First Nation, 51, 52

Storey, K., 117

stranded assets, 104, 293

strategic: as term 251

Strategic Assessment of Climate Change, 241

strategic direction: SEA and, 265–6

strategic environmental assessment (SEA), 3, 232, 238–68, 284, 293; approaches to, 252–4; baseline condition(s), 262; benefits and opportunities, 258–9; in Canada, 244–8; challenges to, 268; characteristics of, 249–50; of climate change, 241; compliance-based, 252, 253, 288; defining, 238–41; design and, 259–68; engagement and, 267–8; evaluating, 267; federal audits, 246, 247–8; follow-up and, 266–7; impact management needs, 266; impacts, opportunities, and risks, 265; monitoring and, 267; as "one-off" activities, 258, 268; opportunities for, 260; origins and evolution, 243–8; policies, plans, and programs (PPPs) and, 238–9, 239–41, 252, 253, 258, 265–6, 267; principles of, 249–54; process components, 259; project EA-like, 252–3, 292; purpose, 260–1; result-based characteristics of, 249, 250; scoping, 261–2; spectrum of, 252, *252*; strategic attributes of, 251–2; strategic direction, 265–6; strategic futures, 253, 293; strategic options, 262–5; strategic transitions, 253–4, 293–4; strategy implementation, 267; timeline of development in Canada, 244–5

strategic options: SEA and, 263–4, 265

strategy: as concept, 251

stress: non-point-source/point-source, 104, 292

study area: local, 83, *83*, 117, 291; regional, 83, *83*, 117, 292–3

subjectivity: significance and, 151, 167

Supreme Court of Canada: on duty to consult, 194

sustainability assessment (SA), 3, 294

sustainability-based criteria: public interest and, 32, 33; significance and, 156–60

sustainable development: role of EA in, 8

Sweden: ecoducts for avoiding wildlife collisions in, 128–9

Swift Current–Webb Community pasture, 62–4

system diagrams, 85, 291

tactical: as term 251

Tahltan First Nations, 142

tailings disposal, 120

Taseko Mines: proposed Prosperity gold-copper mine, 18

Tataskweyak Cree Nation, 197

technical approach: to significance determination, 162–3

Teck Coal, 73; Elk Valley Cumulative Effects Management Framework and, 284; Line Creek Operations Phase II expansion project, 220

Teck Resource Ltd, 142; Frontier oil sands mine, 6

temporal bounding/boundaries, 83–4; CEA and, 228
Tennoy, A., J. Kvaerner, and K. Gjerstad, 121
Terasen Jasper National Park–Mt Robson Pipeline assessment, 112
terms of reference, 67, 294; Indigenous engagement and, 203
Therivel, R. and M.R. Partidário, 239
Therivel, R. and W. Ross, 228
Thompson, M.A., 163
threshold-based approaches: impact significance and, 182
threshold-based prediction, 110, 294
thresholds, 61, 80, 107, 294; for assessment, 60; cautionary, 80, *81*, 287–8; CEA and, 214, 231; critical, 80, *81*, 288; target, 80, *81*, 294
time crowding, 107
time lags, 107
Toro, J. et al., 160
Toro, J., I. Requena, O. Duarte, and M. Zamorano, 154
traditional use studies, 201, 294
Trans Mountain Expansion (TMX) project, 186, 204
Trans Mountain Indigenous Advisory and Monitoring Committee (IMAC), 186
transparency: significance and, 167
transportation projects, 101
treaties, 12
Treaty 5 First Nations, 197
treaty rights, 155, 192, 201
trends analysis, 110; SEA and, 262
Tsay Keh Nay, 158
Tsilhqot'in Nation v. British Columbia, 193
Tsleil-Waututh First Nation, 207

uncertainty: disclosure, 121, 294; impact prediction and, 121–3; level of, 122–3; location of, 122; matrix, 121–3, 294; monitoring and, 178; nature of, 123; scenario, 122, 293; significance determination and, 168; statistical, 122, 293; systemic, 123, 294
understanding: monitoring for, 179
United Kingdom: SEA in, 244
United Kingdom Institute of Ecology and Environmental Management: Guidelines for Ecological Impact Assessment, 163
United Nations 1992 Earth Summit, 243
United Nations Declaration on the Rights of Indigenous Peoples (UNDRIP), xi, 33, 192, 193, 284, 294
United Nations Environment Programme (UNEP), 109; International Resource Panel, 1
United States: environmental assessment in, 2; NEPA in, 2–3, 149, 243, 291; SEA in, 243–4
Upper Bow Basin Cumulative Effects Study, 112, 230
Ur-Energy Inc.: Screech Lake Uranium Exploration Project, 156, 199, 200

usability: monitoring and, 231
US Army Corps of Engineers, 149
US Council on Environmental Quality, 105

valued component–centred approach, 212, *213*, 294
valued components (VCs), 72–80, 294; condition and change analysis, 85–8; examples of condition- and stress-based indicators for, 78; impact prediction and, 94–5; interaction with project actions or activities, 73–7; objectives, 80; pathways, 77, *77*; selecting indicators, 72–3, 77–80
values, 278
VC indicators, 77–80, 294; selecting, 72–3, 77–80
Victor diamond mine, 201–2
Voisey's Bay project, 73
vulnerability: significance and, 154–5

Walker, W.F. at al., 121
Wallington, T., O. Bina, and W. Thissen, 251
War Lake First Nation, 197
water resource projects: climate change and, 102
watersheds: cumulative effects and, 104–5
West Coast Connector Gas Transmission project, 156
Westcoast Connector Gas Transmission Project, 12
West Moberly First Nation, 156
wetlands, 1, 44; LNG Canada Export Terminal wetland compensation strategy, 132–3
Weyerhaeuser, 194
Whabouchi spodumene mine, 179
Wilgenburg, H., 21
wind turbine projects, 60
Wong, L., B.F. Noble, and K. Hanna, 185, 230, 231
Wood, C. and M. Djeddour, 243
Wood, G., 121, 150, 155, 160, 167–8
Wood, G., J. Glasson, and J. Becker, 164–5
Wood Buffalo Environmental Association, 186–7
Wood Buffalo strategic assessment, 284
Woodfibre Liquefied Natural Gas Facility, 100; Squamish Nation EA of, 206–7, 284
Woodstock First Nation, 51
World Bank, 243
World Commission on Environment and Development: *Our common future*, 243

Xay Temíxw land-use plan, 206

Yellowknives Dene First Nation, 175
York Factory First Nation, 197
Yukon: EA in, 24–5; Environmental and Socioeconomic Assessment Act, 23, 24–5, 173; resource development in, 253
Yukon Environmental Socio-economic Assessment Board (YESAB), 24, 25

zone of influence, 83, 294